BIOLOGY AND GEOLOGY OF CORAL REEFS

VOLUME IV: Geology 2

CONTRIBUTORS

Richard E. Chapman

D. C. Green

O. A. Jones

Harry S. Ladd

Alan R. Lloyd

G. R. Orme

J. A. Steers

D. R. Stoddart

H. H. Veeh

BIOLOGY AND GEOLOGY OF CORAL REEFS

EDITED BY

O. A. JONES

Department of Geology
University of Queensland
St. Lucia, Brisbane
Queensland, Australia

R. ENDEAN

Department of Zoology
University of Queensland
St. Lucia, Brisbane
Queensland, Australia

VOLUME IV: Geology-2

ACADEMIC PRESS   New York   San Francisco   London   1977

A Subsidiary of Harcourt Brace Jovanovich, Publishers

ACADEMIC PRESS, INC.
111 Fifth Avenue, New York, New York 10003

United Kingdom Edition published by
ACADEMIC PRESS, INC. (LONDON) LTD.
24/28 Oval Road, London NW1

Library of Congress Cataloging in Publication Data

Jones, Owen Arthur.
    Biology and geology of coral reefs.

    Includes bibliographies.
    CONTENTS: v. 1.   Geology 1.–v. 2.   Biology 1.–
v. 3.   Biology 2.–v. 4.   Geology 2.
    1.   Coral reef biology–Collected works.   2.   Coral
reefs and islands–Collected works.   I.   Endean, R.,
joint author.   II.   Title.
QH95.8.J66   574.909′4′2   72-84368
ISBN 0–12–389604–5

PRINTED IN THE UNITED STATES OF AMERICA

To the Great Barrier Reef Committee, its office-bearers and its many members who have worked unremittingly for fifty years to further our knowledge of the Great Barrier Reef; and to the memory of geologist Professor H. C. Richards, one of its founders, and biologist Professor E. J. Goddard, an early enthusiastic supporter

CONTENTS

1. Types of Coral Reefs and Their Distribution

Harry S. Ladd

2. The Origin of Fringing Reefs, Barrier Reefs, and Atolls

J. A. Steers and D. R. Stoddart

3. The Nature and Origin of Coral Reef Islands

D. R. Stoddart and J. A. Steers

4. Economic Geology and Fossil Coral Reefs

Richard E. Chapman

5. Aspects of Sedimentation in the Coral Reef Environment

G. R. Orme

6. Radiometric Geochronology of Coral Reefs

H. H. Veeh and D. C. Green

7. Some Notes on the Coral Reefs of the Solomon Islands Together with a Reference to New Guinea

O. A. Jones

8. The Great Barrier Reefs Province, Australia

O. A. Jones

Part I. A Summary of the Geology and Hydrology of the
Queensland Coast and the Continental Shelf

Part II. A Bibliography of the Geology and Hydrology of
the Great Barrier Reefs and the Adjacent Queensland
Coast (Including a Number of Titles Brought Together
by the Late Professor Emeritus W. H. Bryan)

9.   The Basement beneath the Queensland Continental Shelf

Alan R. Lloyd

10.   The Coral Sea Plateau—A Major Reef Province

G. R. Orme

LIST OF CONTRIBUTORS

Numbers in parentheses indicate the pages on which the authors' contributions begin.

RICHARD E. CHAPMAN, Department of Geology and Mineralogy, University of Queensland, St. Lucia, Brisbane, Queensland, Australia (107)

D. C. GREEN, Department of Geology and Mineralogy, University of Queensland, St. Lucia, Brisbane, Queensland, Australia (183)

O. A. JONES, Department of Geology, University of Queensland, St. Lucia, Brisbane, Queensland, Australia (201, 205)

HARRY S. LADD, National Museum of Natural History, Smithsonian Institution, Washington, D.C. (1)

ALAN R. LLOYD,* P. T. Geoservices Ltd., Bandung, Indonesia (261)

G. R. ORME, Department of Geology and Mineralogy, University of Queensland, St. Lucia, Brisbane, Queensland, Australia (129, 267)

J. A. STEERS, St. Catherine's College, Cambridge, England (21, 59)

D. R. STODDART, Department of Geography, University of Cambridge, Cambridge, England (21, 59)

H. H. VEEH, School of Earth Sciences, Flinders University of South Australia, Bedford Park, South Australia (183)

* Present address: 48A Jellicoe Street, Toowoomba, Queensland, Australia.

GENERAL PREFACE

This four-volume work (two volumes covering geological and two biological topics) originated from an article on The Great Barrier Reefs of Australia written by the editors and published in the November 1967 issue of *Science Journal*. The prime aim of this treatise is to publish in one source as many as possible of the major advances made in the diverse facets of coral reef problems, advances scattered in a multitude of papers and published in a variety of journals.

Initially a one-volume work was projected, but the wealth of material available led to this four-volume treatise. Two contain chapters on aspects of geomorphology, tectonics, sedimentology, hydrology, and radiometric chronology relevant to coral reefs, and two accommodate articles on pertinent biological topics.

The task of organizing the contributions by some forty-one authors (situated in eight different countries) on forty-six different topics proved a formidable one. We wish to thank the contributors, all of whom have been as cooperative as their teaching and/or other commitments permitted.

We feel that the material presented in these volumes demonstrates that many major advances in our knowledge have been made in recent years. We realize that the treatment of the topics covered is not complete and that in many cases new problems have been brought to light. It is our hope that the volumes will provide a powerful stimulus to further work on all aspects of coral reefs.

The editors, of course, accept overall responsibility for all four volumes. Dr. Jones is mainly responsible for the editing of the two geology volumes; Dr. Endean for that of the two biology volumes.

O. A. JONES
R. ENDEAN

PREFACE TO VOLUME IV: GEOLOGY 2

This, the fourth volume in the treatise "Biology and Geology of Coral Reefs" and the second geology volume, appears four years after the first volume and more than that after the first chapters for it were written. Some authors completely revised their chapters recently, others updated their contributions by citing later references in the text and adding them to their bibliographies, and still others, unfortunately, were unable to find the time to do anything.

There has been a great surge in research in the last few years in all aspects of coral reefs and in the resulting publications. Notable among the publications are the *Proceedings of the First and Second International Symposia on Coral Reefs* comprising more than 150 papers. Such a flood of material increases the difficulties of publishing "up to the minute" papers in works of this type. Fortunately, articles which are two or three years behind the literature do give valuable summaries of work up to the date of writing, quite apart from reporting information which was new and unpublished at that time. The bibliographies are also of considerable importance.

In this volume, Dr. Harry Ladd writes on Types of Coral Reefs and Their Distribution, and Professor J. A. Steers and Dr. David Stoddart on The Origin of Fringing Reefs, Barrier Reefs, and Atolls and on The Nature and Origin of Coral Reef Islands. These constitute three very important chapters, still of great value although they were written in 1970 and 1971, respectively. The findings of the 1973 Royal Society and Queensland Universities Expedition to the northern part of the Great Barrier Reefs on the specialized Low Wooded Islands need to be read in conjunction with Chapter 3 by Stoddard and Steers. The far northern section of the Reef Province is the least known, and the papers arising from that expedition should prove of very great interest.

Another chapter written sometime ago is the one by Dr. R. E. Chapman, which deals with economic deposits associated with fossil coral reefs. More recently written are the two chapters by Dr. G. R. Orme, one dealing with the very important topic of the interrelation of ecology and sedimentation in coral reef complexes and the other with the coral sea plateau, an area of scattered coral reefs, which has until now been almost completely neglected.

In Chapter 6, Drs. H. H. Veeh and D. C. Green discuss the techniques and results of radiometric dating of coral reefs, an area in which there have been major advances in recent years, and in Chapter 7 O. A. Jones rectifies an omission from Volume I, Geology 1, by providing some notes on the coral reefs of the Solomon Islands.

The remaining two chapters deal with the Great Barrier Reef Province. A. R. Lloyd presents a summary of what is known of the geology of the basement upon which the reefs rest and O. A. Jones describes the reefs from a geological, geophysical, and hydrological viewpoint and attempts to provide a complete bibliography on the reefs. The latter is surprisingly lengthy, particularly as it includes papers on the geology of the coastal areas as far west as the main continental divide, areas which have contributed and continue to contribute eroded material to the platform upon which the reefs rest.

O. A. JONES
R. ENDEAN

CONTENTS OF PREVIOUS VOLUMES

1

TYPES OF CORAL REEFS AND THEIR DISTRIBUTION

Harry S. Ladd

I. Introduction—Definition of a Reef

Coral reefs of one sort or another are found in all of the warmer parts of the oceans. On first view, the wave-breaking structures arouse a feeling of amazement, if not bewilderment. Charles Darwin felt such astonishment even more deeply after he had examined the soft and almost gelatinous bodies of the "apparently insignificant creatures" that were responsible for building a solid reef which "day and night is lashed by the breakers of an ocean never at rest" (1842, p. 15). More recently, an oceanographer, after surveying reefs in the Marshall Islands, aptly compared the deep sea atoll to a Gothic cathedral, "ever building yet never finished, infinite in detail yet simple and massive in plan" (Revelle, 1954, p. iii).

The earliest observers of coral reefs who recorded their discoveries were led to speculate about reef origin but these writers did not formu-

1

late concise definitions. For example, Adalbert von Chamisso, who saw many reefs during the years 1815–1819 when he sailed as a naturalist with Otto von Kotzebue, recognized that low islands in the South Seas and the Indian Ocean owed their origin to several species of coral that built the reefs on the tops of mountains lying under water. He described the shape of the reef, its relation to the prevailing wind, the passes, the lagoons, and the islets (Adalbert von Chamisso, 1821, pp. 331–336).

Charles Darwin, who followed Chamisso, wrote extensively about the classification structure, and distribution of coral reefs, but he did not formulate a concise definition. A later student of corals and coral reefs, T. Wayland Vaughan, felt that a definition was essential and offered the following to cover both ancient and existing reefs: "A coral reef is a ridge or mound of limestone, the upper surface of which lies, or lay at the time of its formation, near the level of the sea, and is predominantly composed of calcium carbonate secreted by organisms, of which the most important are corals" (Vaughan, 1911, p. 238; 1919, p. 238). Vaughan included bedded deposits in the category of coral reefs if reef corals were present in sufficient abundance, but he excluded structures in which algae or other noncoral organisms were dominant.

In 1928 Cumings and Shrock proposed the term *bioherm* as a substitute for reef and coral reef. Four years later Cumings (1932, p. 333) used the term reef to include all ridgelike or riblike structures that rose close to the surface of the sea, regardless of mode of origin. He referred again to the term bioherm for organic structures and proposed a new term *biostrome,* to include "purely bedded structures . . . built mainly by sedentary organisms" but without moundlike or lenslike forms. These terms have been only sparingly used by those studying existing reefs.

In 1954 MacNeil systematically reviewed all definitions and made an attempt to define an organic reef. He stated (1954a, p. 389) that such a reef is a rigid structure composed of calcareous skeletons of various organisms, interlocked or cemented together by growth, and of detrital material derived from the breakup of such skeletons, and that the structure maintains its upper surface at or near the level of the sea.

In recent years there has been an increasing awareness of the quantitive importance of sediments in living reefs. All existing reefs—including those that are definitely wave-resisting—seem to be composed in large part of calcareous sediments. Hence there is a growing tendency to use the generic term reef without qualifying adjectives. The term reef, thus applied, covers the entire structure—surface reef, lagoon deposits, and off-reef deposits. This is the *reef complex,* a useful term introduced by Henson (1950, p. 215).

In describing a reef complex there are several terms that are widely used, particularly in the interpretation of ancient reef deposits. The term *off-reef* is used to describe the dipping beds that lie on the seaward side of the reef. The terms *fore-reef, outer slope deposits,* and *reef talus* are essentially synonymous. The term *back-reef,* meaning behind the reef edge, is variously applied to deposits on the reef flat or in the lagoon. Such deposits may include land-derived sediments as well as reef debris carried landward.

In most discussions of coral reefs the major reef types are arranged in the evolutionary sequence suggested by the Darwin theory—fringing reef, barrier reef, atoll—and this arrangement is followed in the present paper. Darwin himself, however, did not follow this order. He started with atolls and, indeed, he insisted that an atoll could be developed without passing through fringing and barrier stages (1842, p. 121).

In this chapter the world distribution of reefs is considered after which the characteristics and distribution of each of the major types, fringing, barrier, and atoll, are discussed. Similar discussions of minor types such as table reefs, faros, microatolls, knolls, patch reefs, and others are then given.

II. Distribution of Reefs around the World

Charles Darwin was the first to make a serious study of the distribution of coral reefs throughout the world and to delimit the areas occupied by the major types. His map, first published in 1842, showed fringing reefs in red, barrier reefs in pale blue, and atolls in bright blue. Darwin had seen only a limited number of reefs during his world cruise and, of necessity, he depended on the observations of others. He studied all available charts and travel reports. In appraising the evidence presented he exhibited excellent judgment and succeeded in developing regional reef patterns. A second and more detailed map showing the distribution of coral reefs around the world was published by Joubin in 1912.

In 1929 Molengraaff published a map of the East Indies, using color to show reef types, much as Darwin had done (Molengraaff, 1930). Fringing reefs (shore reefs) were shown in red and most of the reefs of the area are in this class. Barrier reefs and atolls were grouped together and shown in two colors: those believed to have been controlled by Pleistocene and later oscillations of sea level were shown in blue; those believed to have been controlled largely by diastrophism were shown in green.

A map showing reef distribution in relation to sea surface temperatures in the Indian and Pacific Oceans was later published by Schott (1935). World maps by Ladd appeared in 1950, by Emery *et al.* in 1954, and by Wells in 1957. All the world maps show that the western halves of the Pacific and Indian Oceans are the areas of concentrated reef development.

Darwin's reef studies were tied to lands that rose above sea level. Very little was known in his day about the surface of the earth below sea level. Only in recent years has a greatly increased knowledge of submarine topography given a better understanding of the apparently erratic distribution of reefs around the world.

In attempting to evaluate the relative importance of various factors that control reef distribution, the ecologic controls such as temperature, light, and salinity are, of course, the most important. Another essential requirement for reef development is the availability of a suitable platform for growth. A third consideration that is of major importance in the development of some barriers and many atolls is subsidence. This subsidence must be slow and it may be of long duration.

Temperature is recognized as a major ecologic control in reef distribution. In general, reef growth requires a surface temperature of 70°F (20°C). This relation is clearly shown on world reef maps showing isotherms (Wells, 1957, Pl. 9). Also shown on the map are the bends in the isotherms relative to the continents. On the western sides of the continents, cold waters upwell from below, restricting the zones of potential reef growth. Combined with this is the comparative rarity of suitable bases for reef growth off the western tropical shores of Africa and the Americas. Most reefs are found in the tropics, but there are a few exceptional localities. Midway and Kure, at the northwestern end of the Hawaiian chain, for example, are well-developed algal-coral atolls north of 28° north latitude. Bermuda in the Atlantic, lying more than 32° north of the equator supports marginal reef growth at the present time (Stanley and Swift, 1967). In the western Pacific temperature conditions permit the growth of reef corals in Tokyo Bay at a latitude of more than 35° north, but no reefs are developed (Wells, 1954).

III. Major Types—Criteria for Recognition

Various criteria have been used in the classification of reefs—morphology, evolution, size, depth of surrounding water, and relationship to nearby land. Of these, the most satisfactory as far as the major types are concerned is morphology. In most places fringing reefs can be separated from barrier reefs and atoll reefs are usually easily distinguished from the other major types. Along some coasts, however, lagoons are narrowed progressively and the offshore barrier becomes a fringing structure. In other areas, reefs that morphologically resemble atolls are found to have one or more volcanic islets in their lagoons; such islets, by definition, disqualify the reefs from the atoll category.

The recognition of three major morphologic types does not mean that all reefs are stages in an evolutionary sequence though, theoretically at least, it is possible for one major type to evolve into another. Reef development is dependent upon suitable ecologic conditions. As Wells states, ". . . organic-reef communities build reefs whose shape and situation are guided by hydrologic, meteorologic, and pre-existent geomorphic controls." The ecologic niches and organic associations of one major reef type do not differ significantly from those found in another. Cross sections of atoll reefs may be compared directly with equivalent sections of barrier or fringing reefs (Wells, 1957, p. 625).

Using morphology alone—size, shape, and relation to nearby land, if any—it is possible, in most cases, to identify reef types. Thus, fringing reefs are commonly developed along the shores of volcanic islands or of islands composed of older, now elevated, limestones. Barrier reefs may occur off volcanic or limestone coasts, or off steep or flat lying coasts, and the intervening lagoon may be wide or narrow.

A. Fringing Reefs (Shore Reefs)

Fringing reefs develop near shore in all well-lighted marine waters wherever temperature and oxygen supply are favorable and where there is some sort of firm bottom. Ideal sites in the tropics are the rocky shores of young volcanic islands and the shores of high islands composed of other types of hard rock. To these areas the sea brings oxygen, nutrient salts, and some food and there is a minimum of fresh water that would reduce normal salinity and bring in undesirable terriginous debris. The fringing reef will flourish in a narrow zone if the submarine slope is steep, and in a wider zone if the shallows near the shore are part of a submerged terrace or platform. Actually, most tropical shores are bordered by an intertidal platform of some kind and this platform may be partly or almost completely veneered with growing coral and other reef organisms. In either case, the structure qualifies as a fringing reef. Similar reefs are also commonly developed along the shores of lagoons.

Drill holes and other types of excavations close to shore on existing fringing reefs and examinations of elevated fringing reefs have indicated that most shore reefs are thin, in many instances they are merely veneers of reef organisms on a platform of nonreef origin. A fringing reef, however, may acquire appreciable thickness if it is developed along a coast that is slowly subsiding relative to sea level. Examples have recently been described from the island of Oahu in Hawaii. Lum and Stearns (1970) described a reef and associated limestones more than 100 feet in thickness in drill holes in the Wainimalo area near the windward coast; another drill hole, near Pearl Harbor on the south coast, passed through more than 1,000 feet of reef limestone and lagoonal sediments (Resig, 1969).

The longest stretch of reef in the world is the fringing reef that adjoins the shores of the Red Sea. This shore reef if straightened into a single line would exceed 2500 miles. One reason for the health of this long reef is the aridity of the climate. There are no streams to bring in terrigenous sediment or damaging floods of fresh water.

Though fringing reefs are best developed on rocky costs, they also may appear as discontinuous patches on deltas and other depositional flats where conditions are less favorable. Many types of reef-building coral do not thrive in the muddy waters that overlie depositional flats but other corals, such as certain species of *Porites,* grow in turbid waters. For example, waters covering the reef flats at points near the mouth of the Mba River on the island of Viti Levu in Fiji are often heavily laden with clay and silt. At certain times the waters become so murky that visibility of the bottom is limited to 3 feet—yet living colonies of *Porites* cover much of the flat's surface, forming a discontinuous fringing or shore reef. This may be a somewhat unusual situation. Squires (1962) reported briefly on coral growth near the mouth of the still larger Rewa River on Viti Levu. He noted the limits imposed on coral growth by the presence of suspended sediment and the fluctuations in salinity caused by the large volumes of fresh water. No fringing reef was noted but he found a faunal gradient that developed as the effectiveness of sediment and fresh water decreased seaward.

B. BARRIER REEFS

Barrier reefs are linear structures separated from the land by a lagoon. Most barriers rise from a platform or a terrace. In many places the reef lies close to the edge of the platform while in others it is well back from the edge, the platform continuing seaward as an off-reef terrace. Barriers that surround islands are annular and the size of the encircled islands varies greatly. In cases where the nonreef island is exceedingly small the barrier is, in all but name, an atoll. The distance from the barrier to the shore is likewise a variable feature. As the lagoon narrows, the barrier may become a fringing reef. Double barriers are known, the two elongate reefs being separated by a second lagoon.

The Great Barrier Reef off the Queensland coast of Australia is the largest aggregation of reefs in the world. It stretches from Torres Strait in the north to the Swain Reefs in the south, a distance of more than 1200 miles. It varies in width from 10 to nearly 200 miles and covers an area of more than $80,000^2$ miles. Though often referred to in the singular it is actually composed of many reefs of varying size and shape. Between the surface reefs are wide areas of blue lagoonal waters. Observable differences in morphology coupled with information obtained by drilling clearly indicate that the area has had a long and complex geological history. In 1968 W. G. H. Maxwell published an Atlas of the Great Barrier Reef, a volume that contains numerous detailed maps and photographs. Many of the photographs were taken from the air, others from underwater.

Fairbridge (1950, p. 338) referred to the well-developed zone of *ribbon reefs* that festoon the seaward edge of parts of the platform that supports the Great Barrier. The reefs do indeed resemble ribbons when seen from the air or on a map. Actually, the reef that forms the rampart is 1000–1500 feet in width and the broad scallops in the ribbon are 2–15 miles in length. As noted by Fairbridge, and confirmed by the author's observations, the reefs at the end of each scallop curve into the lagoon like the horns of a sand spit. The ribbon reefs of the Great Barrier have apparently never been closely examined on the ground. Many scallops show two distinct margins, one below the surf line, the other above. Individual scallops are separated by wide passes.

Second only to the Great Barrier is the reef that lies off New Caledonia. As a barrier, the reef parallels the northeast coast and extends northwestward beyond the islands for a total distance of 400 miles. The enclosed lagoon varies in width from 1 to 8 miles; its depth ranges up to 50 fm but there are shoals and reef patches. At its northwest end the reef reverses, forming a broad loop and continuing to the southeast to form a barrier off the southwest coast.

A third barrier that is comparable to the two already mentioned is the Great Sea Reef lying north of the two main islands of Fiji, Viti Levu and Vanua Levu. The reef runs roughly east-west for a distance of 165 miles. Its western end partly encloses the historically famous Bligh Water, the main reef extending from near the Yasawa Islands eastward to near the

eastern tip of Vanua Levu where it becomes a fringe. This relatively in-
accessible and little-known structure has a maximum width of 3 miles
and off Viti Levu it lies as much as 60 miles from the shore. The enclosed
lagoon, now partly charted, contains many small reefs.

C. Atolls

Atolls are annular reefs that develop at or near the surface of the sea.
They are commonly pictured as circular or subcircular, though it has
long been recognized that a few depart widely from such a plan. Stoddart (1965) who studied the shape of 99 existing atolls demonstrated
the fundamental homogeneity of shape.

The exact outline of the ring that encircles a lagoon without pro-
jecting land may be influenced by the shape of the foundation on which
the reef rests, but the annular character of the reef is explained primarily
by the ecologically favorable positions of the builders that settle near the
margins of the submarine foundation. Factors controlling growth of the
reef organisms include water depth, agitation, oxygen, and supplies of
food and nutrient salts.

For purposes of description, atolls may be divided into two groups:
those that rise from the deep sea and those that are found on the continental shelf. In most cases the differences between the two types are
essentially those recognized by Patrick Marshall (1931) when he divided
coral reefs into two classes—rough-water and calm-water types.

1. Deep-Sea Atolls

Deep-sea atolls are isolated structures and they vary considerably in
size. Small rings, usually without reef islets, may be less than a mile in
diameter, but many atolls have a maximum diameter exceeding 20 miles
and bear a dozen or more islets. Kwajalein in the Marshall Islands in
the Pacific and Suvadiva in the Maldives in the Indian Ocean are the
largest existing atolls, each covering an area of more than 700^2 miles.

The major factor controlling the distribution of atolls in the deep
oceans seems to be submarine vulcanism, a process that provides the
numerous isolated cones that rise two miles or more above the ocean
floor and that individually or collectively may subside at a slow rate,
permitting reef upgrowth. Theoretically, at least, atolls can develop
without the subsidence that permits thick accumulations of reef rock but
in all cases where deep sea atolls have been tested by the drill (Fig. 1)
subsidence has occurred. All such tests have been made in the Pacific.

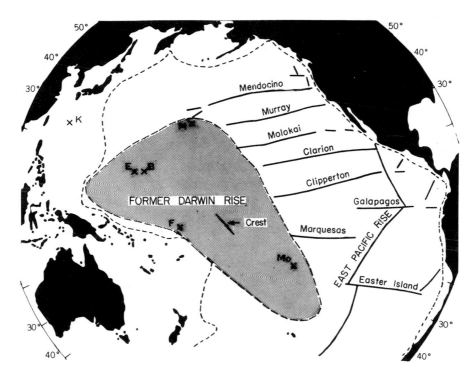

Fig. 1. Map of the Pacific Ocean basin showing major tectonic features, the former Darwin Rise and sites of deep drill holes on atolls. Explanation: ———— crest of rise; – – – approximate boundary of rise; ——·—— Andesite line marking structural boundary of basin. Deep drilling sites: B, Bikini; E, Eniwetok; F, Funafuti; K, Kita-daito-jima; M, Midway; Mo, Mururoa. Modified after Menard (1964, with permission of McGraw-Hill Book Company), Menard (1967), and Ladd *et al.* (1970).

No Indian Ocean atoll has yet been drilled. Deep water atolls, comparable to those of the Pacific, occur in the Caribbean (Milliman, 1969) but only shelf atolls have been drilled in that area (Fosberg, 1969).

As previously mentioned, the general distribution of atolls was indicated on Darwin's 1842 map. In 1953 Bryan issued a check list of about 400 atolls and in 1958 Cloud published a map showing the distribution of some 330 known and reported atolls around the world. The Pacific part of Cloud's map was republished by Menard and Ladd in 1963 (Fig. 7) and another map in the same paper (Fig. 2) showed atolls outside the Pacific.

The distribution of atolls in the Pacific was given a credible explanation by Menard in 1964 when he outlined a broad area in the central and southwestern part of that ocean and recognized it as a rise that had subsided, a vast bulge of the earth's mantle measuring 6000×2500 miles. The structure was thought to have existed in late Mesozoic time and to have subsided during parts of the Cenozoic. Hundreds of volcanoes, many closely spaced over parts of the area were, in some instances, arranged linearly and ridges were built by overlapping extrusions. Many of the volcanoes formed islands and, when eruptive action ceased, were cut to platforms just below sea level. When the postulated rise began to subside, many of the truncated islands formed guyots, and many others served as the foundations of atolls. Menard appropriately named this area, where a large percentage of the world's atolls are found, the Darwin Rise. It is shown in Fig. 1.

The explanation summarized above is plausible as an overall explanation of the distribution of Pacific atolls. However, as Menard himself recognized, the postulated subsidence was not at a uniform rate in all parts of the area nor was it continuous throughout the Cenozoic. Geophysical operations have indicated that the sedimentary sections beneath atolls vary considerably in thickness and this has been confirmed by deep drilling. Such drilling also has shown that long periods of slow subsidence were interrupted by periods when atolls stood hundreds of feet above the sea. During these emergent periods the reef rock was leached and recrystallized under atmospheric conditions. The emergent periods were long enough, in some instances, to permit the establishment of a land flora and settlement by high island land snails (see Chapter 3, Volume I, Geology 1).

Some who studied coral reefs believed that the atoll was an unique type developed in existing oceans under the peculiar conditions foisted upon the world by Pleistocene glaciation. Such a view is not widely held today. Pleistocene changes in sea level did cause extensive modifications in the upper levels of existing reefs, but some of today's atolls were in existence as reefs long before the Pleistocene. These atolls and other types of ancient reefs will be described in other parts of the present work. The major types of reefs here recognized appear to have been in existence in Paleozoic times. The builders have changed radically but the reef forms are little altered.

Most atolls are found within the Trade Wind belts. These prevailing winds and the waves and currents engendered by them are primarily responsible for the changes that may be noted in the atoll reef as it is followed from the windward edge around the ring to leeward. Grooves and buttresses form the "toothed edge" of the windward side. The chief cause of grooves and buttresses—be it growth or erosion—is favored by exposure to strong and persistent surf. The grooved zone is usually fol-

lowed, in the Pacific at least, by an algal ridge. This ridge is a broad elevation and along most of the windward side of the atoll it is inconspicuous even at low tide. Where islands are present on the reef flat,
however, they interfere with cross reef circulation, the water level is
raised locally, and the marginal algal ridge becomes a cuesta whose
steeper slope faces the waves. This structure may be very conspicuous
at times of low tide on atolls with high tidal ranges. Zonation that parallels the reef front is better developed windward than leeward and the
reefs that border the lagoon of the atoll have their own characteristic
features. Reefs on the lee sides of large atoll lagoons face waves that
have a fetch of 20 miles or more. These reefs take on some of the characteristics of outside windward reefs.

In the North Pacific, three deep-sea atolls—Kure, Midway, and Pearl
and Hermes at the northwestern end of the Hawaiian chain—lie outside
the tropics (Gross *et al.*, 1969). In the South Pacific a half dozen are
shown on Cloud's map (1958). None of these has been studied in detail
but in 1962 17 Tongans were marooned for 14 weeks on one of them, the
Minerva Reef south of Fiji, before most of them were dramatically res-
cued (Ruhen, 1963).

There are a fair number of atolls in the southwest Pacific that lie outside the boundaries of the Darwin Rise in Fiji, New Caledonia, the
Solomon and Admiralty Islands, the islands of Indonesia, and the South
China Sea. The Mbukatatanoa or Argo Reefs in eastern Fiji may be cited
as examples. The reefs enclose two irregularly shaped lagoons totaling
20 miles in length. Included in the smaller northern lagoon are two
points of rock called Bacon's Isles. One of these islands is basaltic. To
those who favor genetic terms this is an "almost atoll." The reefs and
the enclosed lagoons have not been closely examined but, after studying
some of the smaller but equally irregular reefs that encircle islands
nearby (Oneata, Aiwa, Lakemba) (Ladd *et al.*, 1945), I doubt whether
the Argo reefs have passed through earlier fringing and barrier stages
and that any great thickness of limestones would be encountered by a
drill hole in the lagoon area.

Atolls are not as numerous in the Indian Ocean as they are in the
Pacific and their distribution appears erratic. Recently, however, the configuration of the floor of the Indian Ocean has become better known
(Heezen and Tharp, 1964, 1967) and it appears that atolls have developed on a variety of foundations. The peculiar distribution pattern is
now more understandable. The greatest concentration of atolls form a
north-south string that starts in the Chagos to the south and, after a wide
gap, continues through the Maldives and the Laccadives. These atolls

rise from a submarine plateau that lies to the west of a north-south trench, the Chagos Trench. Other well-known atolls—the first closely examined by Darwin—lie far to the east. This lesser concentration, the Cocos-Keeling atolls, arises from a poorly aligned series of submarine mountains. Other deep sea atolls are found near the western border of the Indian Ocean, in the Seychelles, and in the Amirante Islands with one or two others east of Madagascar. This is an area of complicated geologic structure.

Caribbean atolls, as Milliman pointed out (1969), have been considered "pale images" of their counterparts in the Indo-Pacific region. He believed the comparison to be a fair one as regards the reefs of the northern Caribbean, but from the southwestern part he described four atolls rising from depths in excess of 1000 m that can be favorably compared to Indo-Pacific structures. The four Caribbean atolls lack a true algal ridge but there is a *Millepora* zone to windward that is emergent at low tide and is, in effect, the ecologic equivalent of the algal ridge.

2. Shelf Atolls

Shelf atolls are found in many parts of the world. Near Australia they rise from the Great Barrier platform on the northeast and from the continental shelf off the northwest coast (Teichert and Fairbridge, 1948). Others are found in Indonesia, the Caribbean Sea, and the Gulf of Mexico. Some small atolls that rise from the slopes of larger island pedestals also belong in this category.

As MacNeil pointed out (1954a, p. 396) special names have been proposed for shelf atolls to separate them from deep sea atolls. Davis (1928, p. 19) called all reefs lying back from the edge of the continental shelf "bank reefs" and, depending on the presence or absence of islands, he named them "bank barriers" and "bank atolls." Small atolls located within larger lagoons were called "lagoon atolls" (Davis, 1928, p. 15).

IV. Minor Types

Many terms have been applied to the almost bewildering variety of small reefs that are associated with or form a part of one of the three major types already described. Many of these have been referred to recently by Stoddart in a brief review of reef terminology (1969, pp. 452–453).

A. Table Reefs

This term was proposed by Tayama (1935) for small isolated reefs that rise from the deep sea to intertidal levels but do not possess a lagoon. He listed 16 examples of sea level reefs of this type in Palau, and the Caroline and Marshall Islands. Tayama also mentioned islands that he regarded as elevated table reefs, noting that submarine examples also occurred in the same area.

B. The Faro

The faro, or atollon, was described by Gardiner (1902, p. 282) as a small ring-shaped basin with small lagoon. Numerous faros were found on the reefs of larger atolls in the Indian Ocean (Gardiner, 1902, pp. 289, 290, and 295). Similar reefs have since been located in other areas, some occurring on the rims of barriers as well as on atoll reefs. The well-defined lagoons of faros rarely exceed 10 fm in depth.

Newell and Rigby (1957, p. 36) briefly described Bahaman structures that appeared similar to faros. These were small patch reefs, 10–200 m across with dark rings or fringes of gorgonians upon a rim of stony corals. The central area was characteristically more or less dead, slightly depressed, and covered with coral fragments.

Faros have been described from the barrier of Mayotte in the Indian Ocean and from the barrier and the lagoon of New Caledonia in the Pacific (Guilcher, 1965). Some of the crescentic reefs and atolls of the Great Barrier also appear to be faros. In 1968, during a visit to Heron Island on the Great Barrier, Peter Woodhead, then Director of the Research Station, showed me numerous farolike coral rings 25 ft or more in diameter that rise abruptly 10–15 ft above the sand floor of Heron's lagoon. Small heads of a variety of living corals were seen but the large masses making up the doughnutlike structures were dead. These unusual reefs are numerous and appear clearly on enlarged aerial photographs of the lagoon.

C. Microatolls

From small farolike structures it is but another step downward to the *microatoll*. These are subcircular masses of one or more living corals. They measure 3–25 ft or more in diameter and develop on reef flats whose general surface lies below low tide. As in true atolls, there is a concentration of living forms around the periphery, although organisms, including algae, may grow sporadically over inner parts of the structure. The upward growth of microatolls is limited by low-tide level but they appear to expand laterally and, as they coalesce, may form a new reef surface (Ladd *et al.*, 1950, p. 415).

Hoskin describing the Alacran Reef in the Gulf of Mexico (1963, p. 27) noted that typical microatolls range in size from single heads to very large and complex structures. He called the largest ones patch reefs and noted that the patch reefs were usually found in deeper water than the smaller microatolls.

D. KNOLLS

Isolated mounds covered by living coral are numerous in the lagoons of both atolls and barrier reefs and are commonly referred to as *coral knolls,* a term used by Darwin (1842, pp. 41 and 91–95). The knolls may be very small—less than 10–50 ft across and 3–20 ft high—but larger ones are numerous, some from several hundred ft to more than a mile in width at the base, rising nearly to the sruface of the water from even the deepest parts of the lagoon. The 180,000 soundings made in Eniwe-tok's 24-mile lagoon revealed more than 2000 coral knolls. The tops of a number of knolls that have been examined by divers are somewhat ir-regular and are largely covered by a rich growth of living corals. Slopes vary, but are generally less than 45° and appear to be covered by sand and coarser debris (Ladd *et al.*, 1950, pp. 421–423).

MacNeil suggested (1954b) that knolls were basically erosion rem-nants formed when the sea stood at a lower level. This may be the ex-planation for the location of many knolls, but others may be due entirely to growth. No such structure has yet been drilled, but Nesteroff (1955a, p. 16) used TNT to remove the tops of knolls and found the radial struc-ture of coral growth. He attributed all knolls in the Red Sea area to dif-ferential reef growth (1955b).

The terms *pinnacle* and *coral pinnacle* have long been applied to areas of coral growth in a lagoon. There are spire-shaped masses of coral in some lagoons, masses that rise vertically or even overhang their bases. They are composed of one or more colonies of living coral and the term pinnacle is quite properly applied to them. Many lagoon structures that resemble pinnacles in cross sections drawn with highly exaggerated ver-tical scale have, in reality, more gentle slopes and should be called knolls.

E. PATCH REEFS

Small sea level reefs that rise from submarine shelves or from the floors of lagoons are loosely grouped as *patch reefs*. In the Great Barrier

area Hill (1960, p. 412) referred to patch reefs inside the barrier, point-
ing out that many of them have the shape of a horse's hoof with the
convexity facing the prevailing trade winds. The windward edge may
dry at low tide. There may be an islet made of bare sand or the islet
may be wooded. Some islets are partly protected by the development of
layers of beach rock. To leeward the lagoon is shallow (green water)
and bears a characteristic mottled pattern of coral growth. In deeper
areas of the lagoon and in the area beyond its margin there are large
coral heads or combinations of heads, known locally as "bommies."
There are, however, many variations in the basic outline of such cuspate
patch reefs; in one direction the variations lead to the shelf atoll (Fairbridge, 1950).

The patch reefs (shelf reefs) of the Great Barrier were studied in
some detail by Maxwell in compiling his comprehensive Atlas of the
Great Barrier Reef (1968). In a later paper, coauthored with J. P. Swinchatt (1970), the development of reef types on the Queensland shelf is
discussed and ideas of development are illustrated in a diagram here
reproduced as Fig. 2 [earlier versions of this arrangement were published by Maxwell in the Atlas (Fig. 65) and in 1969 (Fig. 66)]. The
diagram represents a supposed evolutionary sequence, starting at the top
with an embryonic colony. The authors expressed the belief that variations in reef shape were controlled by differential growth in response to
bathymetric and hydrologic influences. "Where these are uniform around
the reef, symmetrical growth results. Where favorable conditions are
restricted to one side of the reef, elongation, cuspation, and closing result. Unfavorable conditions cause 'resorption.'" The Maxwell-Swinchatt
diagram (Fig. 2) shows a dozen types of shelf reefs plus the end step,
the "resorbed reef." In the diagram and in aerial photographs (Maxwell,
1968, Figs. 64, 104, and 105) the "resorbed" structures with their embayed
margins and dispersed reef segments certainly suggest reefs that have
passed their peak. Their moth-eaten appearance is not fully understood
although Maxwell and Swinchatt stated that some reefs of this type may
result from erosion of an older reef during an early low stand of the sea
and failure to respond to a later sea level rise. This seems to be a plausible
explanation for some "resorbed" reefs but it is not certain that all such
reefs had a similar history.

F. OTHER TYPES

A variety of genetic terms have been applied to coral reefs. Davis
(1928) used almost-atoll, Tayama (1952) used "almost-atoll," "almost
barrier reef," and "almost-table reef." He looked upon these as intermedi-

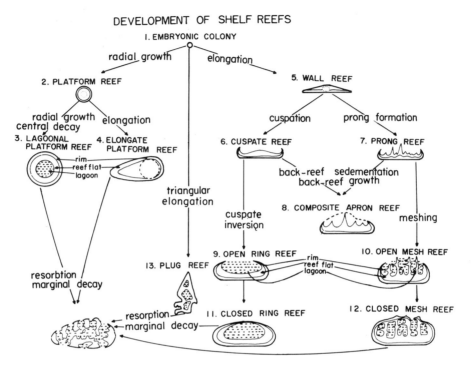

Fig. 2. Reef types of the Queensland shelf. After Maxwell and Swinchatt (1970).

ate stages: the almost-barrier a stage between fringing and barrier (the lagoon shallow and narrow), the almost-atoll a stage between barrier and atoll, and the almost-table a stage intermediate between atoll and table (the lagoon small and shallow). Tayama also used the term "apron-reef" as an initial stage of fringing reef; it is discontinuous and covers a small area. If a term is needed for an embryonic fringing reef, this one may be useful.

MacNeil (1954a, p. 395) cited the terms "near atoll" and "semiatoll" that have been used for atolls with very large reefless segments. He noted that some tables were elongated and curved, or U-shaped, approaching atolls in outline. He proposed that such reefs be known as *curved table reefs.*

Genetic terms such as those cited above may be regarded as undesirable because no one has yet demonstrated that any given reef has actually passed through all three of the classical stages—fringing–barrier–atoll.

Among the reef terms coined by W. M. Davis (1916) were the *extin-*

guished reef and the *resurgent reef.* It seems to this author that these genetic terms, like those of the *almost* series, are undersirable. Neither an extinguished nor a resurgent reef can be identified with certainty until the reef's history and that of nearby islands has been worked out. This had not been done when the terms were applied by Davis (1916) to various Fijian reefs.

Acknowledgment

I am indebted to Dr. Joshua I. Tracey, Jr., of the U.S. Geological Survey for his critical review of the present manuscript.

References

Bryan, E. H., Jr. (1953). Check list of atolls. *Atoll Res. Bull.* No. 19, pp. 1–38.

Chamisso, Adalbert von (1821). "A Voyage of Discovery into the South Sea and Bering's Straits for the Purpose of Exploring a Northeast Passage Undertaken in the years 1815–19 at the Expense of His Highness the Chancellor of the Empire, Count Romanzoff, in the Ship *Rurick,* Under the Command of the Lieutenant in the Russian Imperial Navy, Otto von Kotzebue," Weiman. Remarks and Opinions of the Naturalist, Vol. 3, pp. 331–336. Longman, Hurst, Rees, Orme, and Brown, London (Engl. transl. by H. E. Lloyd).

Cloud, P. E., Jr. (1958). Nature and origin of atolls, a progress report. *Proc. Pac. Sci. Congr., 8th, 1953* Vol. 3A, pp. 1009–1023.

Cumings, E. R. (1932). Reefs or bioherms? *Geol. Soc. Am. Bull.* 43, 331–352.

Cumings, E. R., and Shrock, R. (1928). Niagaran coral reefs of Indiana and adjacent states and their stratigraphic relations, *Geol. Soc. Am. Bull.* 39, 579–620.

Darwin, C. (1842). "On the Structure and Distribution of Coral Reefs." 278 London.

Davis, W. M. (1916). Extinguished and resurgent coral reefs. *Proc. Natl. Acad. Sci. U.S.A.* 2, 466–471.

Davis, W. M. (1928). The coral reef problem. *Spec. Publ. Am. Geogr. Soc.* 9, 1–596.

Emery, K. O., Tracey, J. I., Jr., and Ladd, H. S. (1954). Geology of Bikini and nearby atolls. *U.S., Geol. Surv., Prof. Pap.* 260-A, 1–265.

Fairbridge, R. W. (1950). Recent and Pleistocene coral reefs of Australia. *J. Geol.* 58, No. 4, 330–401.

Fosberg, F. R. (1969). Island news and comment. *Atoll Res. Bull.* No. 126, p. 5.

Gardiner, J. S. (1902). The formation of the Maldives. *Geogr. J.* 19, 277–301.

Gross, M. G., Milliman, J. D., Tracey, J. I., Jr., and Ladd, H. S. (1969). Marine geology of Kure and Midway Atolls, Hawaii: A preliminary report. *Pac. Sci.* 23, No. 1, 17–23.

Guilcher, A. (1965). Coral reefs and lagoons of Mayotte Island, Comoro Archipelago, Indian Ocean, and of New Caledonia, Pacific Ocean. *Proc. Symp. Colston Res. Soc.* 17, 21–44.

Heezen, B. C., and Tharp, M. (1964). Physiographic diagram of the Indian Ocean, with descriptive sheet. *Geol. Soc. Am.*

Heezen, B. C., and Tharp, M. (1967). Indian Ocean floor. Natl. Geogr. Soc. bathymetric map.

Henson, F. R. S. (1950). Cretaceous and Tertiary reef formations and associated sediments in Middle East. *Bull. Am. Assoc. Pet. Geol.* **34**, No. 2, 215–238.

Hill, D. (1960). The Great Barrier Reefs. *J. Geol. Soc. Aust.* **7**, 412–413.

Hoskin, C. M. (1963). Recent carbonate sedimentation on Alacran Reef, Yucatan, Mexico. *N.A.S.—N.R.C., Publ.* **1089**, 1–160.

Joubin, L. (1912). Bancs et récifs de coraux (madrépores). *Ann. Inst. Oceanogr. (Paris)* **4**.

Ladd, H. S. (1950). Recent reefs. *Bull. Am. Assoc. Pet. Geol.* **34**, No. 2, 203–214.

Ladd, H. S., Hoffmeister, J. E., Alling, H. L., Crickmay, G. W., Sanders, J. W., Jr., Cole, W. S., Clark, H. L., Pilsbry, H. A., and Rathbun, M. J. (1945). Geology of Lau, Fiji. *Bull. Bishop Mus., Honolulu* **181**, 1–399.

Ladd, H. S., Tracey, J. I., Jr., Wells, J. W., and Emery, K. O. (1950). Organic growth and sedimentation on an atoll. *J. Geol.* **58**, 410–425.

Ladd, H. S., Tracey, J. I., Jr., and Gross, M. G. (1970). Deep drilling on Midway Atoll. *U.S., Geol. Surv., Prof. Pap.* **680-A**, 1–22.

Lum, D., and Stearns, H. T. (1970). Pleistocene stratigraphy and eustatic history based on cores at Waimanalo, Oahu, Hawaii. *Geol. Soc. Am. Bull.* **81**, 1–16.

MacNeil, F. S. (1954a). Organic reefs and banks and associated detrital sediments. *Am. J. Sci.* **252**, 385–401.

MacNeil, F. S. (1954b). The shapes of atolls: An inheritance from subaerial forms. *Am. J. Sci.* **252**, 402–427.

Marshall, P. (1931). Coral reefs—rough-water and calm-water types. *Rep. Great Barrier Reef Comm.* **3**, 64–72.

Maxwell, W. G. H. (1968). "Atlas of the Great Barrier Reef." Am. Elsevier, New York.

Maxwell, W. G. H. (1969). The structure and development of the Great Barrier Reef. *In* "Stratigraphy and Paleontology, Essays in honour of Dorothy Hill" (K. S. W. Campbell, ed.), pp. 353–374. Australian National University, Canberra.

Maxwell, W. G. H., and Swinchatt, J. P. (1970). Great Barrier Reef: Regional variation in a terrigenous-carbonate province. *Geol. Soc. Am. Bull.* **81**, 691–724.

Menard, H. W. (1964). "Marine Geology of the Pacific." McGraw-Hill, New York.

Menard, H. W. (1967). Extension of northeastern-Pacific fracture zones. *Science* **155**, 72–74.

Menard, H. W., and Ladd, H. S. (1963). Oceanic islands, seamounts, guyots and atolls. "The Sea," Vol. 3, pp. 365–387. Wiley, New York.

Milliman, J. D. (1969). Four southwestern Caribbean atolls: Courtown Cays, Albuquerque Cays, Roncador Bank and Serrana Bank. *Atoll Res. Bull.* No. 129, pp. 1–26.

Molengraaff, G. A. F. (1930). The coral reefs in the East Indian Archipelago, their distribution and mode of development. *Proc. Pac. Sci. Congr. 4th, 1929* Vol. IIA, pp. 55–89; Vol. IIB, pp. 989–1021.

Nesteroff, V. D. (1955a). Les Récifs coralliens du Banc Farsan Nord (Mer Rouge). *Result. Sci. Camp. "Calypso" Ann. Inst. Oceanog.* (Monaco), **30**, 7–53.

Nesteroff, V. D. (1955b). Quelques résultats géologiques de la campagne de la "Calypso" en Mer Rouge (1951–1952). *Deep-Sea Res.* **2**,, 274–283.

Newell, N. D., and Rigby, J. K. (1957). Geological studies of the Great Bahama Bank. *Soc. Econ. Paleontol. Mineral., Spec. Publ.* **5**, 15–79.

Resig, J. M. (1969). Paleontological investigations of deep borings on the Ewa Plain, Oahu, Hawaii. *Hawaii Inst. Geophys.* **HIG-69-2**, 1–99.

Revelle, R. (1954). Geology of Bikini and nearby atolls. *U.S., Geol. Surv., Prof. Pap.* **260-A,** Foreword, i-vii.

Ruhen, O. (1963). "Minerva Reef." Little, Brown, Boston, Massachusetts.

Schott, G. (1935). "Geographie des Indischen und Stillen Ozeans." C. Boysen, Hamburg.

Squires, D. F. (1962). Corals at the mouth of the Rewa River, Viti Levu, Fiji. *Nature (London)* **195,** 361–362.

Stanley, D. J., and Swift, D. J. P. (1967). Bermuda's southern aeolianite reef tract. *Science* **157,** 677–681.

Stoddart, D. R. (1965). The shape of atolls. *Mar. Geol.* **3,** 369–383.

Stoddart, D. R. (1969). Ecology and morphology of Recent coral reefs. *Biol. Rev. Cambridge Philos. Soc.* **44,** 433–498.

Tayama, R. (1935). Table reefs, a particular type of coral reef. *Proc. Imp. Acad. (Tokyo)* **11,** No. 7, 268–270.

Tayama, R. (1952). "Coral reefs in the South Seas," Bull. Hydrogr. Off., Vol. 2 (translation prepared by Engineer Intelligence Div., U.S. Army, 1955).

Teichert, C., and Fairbridge, R. W. (1948). Some coral reefs of the Sahul Shelf. *Geogr. Rev.* **38,** No. 2, 222–249.

Vaughan, T. W. (1911). Physical conditions under which Paleozoic coral reefs were formed. *Geol. Soc. Am. Bull.* **22,** 238–252.

Vaughan, T. W. (1919). Fossil corals from Central America, Cuba, and Porto Rico, with an account of the American Tertiary, Pleistocene, and Recent coral reefs. *U.S., Natl. Mus. Bull.* **103,** 1–524.

Wells, J. W. (1954). Recent corals of the Marshall Islands. *U.S., Geol. Surv., Prof. Pap.* **260-I,** 385–486.

Wells, J. W. (1957). Coral reefs. *Treatise Mar. Ecol. Paleoecol.* **1,** Mem. **67,** Geol. Soc. Am. 609–631.

2

THE ORIGIN OF FRINGING REEFS, BARRIER REEFS, AND ATOLLS

J. A. Steers and D. R. Stoddart

I. Introduction

Since the first scientific exploration of Pacific Ocean atolls in the eighteenth century, reef workers have theorized on the origins, foundations and history of these and other reef formations. In the absence of data on reef foundations and bathymetry, much early speculation was hypothetical and deductive. Theories were proposed by both geologists and biologists, and each group from time to time denied the other's competence to understand the true nature of the problem. Historical accounts of early theories have been given by Böttger (1890) and Günther (1911), and later nineteenth century ideas have been exhaustively reviewed by Langenbeck (1890) and by W. M. Davis (1928) in "The Coral Reef Problem."

Of the general theories proposed, the two of greatest generality and explanatory power are Charles Darwin's Subsidence Theory and R. A. Daly's Theory of Glacial Control. Advocates of these theories have engaged in a long and at times acrimonious debate, in which Davis played a major part, but which we can now recognize to have been largely based on false premises. It must be emphasized that Darwin's was essentially a theory of the *structure and origins* of coral reefs, while Daly's sought to explain particular features of *surface morphology*. Most of the discussion of "the coral reef problem" has been clouded by the failure to understand this distinction, a failure illustrated by the claims frequently made that the two theories were mutually exclusive in their basic propositions. Once the distinction is recognized, it is not surprising that the search for evidence of subsidence in minor and ephemeral features of modern reefs has been fruitless, or that Daly's theory, while partially successful, failed to explain the deep foundations of atolls. Much confusion stemmed from an inadequate understanding of the length and complexity of the Pleistocene glaciations and of their effects in the reef seas, and it is now clear that no reef theory can be complete without taking these into account. However, it is also apparent that many reefs are of such magnitude and antiquity that Pleistocene events alone cover only part of their history. It is now possible to outline a unified reef theory, based on reef studies over the last 40 years. It is interesting to note that the last major deductive treatment of reefs, in Davis's "Coral Reef Problem," appeared in the same year that the Great Barrier Reef Expedition of 1928–1929 initiated a new phase in the careful field study of living coral reefs, an approach pioneered by the Royal Society Funafuti investigations of 1896–1898. After 1928 the "home study of coral reefs" (Davis, 1914) ceased to be a useful exercise.

In this chapter three main theories of reef development are reviewed: the Subsidence, Glacial Control, and Antecedent Platform Theories. The implications of each, the problems inherent in their original formulation, and the ways in which they must be modified to take account of recent work are briefly discussed. Finally, the structural and morphological evidence that a unified theory must explain is considered, and an outline of this more comprehensive view is presented.

II. Subsidence Theory

A. DARWIN'S SUBSIDENCE THEORY

Darwin was impressed during the voyage of the *Beagle* by geological evidence of uplift in the South American Andes and by his personal involvement in the Chilean earthquake of February 1835. He hypothesized that such uplift must necessarily be balanced by a corresponding crustal subsidence, and before leaving South America he formed the view that one result of such subsidence would be the transformation of oceanic volcanic islands into atolls (Fig. 1). From South America he sailed through the Tuamotus to Tahiti, where from the high volcanic slopes he viewed the neighboring reef-encircled island of Moorea, comparing the island, lagoon and reef to an engraving in a frame. He recognized that if the mountain were removed an atoll would remain, and subsidence provided the mechanism:

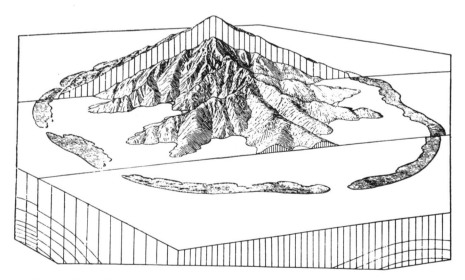

Fig. 1. Block diagram showing the formation of fringing, barrier, and atoll reefs according to Darwin's Subsidence Theory. From Davis (1928).

. . . if we imagine such an Island, after long successive intervals to subside
a few feet, in a manner similar, but with a movement opposite to the continent of S. America; the coral would be continued upwards, rising from the
foundation of the encircling reef. In time the central land would sink beneath the level of the sea to disappear, but the coral would have completed
its circular wall. Should we not then have a lagoon Island [atoll]?—Under
this view we must look at a lagoon Island as a monument raised by myriads
of tiny architects, to mark the spot where a former land lies buried in the
depths of the ocean (Darwin, 1933, p. 400).

Darwin wrote a full account of his theory in 1835 (published 1962),
first announced it in a paper on regional patterns of uplift and subsidence
(1837), expanded it in his "Journal of Researches" (1839), and gave it
definitive treatment in "The Structure and Distribution of Coral Reefs"
(1842).

The main problem for which Darwin's theory provided an explanation
was that, whereas open-ocean atolls rose abruptly from the deep ocean
floor, it had been shown by Quoy and Gaimard (1824) that reef-building
corals were restricted in their growth to near-surface waters (mostly less
than 20 m). Therefore, corals could not build up reefs from the ocean
floor, but it was unreasonable to assume, as did Lyell (1832, pp. 283–
301), that they were growing on the submerged rims of volcanic craters
occurring at suitable depths. Such widespread accordance of crater rims
would be improbable, while many atolls are much larger than any known
terrestrial volcanic craters. Darwin avoided the difficulty by supposing
that fringing reefs begin growing on the shores of high volcanic islands,
that as the island subsides the fringing reefs grow upwards to form
barrier reefs, and that when the central island finally disappears the encircling barrier reef remains as an atoll.

The simplicity and generality of the theory are remarkable, and Darwin
himself at the end of his life was able "to reflect with high satisfaction"
on "solving the problem of coral islands" (1958, p. 80). The theory was
also flexible. Darwin recognized that subsidence could occur "either a few
feet at a time or quite insensibly," and that there might be "intervening
stationary periods" (1842, p. 99); the slower the subsidence or the longer
the stillstand, the wider the present reefs and the greater the area of
land developed on them. Subsidence, in fact, was a "renovating agency,"
determining the existing form of reefs: without it lagoons would fill, reefs
widen, and featureless reef banks would ultimately form (1842, p. 31).

Darwin recognized that this sequence of development was not in-
evitable. Subsidence could be too rapid for reef growth to maintain the
reef at sea level, and the reef would thus be drowned. Reefs might also
be formed, as Lyell supposed, on the rims of submarine craters, or on

other submarine banks at suitable depths (Darwin's letters to C. Semper, 2 October 1879, and A. Agassiz, 5 May 1881), in which case atolls could be formed without going through an earlier barrier-reef stage (1842, p. 101). But these were special cases lacking the explanatory and predictive power of the Subsidence Theory.

The theory was proposed primarily to account for the features of open-ocean atolls and reef-encircled oceanic volcanic islands. Darwin extended it to include the barrier reefs of continental shores, which he considered to differ in no material way from open-ocean reefs. It was adopted by workers such as Jukes (1847) to explain features as large as the Great Barrier Reefs of Australia. Support was also provided by Dana's work during the Wilkes Expedition of 1838–1842, and by his book "On Coral Reefs and Islands" (1853; later work 1872, 1885).

B. Implications of the Subsidence Theory

Some of the implications of the Subsidence Theory were recognized by Darwin, others by later workers. The major, but then untestable, inference is that atolls and to a lesser extent barrier reefs are underlaid by great thicknesses of shallow water reef limestones, overlying nonlimestone rocks of the subsiding foundation (Darwin, 1842, pp. 48–49). Darwin knew that deep borings could provide a crucial test of his theory, but all early attempts were unsuccessful. The evidence of successful borings is considered in Section V of this chapter.

A less direct but more immediately apparent inference concerns the spatial distribution of different reef types in island groups of different ages. Thus within the Society Islands there is a progression from the fringing and poorly developed barrier reefs of Tahiti in the southeast, through the barrier-encircling dissected volcanic islands of Moorea, Raiatea, Huaheine, and Borabora, to the northwestern atolls of Mopelia and Manuae. A similar trend is seen in the Hawaiian Islands, and to a lesser extent in other Pacific groups (Chubb, 1957). In the mobile continental marginal areas of Melanesia and Micronesia many cases are known of transitions between fringing, barrier, and atoll reefs consequent on tectonic deformation (for example, the eastern Carolines and the Palau Islands: Tayama, 1952). Darwin himself (1842, pp. 119–148) went to much trouble in compiling his map of reef-type distribution to show areas of uplift, stability, and subsidence. In general the largest and best-developed atolls (i.e., the oldest) are more remote from existing high islands and continental land, whereas atolls in the Carolines, Fiji Islands, and the East Indies are smaller and presumably younger (Cloud, 1957,

p. 1,012). Other reef types are also regionally concentrated, e.g., table reefs in the Tobi group (Tayama, 1952). Early speculations on the causes and geophysical implications of subsidence (Molengraaff, 1916) have now been superseded by a new understanding of the structure of the sea floor and of the formation, migration and disappearance of volcanoes on it (Menard, 1964, 1969; Wilson, 1963).

Davis (1928) added two further geological implications: the unconformability of contacts between reef limestones and their foundations, and the role of subsidence in disposing of detritus eroded from high islands. He believed that unconformable contacts could generally be observed in elevated reefs, which could then be shown to have subsided before they were elevated. Often, however, such reefs are so eroded that it is difficult to interpret the contacts, while Hoffmeister and Ladd (1935, p. 655) have shown that on many West Pacific islands the reef limestones are underlaid by foraminiferal and other limestones, so that the contact between basement rocks and limestone has no relevance to the history of the reef itself. There is no doubt that Davis frequently grossly oversimplified the structure and history of elevated reefs to make them fit his view of the Subsidence Theory, as is shown by his accounts of Fiji (cf. Ladd and Hoffmeister, 1945), Tonga (cf. Hoffmeister, 1932), and the southern Cook Islands and Austral Islands (cf. Marshall, 1927, 1929; Hoffmeister, 1930).

Davis's second point is more complex. He believed that on new volcanic islands the shedding of eroded sediment would prevent reef growth, and that coasts would thus be cliffed; large amounts of sediment would be carried down to deeper slopes. Once land erosion was largely completed, corals would begin to grow. He estimated that many eroded volcanic islands had lost 50–100 times as much sediment by volume as could be accommodated by their present lagoons, and 500–1,000 times as much as is represented by present deltas. In view of his underestimation of the time scale involved and his neglect of Pleistocene events, however, the point remains largely academic without detailed studies of particular islands. Rates and volumes of erosion on islands of known ages have been established with precision only in the Hawaiian Islands (Wentworth, 1928; McDougall, 1964), but here coral reefs are weakly developed and barrier reefs are absent.

More convincing evidence for subsidence is provided by morphology. First, Dana (1849) recognized that long-continued subsidence of sub-aerially-eroded landscapes would drown river valleys and produce deeply embayed shorelines. As a physiographer, Davis laid great emphasis on the significance of shoreline embayments thus produced, which he termed

"Dana's Principle" (Davis, 1913). By extension he argued that the small islets of almost-atolls would have gentle, embayed, noncliffed shores (Davis, 1920). Both Guppy (1888) and Gardiner (1898) denied any significance to embayed shorelines, while later workers argued that drowning and formation of embayments could result from the Postglacial rise of sea level (Ladd and Hoffmeister, 1936, p. 89). This overestimates the rate of formation of terrestrial valleys, however, and it cannot explain the widespread evidence of deep sediment-filled valleys, up to 300 m deep, again well documented in the Hawaiian Islands (Stearns and Vaksvik, 1938; Stearns and Chamberlain, 1967; Lum and Stearns, 1970).

Second, it was recognized early that constructional reefs could maintain seaward slopes at angles steeper than those of volcanic material (Gardiner, 1915). These steep slopes have been shown to extend to depths of some hundreds of meters in some cases, well below the limit of Pleistocene sea-level fluctuations (Kuenen, 1933, 1938), and to contrast with the mean slope of about 10° exhibited by constructional volcanic islands (Menard, 1969). It has also been shown that the deeper slopes of reef-crowned islands correspond in form with the slopes of reefless volcanic islands (Robertson and Kibblewhite, 1966).

C. Subsidence and Reef Types

Darwin's recognition of the three main stages in his developmental sequence corresponding to fringing, barrier, and atoll reefs did not exhaust the theoretical range of possible reef types (Darwin, 1842, pp. 102–3). Later workers, notably Tayama (1952), have added new intermediate and end members to complete the sequence of forms. In this section we will consider briefly the theoretical effects of subsidence on particular reef types, postponing the assessment of structural and morphological data to Sections V and VI.

Of the three main types, fringing reefs are the simplest, apparently the least in need of complex explanations, and also the least studied. One of the best known fringing reefs is Hanauma Reef, Oahu, where ^{14}C dating shows that reef growth began during the Holocene transgression at about 6,000 B.P., the present surface of the reef being reached by 3,300 B.P., with a rate of reef growth of 1 m/180 year (Easton, 1969). Darwin considered that fringing reefs are mainly characteristic of stationary or rising coasts, on the grounds that on subsiding coasts they are rapidly converted into barrier reefs. He did, however, allow that once such barrier reefs are growing, "second-generation" fringing reefs may be formed during periods of stillstand, presumably on the headlands of already embayed coasts. Young fringing reefs may be ephemeral features. For example, Stoddart has seen largely dead fringing reefs smothered by red clayey sediments around the eroding cone of Kolombangara, Solomon Islands; and the vicissitudes of fringing reefs around Krakatoa following the explosion of 1883 have been well documented by Umbgrove (1930). Because of their limited age, fringing reefs are narrow as well as thin, although, as Darwin (1842, p. 57) recognized, their dimensions are partly a function of the underlying slope. Moreover, they are often absent on coasts protected by barrier reefs, such as the mainland coasts of Queensland and British Honduras, and it is difficult to see how the present barrier reefs could have been initiated as fringing reefs on such coasts without very different conditions.

While small Holocene fringing reefs present no major problems, it remains true, as Vaughan (1916) has said, that "no evidence has as yet been presented to show that any barrier reef began to form as a fringing reef on a sloping shore, and was converted into a barrier by subsidence." One might indeed expect that with a rising sea level, fringing reefs might simply migrate retrogressively upshore, maintaining contact with the coast, rather than building vertically to form a barrier. The transitional stage between fringing and barrier reefs is perhaps the weakest in Darwin's theory.

Three facts may be adduced in favor of barrier formation by subsidence. First, the depth of some enclosed lagoons is greater than the depth in which reef corals will grow: thus one cannot envisage barriers growing vertically from a shelving sea floor, although since most lagoons are within the range of Glacial sea-level movements the argument in favor of subsidence is not wholly convincing. Second, Darwin's theory predicts great vertical thickness of barrier reefs. This can often be inferred on geometric grounds by submarine projection of land slopes, e.g., in the case of the barriers around Gatukai and Vangunu volcanoes, Solomon Islands (Stoddart, 1969b). The thickness of barrier reefs has also been confirmed by boring and by geophysical surveys (Section V). Third, the formation of barrier reefs by local tectonic movements, where subsidence is clearly demonstrated, can often be traced in such active areas as the Palau and Fiji Islands.

It does not follow that all barriers have been formed by simple subsidence, or that barriers cannot be formed in other ways. Darwin anticipated Guppy (1890) in showing how barrierlike reefs could grow at some distance from the shore in gently sloping areas, though such barriers would closely resemble fringing reefs with deep boat channels and could not be compared with the major barrier and lagoon systems. In the case of large barriers, such as the Great Barrier Reefs, it is clear that Jukes's model of simple transverse down-warping must be replaced by a more complex picture of subsidence through longitudinal faulting and warping (Steers, 1929), and this is discussed more fully in Section V. Some barriers, such as that around Tahiti (Agassiz, 1903; Crossland, 1928), have been explained by the outgrowth of a fringing reef followed by solution of the lagoon. Apart from the question of lagoon depth, such interpretations predate our understanding both of Pleistocene events and of the chemistry of calcium carbonate in sea water, and no longer require serious consideration.

Variants on the development of the simple barrier can be readily envisaged. Continued subsidence could convert second-generation fringing reefs into second barriers inside the older barriers. Such double barrier reefs are rare. They have been described from New Caledonia, Vanua Levu, and Mayotte (Guilcher, 1965; Guilcher *et al.*, 1965), but the best-developed in the world (in part a triple barrier) partially surrounds New Georgia, Solomon Islands (Stoddart, 1969b). Continued subsidence affecting second or third generation reefs is an attractive but undoubtedly oversimplified explanation in each case. Davis (1928) explained the general paucity of double barriers by the less luxuriant growth of fringing compared with open-ocean reefs, and their general inability to maintain themselves at sea level during vertical upgrowth. It is also possible that with accelerating subsidence drowned barriers may be formed. Again these are not common, but good examples are known south of Vangunu, Solomon Islands; possibly in northwest Madagascar (Guilcher, 1956); and on the south coast of Tutuila, Samoa (Chamberlin, 1924).

Atolls themselves vary greatly in size, depth of surrounding water, and relationship to volcanic or continental land. Darwin's theory was developed specifically to explain the atolls of the open Pacific, but it has been widely accepted to explain the formation of the smaller atolls of continental-marginal seas (e.g., Kuenen, 1933; Umbgrove, 1947; Stoddart, 1962). The "almost-atoll" stage with a residual nonlimestone island is also quite common (e.g., Truk, Aitutaki, Clipperton). Irregularities in atoll shape can often be most easily explained by inheritance from subaerial island forms, though other explanations, such as landslide scarring, have been put forward for less pronounced irregularities (Fairbridge, 1950b). Since some well-developed double barriers exist, it is perhaps surprising that few double atolls or "nested atolls" (Kuenen, 1951) are found. Possible examples include Nukuoro, Ant, and Taongi Atolls (Tayama, 1952, p. 246), but it requires considerable imagination to see a triple-nested atoll in Kapingamarangi Atoll (Tayama, 1952, Fig. 151). Such forms may have been more common before the erosional events of the Pleistocene. Drowned atolls are, however, common, though few have been described in detail (e.g., Fairbridge and Stewart, 1960, on Alexa Bank). Davis (1928) was puzzled by the concentration of drowned atolls at depths of 60–100 m, and suggested that deeper ones were still to be discovered. Many more drowned atolls, and also guyots at depths of 1–2 km (Hess, 1946; Hamilton, 1956), are now known, and demonstrate that subsidence too rapid for reefs to maintain themselves at the surface has been common during the last 70×10^6 year.

Both Davis (1928) and Tayama (1952) have developed a suggestion of Darwin's that sea-level atolls could be characterized as young, mature, or old by such features as width of reefs and depth of lagoon, but there is no doubt that control by Pleistocene events is more important than position in a developmental sequence. Von Lendenfeld's (1896) discussion of the effects of centrifugal and centripetal upgrowth on the surface dimensions of reefs over time ignores the scale and time problems involved and is of little practical significance.

D. PROBLEMS AND ALTERNATIVES

Darwin's theory had no serious rivals for some 40 years after its announcement. Problems arose, however, with the extension of field work, particularly in the western Pacific, a tectonically active area with many uplifted reefs. Semper (1881) in the Palau Islands concluded that fringing, barrier, and atoll reefs had formed in an area of elevation, thus leading him to doubt the generality of Darwin's explanation. Guppy (1886) found similar elevated reefs in the Solomon Islands. Such reefs were thin, and in some cases overlaid pelagic marine deposits. They lacked both the thickness and the unconformable contacts with nonlimestone foundations which the Subsidence Theory required. Guppy (1889) and Wood-Jones (1912, p. 282) also worked on the Cocos-Keeling Atoll in the Indian Ocean, and could find no evidence of contemporary subsidence in the present surface features of the reefs. As a result Guppy (1889, 1890) and many others abandoned Darwin's explanation. Later Hoffmeister and Ladd (1935) extended these arguments in their study of elevated reefs in Fiji and Tonga.

The discovery of the Pacific raised reefs approximately coincided with the recognition of the importance of pelagic sedimentation in the oceans, especially during the *Challenger* Expedition of 1872–1876. Both Rein (1870) in Bermuda and Murray (1880, 1888) more generally proposed that pelagic sediments accumulating on eroded submarine volcanoes could form platforms at suitable depths for reef growth. From work in the Maldives, Gardiner (1903) believed that submarine erosion by currents could form platforms on which first ahermatypic and then, at shallow depths, reef-building corals could grow. Murray (1880) also proposed that once reefs reached the surface lagoons would be formed by solution, though this was more difficult to accept than the pelagic sedimentation theory. Agassiz, drawing on experience in Fiji (1899), the Great Barrier Reefs (1898), and many other parts of the world, emphasized the role of wave erosion in preparing reef foundations, and Wharton (1890, 1897) had similar ideas. Such was the impression made by these theories that Daly (1915, p. 234), in his review of the literature, found that "no case is recorded where a region bearing an atoll or barrier reef has been shown, beyond question, to be now visibly sinking."

Most of these theories were internally inconsistent; many were based on supposition alone; some, notably Agassiz's, resulted from astonishingly unreliable field observations; and even the most satisfactory were of purely local application. Davis (1928) has shown how widely and readily the Subsidence Theory was abandoned in favor of these new ideas, but he also had no difficulty in demolishing their arguments. The idea which lasted longest, that of lagoon formation by solution (Gardiner, 1930, 1931), could not survive the demonstration that tropical seawater is chemically incapable of dissolving limestone on this scale (Cloud, 1965), and that lagoons are sites of sedimentation not erosion.

The controversy which came to surround the "coral reef problem" as a result of these divergent ideas did however emphasize the need for more careful and detailed field studies and for direct testing of alternative theories, for example, by deep drilling on an atoll.

III. Glacial Control Theory

A. Daly on Glacial Control

Of the alternatives to subsidence, R. A. Daly's theory of "glacial control" was argued in greatest detail and possessed the greatest generality. In a long series of papers (Daly, 1910, 1915, 1916a,b, 1917, 1919, 1948) and several books (especially "The Changing World of the Ice Age," 1934), Daly considered the effects of glacial fluctuations of the sea level on coral reefs and tropical coasts and islands, arguing that the foundations of modern reefs were formed during Pleistocene low stands of the sea. He based his theory, partly anticipated by Tylor (1872), Penck (1894), and others, on two main facts: first, the general existence of level lagoon floors, already stressed by Wharton (1897); and second, their approximate accordance at depths of 50–90 m (tabulation in Daly, 1915, pp. 187–192). Such uniformity could not be ascribed to sedimentation, through aggradation could explain the shallower lagoon floors of smaller and narrower atolls and barrier reefs.

The accordant platforms were formed, according to Daly, by low-level abrasion during glacial times, when reef corals were no longer able to protect coastlines because of lowered water temperatures and increased turbidity (Fig. 2). Daly's emphasis changed so much as he developed the theory that it is not possible to give a complete account of his views (see the critique by Ladd and Hoffmeister, 1936). Initially he ascribed great power to low-level marine erosion; by means of analogy with rates of retreat of the Chalk cliffs of Dover (9–100 cm/year) he argued (Daly, 1910, p. 305; 1915, p. 181) that an atoll the size of Great Chagos Bank, 95 × 145 km, could be beveled in 50,000 years. If this were so, even the biggest atolls, such as Kwajalein, Rangiroa, and Suvadiva, could be so leveled. Under this rather extreme view barriers and atolls surrounding flat-floored lagoons were the results of rather peculiar Pleistocene conditions and were probably not found in Preglacial times. Later he laid less emphasis on marine erosion, and spoke of the smoothing (Daly, 1915) and sandpapering (Daly, 1919) of surfaces weathered and largely lowered by erosion in Preglacial times.

Once platforms were formed, corals recolonized them as the sea level rose, and for ecological reasons grew more strongly on the rims than in

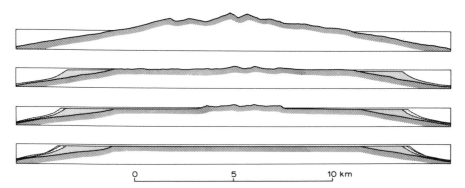

Fig. 2. Stages in the formation of an atoll according to Daly's Glacial Control Theory. The stages show a normal volcanic island; the island largely peneplained, with an encircling embankment of detritus; the island extensively benched by waves, with a wider detritus embankment; and the island completely truncated by wave abrasion. Source: Daly (1915, pp. 160–161).

the centers of the banks. In the case of atolls, Daly still had to explain the existence of so many cones rising from the deep ocean floor and beveled at exactly the right depth for coral growth, but he found support for the rapidity and efficiency of the beveling of volcanic islands in the observed rapid marine erosion of the pyroclastic Falcon Island, Tonga (Hoffmeister *et al.*, 1929), which was rapidly reduced to depths of 10–20 m within a few years.

It is unfortunate that Daly's theory was seen as "diametrically opposed to the theory of upgrowing reefs on subsiding foundations" (Davis, 1928, p. 89), since it quite clearly is not. It was Daly's misfortune to attempt to describe Pleistocene effects with only rudimentary knowledge of the length of time involved, the magnitude of sea-level fluctuations, and the processes at work, and it is perhaps more surprising that his assumptions were often of the right order of magnitude than that his conclusions can now be shown to be untenable.

B. Davis on the Marginal Belts

In one respect, however, Daly's theory was accepted by the exponents of Subsidence Theory. Davis (1923) rejected it for the central zone of the coral seas, largely because of the apparent absence of cliffs on reef-surrounded islands (though cliffs may be more widespread than Davis supposed: e.g., see Tayama (1952) on the 50–100 m high cliffs of Palau, Truk, and Ponape). He did, however, find islands with straight coasts, plunging cliffs, and coral reefs in a narrow zone between the strongly cliffed reefless islands of the cool seas and the uncliffed reef-encircled islands of the Tropics. He ascribed these cliffed coasts of the "Marginal Belts" to low-level abrasion as the reef zones contracted and the sea level fell during the Glacial periods. The main examples of islands in the Marginal Belts included the Hawaiian and Marquesas Islands in the Pacific and the Lesser Antilles in the Atlantic. In the central reef seas Davis argued on the basis of the coastal landforms that the reefs could not have been sufficiently inhibited to have exposed the coasts to wave attack and beveling in Glacial times.

C. Problems of Daly's Theory

Daly's theory can be criticized both in terms of basic premises and of its relationship to actual reef features, which were poorly known when he was writing.

The level of platform formation resulting from glacial abstraction of seawater and its gravitational effects was placed at depths of 60–70 m, perhaps extending to 100 m (Daly, 1915, pp. 174 and 182–183). Recent calculations (Donn et al., 1962), based on ice volumes, indicate appreciably greater sea-level falls: to 106–124 m in Wisconsin I time, 115–134 m in Wisconsin II, and 138–160 m at the maximum extent of Pleistocene glaciations. There is some evidence from Australia of regressions to at least −170 m (Veeh and Veevers, 1970; Jongsma, 1970). With shifts of this magnitude Davis's (1928, p. 102) argument that depth of drowned embayments was greater than the sea-level fall proposed by Daly loses some of its force. It is, however, probable that these glacial shifts were superimposed on a general Pleistocene regression from an early Pleistocene (Calabrian) level 180 m above the present level. If this is so, early Pleistocene glacial regressions could still have been well above present sea level, and the relationship between sea levels and Daly's platforms assumes a new complexity. Suggestions made by Johns (1934) and Shepard (1948) that Pleistocene sea-level falls may have been as great as 1–2 km are not now taken seriously.

Daly had only imprecise information on the time available for low level erosion in the Pleistocene. In 1915 (p. 181) he considered erosion to have been fairly continuous between the Kansan and the Wisconsin, a period he estimated as up to 0.9×10^6 year, though he later modified this view. Adequacy of time became less important once he replaced the idea of large-scale Pleistocene beveling by relatively minor smoothing. Current estimates place the duration of the Pleistocene at more than 2×10^6 year, of which only a small proportion can be assigned to full Glacial periods with low sea levels.

The postulates of sediment and temperature control of reef growth during glacial periods is also open to doubt. There is no evidence of excessive water turbidity around modern reefs which have been tectonically elevated above the sea (Kuenen, 1933), nor is any general increase in ocean turbidity during the Pleistocene apparent from ocean floor sediments. It has, moreover, been shown that corals can thrive in relatively turbid water (Marshall and Orr, 1931). Sediment control was probably limited to those areas where discharge from major rivers, especially draining glaciated or periglacial mountainous areas, carried sediment out to the edge of continental shelves. The discharge of the Pleistocene Fly River of New Guinea and its effects on the northernmost Great Barrier Reefs, though not yet studied in detail, provides a good example of such control. Daly estimated the fall in tropical sea temperatures during glacial periods at 5–10°C. Isotopic paleotemperature measurements now available from the tropics of the Caribbean, Indian, and Pacific Oceans all suggest a fall in surface water temperatures of 5–6°C for the equatorial Atlantic and Indian Oceans, 3–4°C for the equatorial Pacific, and 7–8°C for the Caribbean (Emiliani, 1955, 1970; Oba, 1969). Such falls would doubtless lead to contraction of the reef zones and perhaps to some faunistic attenuation, but the reefs themselves would be able to survive over most of the coral seas. Davis had already reached this conclusion on physiographic grounds (Section III,B). Daly's inference from the proposed temperature and sediment control that the reef corals retreated to refuges, mainly in the East Indies, during the glaciations and subsequently spread out to recolonize during Inter- and Postglacial periods is thus difficult to sustain. Since reefs are complex integrated communities which are remarkably similar in structure and composition over large areas, probably indicating long continuity of development, retreat appears unlikely on biological grounds. Daly also proposed that the major faunistic differences between the Indo-Pacific and Caribbean areas might result from glacial attenuation of reef faunas in the latter, but we know from Vaughan's (1919) work that the relict nature of Atlantic fauna dates mainly from the closure of the Central American isthmus in Tertiary times.

Finally, Daly made large assumptions about the nature and rate of operation of processes involved in the benching. His earlier optimism about truncation of banks the size of Great Chagos by marine erosion gave way to an emphasis on the possible importance of easily eroded pyroclastic material in Pacific vulcanism and to the revised view that Glacial erosion only slightly modified forms produced by deep Preglacial weathering and erosion. Such a modification, however, effectively abandons the central tenet of the Glacial Control Theory and reduces it to a mere variant of the Antecedent Platform Theory (Section IV), as indeed Vaughan (1918, p. 224) realized. The processes at work are considered in more detail in Section VII,A.

The primary direct evidence for Daly's theory was in the flatness and accordance of lagoon floors, and Daly presented many profiles of atolls and banks to support his interpretation (Fig. 3). Critics such as Davis (1928) were able to find many examples of unusually deep atoll lagoons, and of submerged banks with great and variable depths, at the same time arguing that where accordance could be demonstrated it could be otherwise explained. The evidence of lagoon-floor morphology is examined in Section VI,A; it is now clear that irregularity and variability are more common than the smoothness and accordance Daly claimed. Davis also made the point that if the lagoon floor were formed by marine abrasion, with inward-moving cliffs, the resulting profile should be shallowly

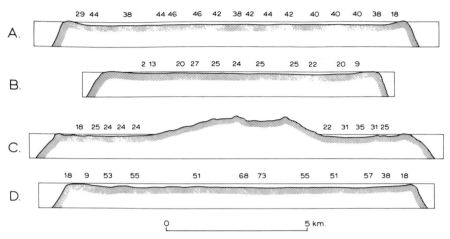

Fig. 3. Profiles of atoll and barrier reef lagoons, and of a submerged bank, show-
ing accordant lagoon depths and smooth floors. A. Funafuti, Ellice Islands; B.
Diego Garcia Atoll, Chagos Archipelago; C. Nairai, Fiji; D. Tizard Bank, China Sea.
Depths are given in meters, vertical exaggeration ×3. Source: Daly (1915, p. 193).

convex upwards, with unconsumed residuals in places; these conditions
are not met in nature, since lagoon floors are generally concave.

Lastly, Daly's theory makes predictions about the age, thickness, and
lateral dimensions of modern surface reefs. The average reef flat width
of about 0.5 km was ascribed by Daly to similarity of age, though clearly
ecological conditions are important controls. Much wider reef flats are
known, there is a systematic difference in width between windward and
leeward reefs, and reefs round shallow lagoons are generally wider than
reefs round deep ones. If reefs were upgrowths from Pleistocene platforms their thickness would be limited by the depth of the platforms.
Daly (1915, pp. 192 and 218) suggested limiting thicknesses of 110 m
for Recent reefs and of 5–25 m for lagoonal sediments. Data on the size
and shape of modern reefs are summarized in Section VIB.

D. Conclusion

Daly's main contribution was to emphasize the importance of
Pleistocene sea-level shifts and to show that they must form part of any
theory of coral reefs. He further clearly showed that modern surface reefs
could only have grown during and since the last transgression of the sea.
His detailed arguments about processes and their results have not survived subsequent work, nor is Daly's understanding of reef morphology

now acceptable. The importance of sea-level movements is not now in question, but we need to revise Daly's views on their effects.

IV. The Antecedent Platform Theory

A. OUTLINE

The Antecedent Platform Theory was formally proposed by Hoffmeister and Ladd (1935, 1944) following work in Fiji and Tonga, but elements of it are apparent in much earlier work, particularly that done in the West Indies. In its most general form the theory states that "any bench or bank . . . that is located at a proper depth within the circumequatorial coral-reef zone can be considered a potential coral-reef foundation and . . . if ecological conditions permit, a reef could grow up to the surface without any change in ocean level" (Hoffmeister and Ladd, 1944, p. 389). The theory resulted from Hoffmeister and Ladd's dissatisfaction with the Subsidence Theory after studying the raised reefs of Fiji and Tonga, and with the inconsistencies and deductive nature of the Glacial Control Theory (Hoffmeister and Ladd, 1935; Ladd and Hoffmeister, 1936).

In their main paper of 1944, Hoffmeister and Ladd classify the antecedent platforms on which reefs may grow in terms of origin—as platforms of erosion, deposition, volcanic eruption, and of tectonic origin. They have little difficulty in showing, mainly from examples in Tonga, Fiji, and the southern Cook Islands, that modern reefs are growing on platforms of diverse origin, without necessarily involving relative change in sea level either by subsidence or glacial control.

Similar suggestions had been previously made by Le Conte (1857) for Florida and by Agassiz (1894, p. 172) in the West Indies, interpreting the present surface reefs as comparatively thin veneers on basements of older and in some cases nonreef origin. Guppy (1884) reached a similar conclusion in the Solomon Islands. Darwin (1842, p. 101) had clearly stated that reefs might grow wherever suitable banks occurred. The most complete evidence for this position is provided by work done in the West Indies by Vaughan (1914, 1916, 1918, 1923), who in a series of papers clearly distinguished between the foundations of reefs and either their present surface features or the distribution of modern living reef communities. He clearly showed that in many cases living reefs were less extensive than the foundations or platforms on which they stood (e.g., on the Great Barrier Reefs and the British Honduras barrier reef). Whereas Daly concentrated on explaining the origin of the foundations by glacial control, Vaughan was more concerned with subsequent reef growth upon them. It is thus possible to interpret many of his views in terms of the Glacial Control Theory and the renewal of reef growth during the Holocene transgression. Vaughan's work in the Lesser Antilles was so careful that even Davis (1928, p. 118) was compelled to accept its general outline.

B. Problems

As a general theory of coral reefs the Antecedent Platform Theory is of limited usefulness. First, in Hoffmeister and Ladd's (1944) statement it is proposed in such broad and inclusive terms that it ceases to be a theory at all and is more a descriptive classification of reef foundations. Its predictive power is consequently low, and each reef must be explained in terms of local conditions.

More important, it succeeds in being an explanation only by severely limiting the range of problems with which it deals. It is most successful in explaining small reefs in such mobile areas as the Caribbean and the West Pacific, where large-scale Darwinian subsidence has probably not occurred. Here it can explain the development of fringing reefs, barrier reefs, and shallow-water shelf atolls such as the Alacran and other Yucatan reefs (Hoskin, 1962; Logan, 1969). It cannot, however, begin to explain the foundations of open-ocean atolls, except to state that surface reefs are thin and overlie older reef foundations. We have seen that it is the history of these foundations which classically formed the "coral reef problem."

It is possible to analyze Hoffmeister and Ladd's foundation types in greater detail, and to show from examples how in some cases subsidence or other explanations are more appropriate (Kuenen, 1950, p. 454). Thus it is difficult to see how pelagic deposition can follow abrasion on cinder cones to form a reef foundation (cf. the remarks on the theories of Murray and others presented in Section II,D). Furthermore, many barrier reefs arise from supposedly antecedent platforms at depths of about 80 m, too great for reef growth, and it is difficult to envisage reef formation on them without relative movement of the sea level.

The Antecedent Platform Theory, as it is at present formulated, is of very limited and local application, applying particularly to small reefs of mobile continental borderlands. Its most useful contribution was to emphasize the thinness and recency of most modern reefs, and it can be viewed as an adjunct rather than an alternative to the theories of Subsidence and Glacial Control.

V. The Evidence of Reef Structure

A. ATOLLS

With the development of deep boring and geophysical techniques the structural implications of the Subsidence Theory could be directly tested. Following early attempts by Belcher (1843) and Moresby (1835) to bore through reefs at Hao in the Tuamotus and in the Maldives, the first major attempt to reach volcanic rock under an atoll was made by the Royal Society at Funafuti in 1896–1898 (Royal Society, 1904). The deepest bore reached a depth of 340 m, the first 194 m being coral limestone and the rest dolomite. Judd (1904) believed the whole to be shallow-water in origin and probably post-Miocene, while others believed that the lower section could be a forereef talus and hence inconclusive as to subsidence. No volcanic rock was reached. Later seismic work at this atoll and at nearby Nukufetau indicated 550–770 m of limestone overlying what was probably volcanic rock (Gaskell and Swallow, 1953).

The first conclusive demonstration of deep subsidence on open-ocean atolls resulted from borings in the Marshall Islands (Fig. 4). Bores at the Bikini Atoll in 1947 reached 410 and 780 m, and showed shallow-water reef limestones with little dolomitization, dating back to the Oligocene (Ladd *et al.*, 1948; Cole, 1954; Wells, 1954). Seismic work in 1946 and 1950 indicated a volcanic basement beneath the limestones, with a least depth of 1,600 m (Dobrin and Perkins, 1954; Raitt, 1954), and volcanic rocks were dredged from the atoll slopes at depths of 1,460–3,660 m in 1950 (Emery *et al.*, 1954). The subsidence thus indicated was finally demonstrated by deep drilling at nearby Eniwetok Atoll in 1951. Hole F-1 penetrated 1,405 m of reef limestone and entered basement rock; hole E-1 went through 1,283 m of reef limestone into 4.3 m of olivine basalt; the drills reached 1,411 and 1,287 m respectively. An earlier Eniwetok drill, to 392 m, was entirely in limestone (Ladd *et al.*, 1953).

Eniwetok Atoll is thus formed by 1.25 km of shallow-water coral limestone, dating back to the Eocene, standing on top of a 3.2-km-high basalt volcano on the ocean floor. The cores have been intensively studied. The bottom of F-1 is in forereef deposits, while E-1 is in reef-rock. The entire limestone column consists of lagoonal or shallow-water reef deposits, mostly laid down in water less than 75 m deep (Todd and Post, 1954; Cole, 1957). Three horizons have been found, at 91, 335, and 847 m, where aragonite has been replaced by calcite, indicating emerg-

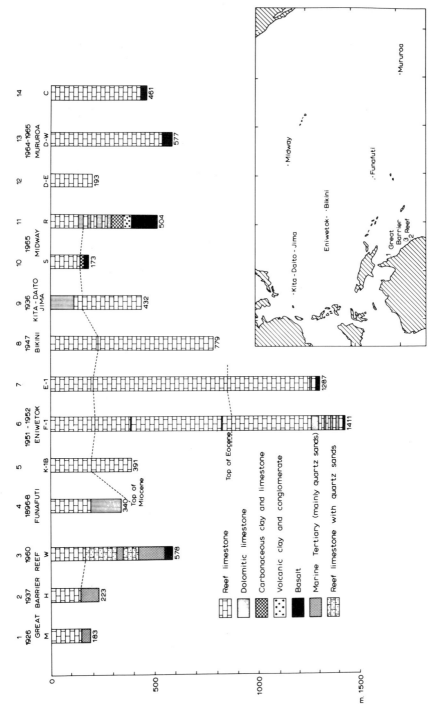

Fig. 4. Composition of cores from deep borings through Pacific reefs. Mainly after Ladd *et al.* (1969), with additions.

ence and solution of the limestone cap, and showing that subsidence has been intermittent (Fig. 5). Pollen and land Mollusca were found in these horizons (Ladd and Schlanger, 1960; Ladd, 1958; Schlanger, 1963). The sediments have been dated paleontologically, giving rates of subsidence of 51.9, 39.6, and 15.2 $m/10^6$ year in the Eocene, Miocene, and post-Miocene, respectively (Cole, 1957). Volcanic rocks have also been dredged from slopes around the Wotje and Ailuk Atolls, and in the Marshalls at 1,265–2,130 m. Geophysical work at Kwajalein, the world's largest atoll, indicates a volcanic basement at depths of 1,006–1,980 m.

More recently, deep drilling has been carried out at the Mururoa

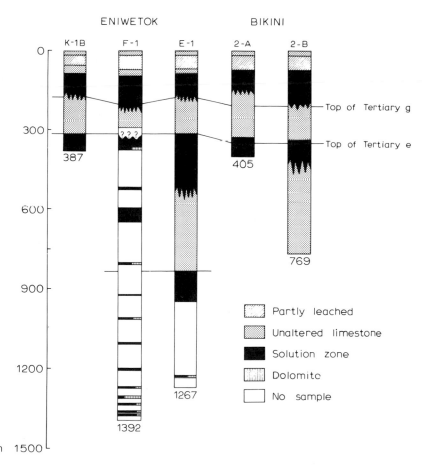

Fig. 5. Diagenetic horizons in cores from Marshall Island coral reefs. After Schlanger (1963).

Atoll, Tuamotus. Basalt was reached in two bores at 438 and 415 m, and was penetrated for 160 m; geophysical work had indicated a volcanic basement at about 400 m. Radiometric dating of the upper part of the limestone column showed remarkable agreement with the chronology in the northern Marshall Islands, 7,000 km away, and similar diagenetic horizons have been found (Lalou et al., 1966; Chauveau et al., 1967). Holes have also been drilled on the Midway Atoll, Hawaiian Islands, penetrating some 120 m of post-Miocene limestone, and reaching basalt at 157 and 384 m, considerably shallower than the seismic predictions of 260–290 and 630–710 m (Ladd et al., 1969; Shor, 1964). As in the Marshalls, the limestone column shows recrystallization below 60 m.

Gravity surveys have also been carried out on the atolls of Manihiki and Rakahanga in the northern Cook Islands, which rise about 4 km from the ocean floor (Robertson, 1968). Seismic refraction work on Manihiki indicates a comparatively thin coral cap, varying from 50 m in the center of the lagoon to about 500 m under the peripheral reefs (Hochstein, 1967).

Borings in other oceans have been less spectacular than in the Pacific. At Bermuda an early bore to 427 m entered lava at 171 m (Pirsson and Vaughan, 1913), and more recently two short bores, to 33 and 43 m, both went into volcanics at 21–24 m (Newman, 1959). The reef cap here is too thin to bear on Darwin's theory. In the Bahamas a bore in 1932 went through 117 m of limestone, mostly dolomitized reef limestone (Field and Hess, 1933). A deep well at Andros Island, Bahamas, has subsequently gone through 161 m of surface-type limestone and then into dolomite, ending in the lower Cretaceous at 4,488 m, with apparently shallow-water facies through the whole column and with successive horizons of solutional diagenesis (Spencer, 1967; Goodall and Garman, 1969). This indicates a rate of subsidence approximately the same as that in the Marshall Islands, but over a much longer span of time and with the production of much larger limestone banks (Dietz et al., 1970; Lynts, 1970). In the Caribbean, recent drilling has reached nonlimestone rocks beneath a limestone cap at 959 m beneath Glover's Reef and 1,219 m under the Turneffe Atoll, British Honduras.

Therefore, open-ocean drilling leaves no doubt of the correctness of Darwin's basic postulate for the formation of atolls. As far as explaining the existence and location of atolls in the open ocean is concerned, the "coral reef problem" may be regarded as solved, although this is not to imply that the Subsidence Theory is equally successful in explaining the surface features of the reefs themselves.

Two elevated atolls have also been drilled: Kita-Daitō-Jima in

1934–1936, and Maratoea, Borneo. At Kita-Daitō-Jima the section consisted of Upper Oligocene-Pleistocene calcareous sands and limestones, dolomitized in the upper 104 m, and clearly indicating subsidence before the uplift (Hanzawa, 1940; Sugiyama, 1936; Ota, 1938). The Maratoea bores went through 189 m of coral limestone, then 84 m of reef detritus, and finally into alternating coral and noncoral limestone. No dates are available, but subsidence is indicated by the depth of the section (Kuenen, 1947). The bores on these raised atolls are of interest because they are both located in unstable continental marginal areas where it has often been argued that Darwin's theory does not apply.

B. Barrier and Fringing Reefs

For barrier and fringing reefs, particularly those associated with continental coasts rather than volcanic islands, the problems are more complex. In the case of the Great Barrier Reefs, Steers (1929) and Fairbridge (1950a) have suggested that the reefs overlie complex structures, with longitudinal faulting and warping, rather than the simple flexuring proposed by Darwin (1842) and Jukes (1847).

Bores have been put down at Michaelmas Cay in 1926 to 183 m, at Heron Island in 1937 to 223 m, and at Wreck Island in 1959 to 578 m. The Michaelmas Cay bore went through 115 m of coral limestone into quartz sand with Foraminifera; that at Heron Island, 1,130 km away, went through coral limestone into similar quartz sand at 154 m. The upper limestones were shallow-water and probably indicated subsidence (Richards, 1938; Richards and Hill, 1942), though the depth of reef limestone is approximately within the range of Pleistocene sea-level fluctuations. The Wreck Island bore went into quartz sandstone at 121 m, and then through a sequence of sandstones and calcarenites reaching back into the Miocene; basement rocks were reached at 547 m (Traves, 1960). Recent reinterpretation of the Heron and Wreck Island bores shows that they both penetrated the Miocene, with a long break in sedimentation at the end of the Tertiary (A. Lloyd, in Jones and Endean, 1967).

Bathymetric and geophysical studies of the structure of the Coral Sea basin also throw light on the evolution of the Great Barrier Reefs. It is now believed that the Coral Sea began to open in the early Tertiary, with considerable subsidence beginning in the late Oligocene and early Miocene. A widespread unconformity within the basin, from which the reefs rise, is identified as Miocene; from this it is inferred that subsidence in areas of reef upgrowth has taken place at rates of 79–111 m/10[6] year (rather faster than in the Marshall Islands), and at rates of 170–240 m/10⁶ year in the basin proper where there are no surface reefs. It is interesting to note that the atolls outside the main reef are not now thought to cap subsiding volcanoes but to overlie structural upwarps within the basin itself (Ewing *et al.*, 1970; Gardner, 1970).

A record of alternating carbonate and siliceous deposits, similar to those of the Great Barrier Reefs, is also available from the Florida deep boring made in 1896 (Hovey, 1896). The history of this area is broadly similar to that of the Bahama Banks. A new boring on the southern barrier reef of New Caledonia reached basement at 225 m, with three marine and two detrital or aeolian horizons in the reef limestone (Avias and Coudray, 1967); this record also indicates subsidence, with complications resulting from Pleistocene events.

Fringing reefs generally lie on relatively shallow foundations close to the shore, and have a much shorter history than barriers and atolls. Such superficial recent reefs have been described from Samoa (Mayor, 1924; Cary, 1931), the New Hebrides (Baker, 1929), the Solomon Islands (Stoddart, 1969a), the Seychelles (Taylor, 1968; Lewis, 1968), and elsewhere. Mayor (1924) put down several shallow bores on the Utelei fringing reef in Tutuila, Samoa. These reached basement at 20.7, 36.9, and 36.6 m respectively at distances of 61, 175, and 282 m from the shore. A further bore on Aua reef reached basalt at 47.6 m, 156 m from the shore. 25 bores were put down near Pago Pago, Tutuila, in 1941 and, though these showed that the upper 24 m of the reef consisted of calcareous silt and sand with a surface veneer of cemented coral, the details have never been published (Stearns, 1944, p. 1,284). Borings have been made in recent years through near-shore reefs in the construction of airfields, e.g., on Mahé, Seychelles, and on Tahiti, but no details have been published; the Mahé bores show an irregular granitic basement at a shallow depth beneath the fringing reef (C. J. R. Braithwaite, personal communication).

VI. The Evidence of Reef Morphology

Three types of morphological data on modern reefs are relevant in the discussion of reef origins, especially in terms of the Glacial Control Theory. These are the form of lagoons, the size and shape of recent reef bodies, and the presence of terraces on reef slopes. These cannot be treated in detail here, but in this section we will discuss certain features each of which have theoretical implications.

A. Lagoons and Their Features

Daly (1915, pp. 187–191) concluded from an extensive tabulation of maximum depth and "mean depth of deeper part" of barrier and atoll lagoons, drowned atolls and rimless banks that (a) maximum and general lagoon depths increase with lagoon width up to a width of 20 km, i.e., the smaller the lagoon the more rapid the sedimentary fill; (b) lagoons of similar widths usually have similar maximum and general depths; (c) lagoons of drowned atolls and the surfaces of rimless banks are deeper than reef-rimmed lagoons of similar widths; and (d) the maximum depth of atoll lagoons rarely exceeds 91 m, with a range of 60–90 m. Daly believed that such uniformity could not be expected under conditions of subsidence.

More detailed statistical studies of lagoon depths have been made by Yabe and Tayama (1937), Tayama (1952), Emery *et al.* (1954), Nugent (1946), and Wiens (1962). These have confirmed the relationship between depth and size (Fig. 6), but have shown that the range of depths is greater than Daly supposed. Some extreme depths (e.g., 150–200 m in the Moluccas) have been attributed by Kuenen (1933) to rapid subsidence. Tayama (1952, p. 248) has also demonstrated regional

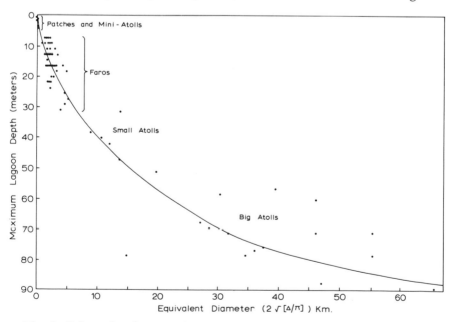

Fig. 6. Relationship between lagoon size and maximum depth in reef patches, faros, and atolls in the Maldive Islands. After Scheer (1972).

variations in maximum lagoon depths: 27 m in the Palau Islands, 41 m in the western Carolines, 63 m in the eastern Carolines, 64 m in the western Marshalls, and 59 m in the eastern Marshalls. Such variations are difficult to explain by the Glacial Control Theory alone. However, the general limiting depth of lagoons of about 70 m closely approximates Daly's limit, in spite of striking anomalies.

The form of lagoon floors, on the other hand, differs considerably from that inferred by Daly and other early workers. According to Darwin, "the greater part of the bottom in most lagoons, is formed of sediment; large spaces have exactly the same depth, or the depth varies so insensibly, that it is evident that no other means, excepting aqueous deposition, could have levelled the surface so equally" (1842, p. 26). Dana (1890, p. 183) found that apart from reef knolls the floors of large lagoons are "nearly uniform," and Gardiner (1915, p. 9) in the Maldives found lagoon floors "almost uniform dead level." Daly (1915) drew many small-scale profiles from hydrographic charts to indicate this flatness.

The supposed uniformity, however, resulted from an inadequate number of soundings, and with the development of continuous echo profiling the irregularity of lagoon floors soon became apparent. Lagoons sounded in detail include Eniwetok, Bikini, Diego Garcia, and Addu (all atolls), and Truk (almost-atoll). All show considerable irregularities superimposed on a steep-sided but not flat-floored basin with a characteristic hypsometric curve (Stoddart et al., 1966). Two kinds of bottom irregularities are important. First, reef knolls and reef patches, of which there are 2,300 in Eniwetok Lagoon, 1,500–2,000 in Raroia, and 900 in Bikini and Glover's Reef. These are either Postglacial reef growths or they are coral-veneered subaerially-eroded limestone pinnacles. The latter interpretation is supported by the number of such knolls which lack actively growing coral at the present time. Excavation of some reef patches in the Red Sea by Nesteroff (1955), however, failed to demonstrate an older limestone core. Second, many lagoons have closed depressions, some of which are irregular (Guilcher, 1963), but others, including the "blue holes" of the Bahamas and the Caribbean, are circular and regular in shape (Agassiz, 1894; Doran, 1955; Stoddart, 1962). The deepest are more than 100 m deep, and in the case of the Lighthouse Reef atoll, British Honduras, closely resemble the karstic *cenotes* of the exposed limestone platform of the Yucatan Peninsula. Shepard (1970) has found irregular closed depressions to a depth of more than 70 m, with hills and basins with a relief of 10–20 m, in Truk and other West Pacific lagoons. He terms the bottom irregularity "astonishing."

Under Daly's theory, pinnacles and knolls are necessarily of Postglacial growth on glacially prepared platforms, though bevels and accordant summits suggest a more complex history for many. Enclosed hollows, particularly of the "blue hole" type, cannot be explained by the Glacial Control Theory as set out by Daly. Thus, while the general similarity of lagoon depths strongly suggests a connection with glacial regressions, the details of lagoon topography show that Daly's processes cannot have been important.

B. Size and Shape of Recent Reefs

As Daly recognized there is approximate equality in the width of modern reef flats. He argued (1915, p. 219) from Gardiner's data on coral growth rates that such existing reefs could be formed during Post-glacial times. Tayama (1952, p. 250) quotes a mean reef-flat width of 500 m, with considerable local (ecological and structural) but minor regional variability. Reef widths strongly suggest a similar history for Recent reefs. In aberrant cases unusual widths have been explained by different rates of subsidence or stillstand (Kuenen, 1933, pp. 112–113). Daly's theory also implies a limit on the total volume of Recent reefs, which should be less than 110 m thick and occupying only a small percentage of the total platform area. More recent data suggest that such a thickness is much overestimated: Holocene and late Pleistocene reefs, defined by radiometric dating, have been shown to be only a few to a few tens of meters thick, and older reefrock still outcrops on the surface of many reef-flats, e.g., in the Tuamotus. Some quantitative data on the magnitude of recent reef growth are given by the British Honduras barrier reef (Fig. 7). Here the present sea-level barrier is a narrow ribbon reef standing on the seaward edge of a much wider submerged reef, which itself margins the edge of the coastal shelf. This shelf increases in depth from north to south, from 30 to 75 m, but as it does so the volume of reef material on its edge remains approximately constant at $75–92 \times 10^3$ m^3 per m of reef length. Since the underlying shelf gets deeper, however, so the width of the main shelf-edge barrier decreases, from 5.5 to 1.5 km. The smaller Recent ribbon-reefs are 460 m wide and approximately 3.7 m thick; they maintain a constant volume of 1,700 m^3 per m of reef length (Stoddart, 1963, pp. 84–88).

The small size of modern reefs, as compared to that of older reefs, is well shown in atolls near the latitudinal limits of the reef seas. Here modern reefs generally stand well back from the edges of the reef platforms from which they rise (e.g., Midway Atoll, Bermuda), in contrast with the more vigorous reefs of warmer seas.

Passes through linear reefs also have theoretical significance. Tayama (1952, p. 248) has shown a general relationship between reef length and

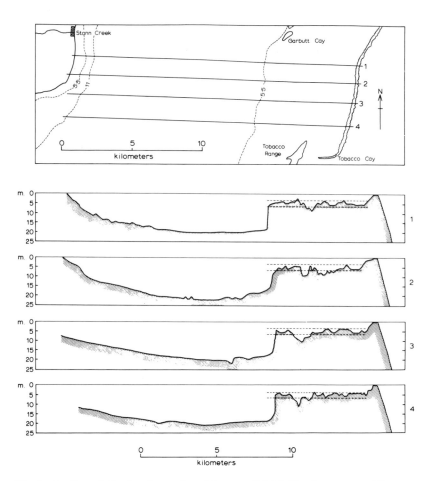

Fig. 7. Profiles of the barrier reef and lagoon, British Honduras coast. The present narrow surface reefs rise at the seaward edge of a much wider reef 5–7 m deep. After Stoddart (1963).

number of passes in the west Pacific; he has also shown that most passes through atoll reefs are on the leeward side. Passes through fringing reefs are located at entrances to valleys, and in many cases barrier reef passes are located opposite valley mouths. In such cases the gap is probably related to discharge during low sea levels rather than to ecological control by freshwater outflow at the present day. Wiens (1962, pp. 32–35) has extended Tayama's statistical work. He has shown considerable regional differences in frequency of passes (high in the Marshalls, where Kwajalein has 28, and low in the Tuamotus, where many atolls have no passes at all). In the northern Maldives the passes are so wide and frequent that the peripheral reef is effectively reduced to a series of scattered patches. Passes are of theoretical interest for the following reason: if subsidence alone were responsible for the surface form of reefs, one would expect that once coral growth was interrupted to form a pass it could not be resumed within it because the floor of the pass would be carried to increasingly greater depths, as the surrounding reef grew upwards. Hence extremely deep passes could be formed. Such deep passes are not, however, found. Most passes are no deeper than the adjacent lagoon floor. They are, therefore, usually interpreted as representing sections of the reef rim where growth has not occurred in Postglacial times, and their limited and often accordant depths are seen as an argument for glacial control (Kuenen, 1951).

C. SUBMERGED TERRACES ON CORAL REEFS

Increased bathymetric information, and radiometrically determined ages of discontinuities in limestone cores from atolls, demonstrate the existence in the reef areas of terraces or benches at levels shallower than the main lagoon floor "platform" of Daly. Such terraces or ledges were known to Darwin (1842, p. 30) and were attributed by him to discontinuous subsidence and the growth of corals on the outer edges of drowned reef flats (1842, pp. 103–104).

Dating of corals from the upper 100 m of the Eniwetok cores (Thurber *et al.*, 1965) shows a marked hiatus between corals less than 10 m deep, dating at less than 6,000 years, and those more than 48 m deep, with ages of $100-130 \times 10^3$ years. The gap in deposition between -12 and -20 m suggests emergence of the reef rim during the last glaciation; the cluster of dates from the last interglacial below 20 m indicates that the gross reef topography is older than the period of modern reef growth, and older than the last Glacial low sea level. Similar dates have been reported in the Bahamas and the Florida Keys, indicating reef growth near present sea level in the last Interglacial ($\sim 80-150 \times 10^3$ years B.P.), with emergence and nondeposition during the last glaciation ($\sim 5-80 \times 10^3$ years B.P.) (Broecker and Thurber, 1965). Cores at Mururoa, Tuamotus, show a similar discontinuity at about -6 m, with no corals dating between 8.8 and 100×10^3 years (Lalou *et al.*, 1966; Chauveau *et al.*, 1967). Many slightly elevated reefs also date from the last Interglacial times (see summary in Stoddart, 1969c, 1971).

This well-marked depositional discontinuity, either at shallow depths or outcroppings, may also be related to shallow terraces now widely reported from reef areas (Newell, 1961). A terrace at 14.5–18 m depth has been found on the seaward slopes of reefs in the Carolines, Marshalls, Tuamotus, and Bahamas, and a shallower terrace is found in some cases within atoll lagoons. Such terraces have recently been studied in detail in Bermuda (Stanley, 1970).

Both the terraces and the limestone discontinuities lie within the range of Pleistocene sea-level changes, and they cannot be simply related to the known sea-level fluctuations of more than 100 m. However, their existence does demonstrate conclusively that erosional beveling of reefs did not take place, at least during the last glaciation, in the manner that Daly supposed.

VII. An Integrated Theory of Coral Reefs

From this review, two clear conclusions emerge: (1) in the case of open-ocean atolls, and to a less certain extent in the case of reefs in mobile areas of the continental margins, Darwin's Subsidence Theory appears to provide an adequate and tested explanation of the development of the main reef types; (2) although Pleistocene fluctuations of the sea level are now well documented, and must have had considerable effects in the reef seas, the processes envisaged by Daly were not capable of forming the features he thought characteristic of coral reefs. Though Daly's processes did not operate in the way he described, it is nevertheless clear that Pleistocene events have been of crucial importance in forming the present surface features of reefs, and we need to specify more precisely the nature of "glacial control" which took place. The opportunity to do so is given by the study of reefs now tectonically upraised, which can serve as a model of the reefs formerly universally exposed by the Glacial falls in sea level.

A. Processes on Emergent Reefs

Two main processes act on emergent reefs: marine erosion, especially at intertidal levels, and rainwater solution. Although Kuenen (1950, p. 462) believed that erosion during Glacial low sea levels was capable of "complete or far advanced leveling of all emerged reef rock," as Daly argued, measurement of present day intertidal notching of limestones give maximum rates of only 0.1 cm/year (Hodgkin, 1964; Trudgill, 1970). The Glacial periods were not long enough for substantial beveling to occur by such a process, especially when it is realized that once notch depths exceed 15–20 m, the overlying limestone collapses and notching must begin again. Newell (1960) reached a similar conclusion on other grounds.

Estimates have also been made of rates of freshwater solution of emerged limestones, based on the solubility of calcite at different temperatures and with differing amounts of rainfall. These estimates, for Florida, Bermuda, and Aldabra (Ginsburg, 1953; Land *et al.*, 1967; Stoddart *et al.*, 1970) suggest orders of magnitude of solution of 1–2 cm/10^3 year. Because of mineralogical and textural variability in the reef limestones and inadequate knowledge of Pleistocene rainfalls, these estimates can be no more than approximate, but they do suggest extremely slow rates of karst erosion on exposed reefs. Direct measurements of surface erosion on the Aldabra Atoll give rates of approximately 10 cm/10^3 year (Trudgill, 1970), more rapid than the above calculations indicate but still extremely slow, when it is remembered that glacial regressions exposed limestone islands up to about 150 m high and in extreme cases up to 100 km in diameter. Subaerial erosion alone would take not less than 1.5×10^6 year to reduce such islands to sea level, a time in excess of the total extent of the full Glacial periods.

B. Forms on Emergent Reefs

We thus conclude that, rather than reefs being beveled by low-level erosion, old reefs survived the Glacial periods with only superficial modification of form (though with considerable diagenetic changes in the limestones), and that modern reef growth took place on old limestone surfaces, furrowed by subaerial erosion, as sea level rose at the rate of 1 m/100 year from 17,000 to 8,000 B.P., stabilizing at about 5000 B.P. Modern reef growth thus veneers an inherited topography, and to this extent the Antecedent Platform Theory for modern reefs must be accepted.

Several workers have suggested that karst erosion forms developed during low sea levels are reflected in modern reef topography, and that this could form the basis of a new Glacial Control Theory. Hoffmeister and Ladd (1935, p. 663) observed the widespread basining of emerged limestones in the Lau Islands, Fiji, the basining being developed "regardless of the structure, age, or organic composition of the limestone." MacNeil (1954) proposed that such basining could give rise to the main features of atolls, particularly the central lagoon. Further evidence for such erosion came from the study of emerged limestones on Okinawa (Saplis and Flint, 1949; Flint *et al.*, 1953). It is also possible that similar basining takes place on emerged fringing and barrier reefs, between the reef limestones and the nonlimestone rocks, and there has been much discussion over the development of such features in the Cook and Austral Islands (especially Mangaia and Rurutu: Marshall, 1929; Hoffmeister, 1930). If such eroded features are subsequently colonized by corals, we have a possible mechanism for the transition from fringing to barrier reefs, though it cannot be a general explanation because the depth of the solutional depression is much less than that of most barrier lagoons. Hoffmeister and Ladd (1945) have produced interior basining and marginal furrowing of limestone blocks experimentally subjected to chemical erosion.

There have been few detailed studies of subaerial reef limestone erosion under different climatic conditions. In particular the development of cenotes, uvalas, and other larger karst features appears to be quite restricted. Caves and cenotes are described from raised reefs in the Trobriand Islands (Ollier and Holdsworth, 1968, 1969), but elsewhere they are less common. In particular, the karst theory needs to explain the main fact in favor of Daly's original theory, the approximate equality of lagoon depths, which, while more variable than Daly supposed, are nevertheless remarkably uniform when viewed against the total vertical relief of features such as atolls, which rise several thousands of meters from the ocean floor.

VIII. Conclusion

The general acceptance of Darwin's Subsidence Theory and the restatement of the Glacial Control Theory in terms of karst rather than marine erosion, as outlined in this chapter, develops the views of Kuenen (1947), Tayama (1952), and other reef workers, particularly MacNeil (1954). It must, however, be emphasized that while the Tertiary history of a number of reefs is now well known, as a result of deep boring, in no case do we yet have a thorough understanding of the changes resulting in a reef from the recurrent sea-level changes of the Pleistocene. We need to know a great deal more about the shallow geology and structure of reefs, to determine their Pleistocene history, on the one hand, and, on the other, to learn more of processes and rates of erosion on modern elevated reefs which simulate the conditions of the Glacial periods. We need to know more precisely the interrelationships between rates of subsidence ($\sim$1.5–5 or 10 cm/10^3 year), of surface solution ($\sim$1–10 cm/10^3 year), and of marine erosion ($\sim$100 cm/10^3 year over short periods), under different climatic and lithological conditions. We also need more precise estimates of the rates of reef growth in Holocene times, and we need to know why in many areas corals have failed to colonize apparently suitable foundations. Only with such information can a comprehensive integrated theory be developed.

IX. Addendum

Since this chapter was completed, much more attention has been given to the implications of the theory of plate tectonics for the development of coral reefs, as briefly sketched by Stoddart (1976). There has been substantial confirmation of the main argument presented in this chapter: that reef morphology is greatly influenced by karst-erosion features, notably in important papers by Purdy (1974a,b). Some further points are added in a review by Stoddart (1973).

References

Agassiz, A. (1894). *Bull. Mus. Comp. Zool.* **26**, 1.
Agassiz, A. (1898). *Bull. Mus. Comp. Zool.* **28**, 95.
Agassiz, A. (1899). *Bull. Mus. Comp. Zool.* **33**, 1.
Agassiz, A. (1903). *Proc. R. Soc. London* **71**, 412.
Avias, J., and Coudray, J. (1967). *C. R. Hebd. Seances Acad. Sci.* **265**, 1867.
Baker, J. R. (1929). "Man and Animals in the New Hebrides." Routledge, London.
Belcher, E. (1843). "Narrative of a Voyage Round the World, Performed in Her Majesty's Ship Sulphur, During the Years 1836–1842," 2 vols. Henry Colburn, London.
Böttger, L. (1890). *Z. Naturwiss.* **63**, 241.
Broecker, W. S., and Thurber, D. L. (1965). *Science* **149**, 58.
Cary, L. R. (1931). *Pap. Tortugas Lab.* **27**, 53.
Chamberlin, T. C. (1924). *Pap. Dept. Mar. Biol. Carnegie Inst. Washington* **19**, 147.
Chauveau, J.-C., Deneufbourg, G., and Sarcia, J. A. (1967). *C. R. Hebd. Seances Acad. Sci.* **265**, 1113.
Chubb, L. J. (1957). *Geol. Mag.* **94**, 221.
Cloud, P. E., Jr. (1957). *Proc. Pac. Sci. Congr., 8th, 1953* Vol. 3, p. 1009.
Cloud, P. E., Jr. (1965). *In* "Chemical Oceanography" (J. P. Riley and G. Skirrow, eds.) Vol. II, pp. 127–158. Wiley, New York.
Cole, W. S. (1954). *U.S., Geol. Surv., Prof. Pap.* **260-O**, 569.
Cole, W. S. (1957). *U.S., Geol. Surv., Prof. Pap.* **260-V**, 743.
Crossland, C. (1928). *J. Linn. Soc. London* **36**, 577.
Daly, R. A. (1910). *Am. J. Sci.* [4] **30**, 297.
Daly, R. A. (1915). *Proc. Am. Acad. Arts Sci.* **51**, 155.
Daly, R. A. (1916a). *Am. J. Sci.* [4] **41**, 153.
Daly, R. A. (1916b). *Proc. Natl. Acad. Sci. U.S.A.* **2**, 664.
Daly, R. A. (1917). *Scientia, Ser. Biol. Paris* **22**, 188.
Daly, R. A. (1919). *Am. J. Sci.* [4] **48**, 136.
Daly, R. A. (1934). "The Changing World of the Ice Age." Yale Univ. Press, New Haven, Connecticut.
Daly, R. A. (1948). *Am. J. Sci.* **246**, 193.
Dana, J. D. (1849). "U.S. Exploring Expedition: Geology." C. Sherman, Philadelphia.

Dana, J. D. (1853). "On Coral Reefs and Islands." Putnam, New York.

Dana, J. D. (1872). "Corals and Coral Islands," 1st ed. Sampson Low, London (2nd ed., 1874; 3rd ed., 1890).

Dana, J. D. (1885). *Am. J. Sci.* [3] **30**, 89, and 169.

Dana, J. D. (1890). "Characteristics of Volcanoes, with Contributions from the Hawaiian Islands." Dodd, Mead and Company, New York.

Darwin, C. R. (1837). *Proc. Geol. Assoc.* **2**, 552.

Darwin, C. R. (1839). "Journal of Researches into the Geology and Natural History of the Various Countries Visited by H. M. S. Beagle." Henry Colburn, London.

Darwin, C. R. (1842). "On the Structure and Distribution of Coral reefs." Smith Elder & Co., London (2nd ed., 1876; 3rd ed., 1889. Reprinted by Cambridge Univ. Press, London and New York, 1962).

Darwin, C. R. (1933). "Charles Darwin's Diary of the Voyage of H. M. S. Beagle" (edited from the Ms. by Nora Barlow). Cambridge Univ. Press, London and New York.

Darwin, C. R. (1958). "The Autobiography of Charles Darwin 1809–1882." Collins, London.

Darwin, C. R. (1962). *Atoll Res. Bull.* **88**, 1.

Davis, W. M. (1913). *Am. J. Sci.* [4] **35**, 173.

Davis, W. M. (1914). *Bull. Am. Geogr. Soc.* **46**, 561, 641, and 721.

Davis, W. M. (1920). *Nature (London)* **105**, 292.

Davis, W. M. (1923). *Am. J. Sci.* [5] **6**, 181.

Davis, W. M. (1928). "The Coral Reef Problem." Am. Geogr. Soc., New York.

Dietz, R. S., Holden, J. C., and Sproll, W. P. (1970). *Geol. Soc. Am. Bull.* **81**, 1915.

Dobrin, M. B., and Perkins, B. (1954). *U.S., Geol. Surv., Prof. Pap.* **260-J**, 487.

Donn, W. L., Farrand, W. R., and Ewing, M. (1962). *J. Geol.* **70**, 206.

Doran, E. (1955). *Univ. Tex. Publ.* **5509**, 1.

Easton, W. H. (1969). *Spec. Pap. Geol. Soc. Am.* **121**, 86.

Emery, K. O., Tracey, J. I., Jr., and Ladd, H. S. (1954). *U.S., Geol. Surv., Prof. Pap.* **260-A**, 1.

Emiliani, C. (1955). *J. Geol.* **63**, 538.

Emiliani, C. (1970). *Science* **168**, 822.

Ewing, M., Hawkins, L. V., and Ludwig, W. J. (1970). *J. Geophys. Res.* **75**, 1953.

Fairbridge, R. W. (1950a). *J. Geol.* **58**, 330.

Fairbridge, R. W. (1950b). *Geogr. J.* **115**, 84.

Fairbridge, R. W., and Stewart, H. B., Jr. (1960). *Deep-Sea Res.* **7**, 100.

Field, R. M., and Hess, H. H. (1933). *Trans. Am. Geophys. Union,* 234.

Flint, D. E., Corvin, G., Dings, H., Fuller, W. P., McNeil, F. S., and Saplis, R. A. (1953). *Geol. Soc. Am. Bull.* **64**, 1247.

Gardiner, J. S. (1898). *Proc. Cambridge Philos. Soc.* **9**, 417.

Gardiner, J. S. (1903). *Am. J. Sci.* [4] **16**, 203.

Gardiner, J. S. (1915). *Geogr. J.* **45**, 202.

Gardiner, J. S. (1930). *Proc. Linn. Soc. London* p. 65.

Gardiner, J. S. (1931). "Coral Reefs and Atolls." Macmillan, New York.

Gardner, J. V. (1970). *Geol. Soc. Am. Bull.* **81**, 2599.

Gaskell, T. F., and Swallow, J. C. (1953). *Occas. Pap. Challenger Soc.* **3**, 1.

Ginsburg, R. N. (1953). *J. Sediment. Petrol.* **23**, 85.

Goodall, H. G., and Garman, R. K. (1969). *Am. Assoc. Pet. Geol. Bull.* **53**, 513.

Guilcher, A. (1956). *Ann. Inst. Oceanogr.* **33**, 65.

Guilcher, A. (1963). *Bull. Assoc. Geog. Fr.* 314–315, 2.

Guilcher, A. (1965). *Exped. Fr. Recifs Coralliens Nouv.-Caledonie* 1 (2), 113.

Guilcher, A., Berthois, L., Le Calvez, Y., Battistini, R., and Crosnier, A. (1965). "Les récifs coralliens et le lagon de l'île Mayotte (Archipel des Comores, Océan Indien)." ORSTOM, Paris.

Günther, S. (1911). *Sitzungsber. Bayer. Akad. Wiss.* 14, 1.

Guppy, H. B. (1884). *Proc. Linn. Soc. N. S. W.* 9, 949.

Guppy, H. B. (1886). *Proc. R. Soc. Edinburgh* 13, 857.

Guppy, H. B. (1888). *Scott. Geogr. Mag.* 4, 121.

Guppy, H. B. (1889). *Scott. Geogr. Mag.* 5, 281, 457, and 569.

Guppy, H. B. (1890). *Proc. Victoria Inst.* 23, 51.

Hamilton, E. L. (1956). *Mem., Geol. Soc. Am.* 64, 1.

Hanzawa, S. (1940). In *Jubilee Publ. Commem. H. Yabe's 60th Birthday* 2, 755.

Hass, H. (1962). *Atoll Res. Bull.* 91, 1.

Hess, H. H. (1946). *Am. J. Sci.* 244, 772.

Hochstein, M. P. (1967). *N. Z. J. Geol. Geophys.* 10, 1499.

Hodgkin, E. P. (1964). *Z. Geomorphol.* [N. S.] 8, 385.

Hoffmeister, J. E. (1930). *Geol. Mag.* 67, 549.

Hoffmeister, J. E. (1932). *Bull. Bishop Mus., Honolulu* 96, 1.

Hoffmeister, J. E., and Ladd, H. S. (1935). *J. Geol.* 43, 653.

Hoffmeister, J. E., and Ladd, H. S. (1944). *J. Geol.* 52, 388.

Hoffmeister, J. E., and Ladd, H. S. (1945). *Geol. Soc. Am. Bull.* 56, 809.

Hoffmeister, J. E., Ladd, H. S., and Adling, J. L. (1929). *Am. J. Sci.* [5] 18, 461.

Hoskin, C. M. (1962). *N.A.S.—N.R.C., Publ.* 1089, 1.

Hovey, E. O. (1896). *Bull. Mus. Comp. Zool.* 28, 65.

Johns, C. (1934). *Geol. Mag.* 71, 66, 176, and 408.

Jones, O. A., and Endean, R. (1967). *Sci. J.* 3, 44.

Jongsma, D. (1970). *Nature (London)* 228, 150.

Judd, J. W. (1904). *In* "The Atoll of Funafuti," pp. 167–185. Royal Society, London.

Jukes, J. B. (1847). "Narrative of the Surveying Voyage of H. M. S. 'Fly'," 2 vols. Boone, London.

Kuenen, P. H. (1933). *Snellius-Exped.* 5 (2), 1.

Kuenen, P. H. (1938). *C. R. Congr. Int. Geogr., Amsterdam* 2 (2B), 93.

Kuenen, P. H. (1947). *Verh. K. Ned. Akad. Wet. Amsterdam, Afd. Naturk.* 43 (3), 1.

Kuenen, P. H. (1950). "Marine Geology." Wiley, New York.

Kuenen, P. H. (1951). *J. Geol.* 59, 503.

Ladd, H. S. (1958). *J. Paleontol.* 32, 183.

Ladd, H. S., and Hoffmeister, J. E. (1936). *J. Geol.* 44, 74.

Ladd, H. S., and Hoffmeister, J. E. (1945). *Geol. Soc. Am. Bull.* 56, 809.

Ladd, H. S., and Schlanger, S. O. (1960). *U.S., Geol. Surv., Prof. Pap.* 260-Y, 863.

Ladd, H. S., Tracey, J. I., Jr., and Lill, G. G. (1948). *Science* 107, 51.

Ladd, H. S., Ingerson, E., Townsend, R. C., Russell, M., and Stevenson, H. K. (1953). *Bull. Am. Assoc. Pet. Geol.* 37, 2257.

Ladd, H. S., Tracey, J. I., Jr., and Gross, M. G. (1969). *U.S., Geol. Surv. Prof. Pap.* 680-A, 1.

Lalou, C., Labeyrie, J., and Delibrias, G. (1966). *C. R. Hebd. Seances Acad. Sci.* 263, 1946.

Land, L. S., McKenzie, F. T., and Gould, S. J. (1967). *Geol. Soc. Am. Bull.* 78, 893.

Langenbeck, R. (1890). "Die Theorieen über die Entstehung der Koralleninseln und Korallenriffe und ihre Bedeutung für geophysische Fragen." Engelmann, Leipzig.

Le Conte, J. (1857). *Am. J. Sci.* [2] **23**, 46.

Lewis, M. S. (1968). *J. Geol.* **76**, 140.

Logan, B. W. (1969). *Mem., Am. Assoc. Petrol. Geol.* **11**, 129.

Lum, D., and Stearns, H. T. (1970). *Geol. Soc. Am. Bull.* **81**, 1.

Lyell, C. (1832). "Principles of Geology, Being an Attempt to Explain the Former Changes of the Earth's Surface, by Reference to Causes Now in Operation," Vol. 2. John Murray, London.

Lynts, G. W. (1970). *Nature (London)* **225**, 1226.

McDougall, I. (1964). *Geol. Soc. Am. Bull.* **75**, 107.

MacNeil, F. S. (1954). *Am. J. Sci.* **252**, 402.

Marshall, P. (1927). *Bull. Bishop Mus., Honolulu* **36**, 1.

Marshall, P. (1929). *Geol. Mag.* **66**, 385.

Marshall, S. M., and Orr, A. P. (1931). *Sci. Rep. Great Barrier Reef Exped.* **1**, 93.

Mayor, A. G. (1924). *Pap. Dept. Mar. Biol. Carnegie Inst. Washington* **19**, 1.

Menard, H. W. (1964). "Marine Geology of the Pacific." McGraw-Hill, New York.

Menard, H. W. (1969). *J. Geophys. Res.* **74**, 4827.

Molengraaff, G. A. F. (1916). *Verh. K. Akad. Wet. Amsterdam* **25**, 215.

Moresby, R. (1835). *J. R. Geogr. Soc.* **5**, 398.

Murray, J. (1880). *Proc. R. Soc. Edinburgh* **10**, 505.

Murray, J. (1888). *Proc. R. Inst. G.B.* **12**, 251.

Nesteroff, W. (1955). *Ann. Inst. Oceanogr. Monaco* **30**, 1.

Newell, N. D. (1960). *Science* **132**, 144.

Newell, N. D. (1961). *Z. Geomorphol., Suppl.* [N.S.], 3, 87.

Newman, W. S. (1959). *Int. Oceanogr. Congr. Prepr., 1st, 1959* p. 46.

Nugent, L. E., Jr. (1946). *Geol. Soc. Am. Bull.* **57**, 735.

Oba, T. (1969). *Sci. Rep. Tohoku Univ., Ser. 2* **41**, 129.

Ollier, C. D., and Holdsworth, D. K. (1968). *Helictite* **6**, 63.

Ollier, C. D., and Holdsworth, D. K. (1969). *Helictite* **7**, 50.

Ota, Y. (1938). *Contrib. Inst. Geol. Paleontol., Tohoku Univ.* 30, 1.

Penck, A. (1894). "Morphologie der Erdoberfläche," Vol. 2. Engelhorn, Stuttgart.

Pirsson, L. V., and Vaughan, T. W. (1913). *Am. J. Sci.* **136**, 70.

Purdy, E. G. (1974a). *Soc. Econ. Paleontol. Mineral., Spec. Publs.* **18**, 9.

Purdy, E. G. (1974b). *Bull. Ass. Amer. Petrol. Geol.* **58**, 825.

Quoy, J. R., and Gaimard, J. P. (1824). *Ann. Sci. Nat.* **6**, 373.

Raitt, R. W. (1954). *U.S., Geol. Surv., Prof. Pap.* **260-K**, 507.

Rein, J. J. (1870). *Ber. Senckenberg. Naturf. Ges.* **1**, 140.

Richards, H. C. (1938). *Rep. Great Barrier Reef Comm.* 4, 135.

Richards, H. C., and Hill, D. (1942). *Rep. Great Barrier Reef Comm.* 5, 1.

Robertson, E. I. (1968). *N. Z. J. Geol. Geophys.* **10**, 1484.

Robertson, E. I., and Kibblewhite, A. C. (1966). *N. Z. J. Geol. Geophys.* **9**, 111.

Royal Society. (1904). "The Atoll of Funafuti: Borings into a Coral Reef and the Results." Royal Society, London.

Saplis, R. A., and Flint, D. E. (1949). *Geol. Soc. Am. Bull.* **60**, 1974.

Scheer, G. (1972). *Proc. Int. Symp. Corals and Coral Reefs* (Mandapam, 1969), p. 87.

Schlanger, S. O. (1963). *U.S., Geol. Surv., Prof. Pap.* **260-BB**, 991.

Semper, C. (1881). "The Natural Conditions of Existence as they Affect Animal Life." Kegan Paul, Trench, Trubner, London.

Shepard, F. P. (1948). "Marine Geology." Harper, New York.

Shepard, F. P. (1970). *Geol. Soc. Am. Bull.* **81**, 1905.

Shor, G. G. (1964). *Nature (London)* **201**, 207.

Spencer, M. (1967). *Am. Assoc. Petrol. Geol. Bull.* **51**, 263.

Stanley, D. J. (1970). *Z. Geomorphol.* [N. S.] **14**, 186.

Stearns, H. T. (1944). *Geol. Soc. Am. Bull.* **55**, 1279.

Stearns, H. T., and Chamberlain, T. K. (1967). *Pac. Sci.* **21**, 153.

Stearns, H. T., and Vaksvik, K. N. (1938). *Bull. Hawaii Div. Hydrogr.* **4**, 1.

Steers, J. A. (1929). *Geogr. J.* **74**, 232 and 341.

Stoddart, D. R. (1962). *Atoll Res. Bull.* **87**, 1.

Stoddart, D. R. (1963). Ph.D. Thesis, Cambridge University.

Stoddart, D. R. (1969a). *Philos. Trans. R. Soc. London, Ser. B* **255**, 355.

Stoddart, D. R. (1969b). *Philos. Trans. R. Soc. London, Ser. B* **255**, 383.

Stoddart, D. R. (1969c). *Biol. Rev. Cambridge Philos. Soc.* **44**, 433.

Stoddart, D. R. (1971). *Symp. Zool. Soc. London* **28**, 3.

Stoddart, D. R. (1973). *Geography* **58**, 313.

Stoddart, D. R. (1976). *Micronesica* **12**, 1.

Stoddart, D. R., Davies, P. S., and Keith, A. C. (1966). *Atoll Res. Bull.* **116**, 13.

Stoddart, D. R., Taylor, J. D., Fosberg, F. R., and Farrow, G. E. (1970). *Philos. Trans. R. Soc. London, Ser. B* **260**, 31.

Sugiyama, T. (1936). *Publ. Geol. Paleontol. Inst. Tohoku Univ.* **25**.

Tayama, R. (1952). *Bull. Hydrogr. Off. (Dep.) Tokyo* **11**, 1.

Taylor, J. D. (1968). *Philos. Trans. R. Soc. London, Ser. B* **254**, 129.

Thurber, D. L., Broecker, W. S., Blanchard, R. L., and Potratz, H. A. (1965). *Science* **149**, 55.

Todd, R., and Post, R. (1954). *U.S., Geol. Surv., Prof. Pap.* **260-N**, 547.

Traves, D. M. (1960). *In* "The Geology of Queensland" (D. Hill and A. K. Denmead, eds.), pp. 369–371. University Press, Melbourne.

Trudgill, S. (1970). *Area 1970* No. 3, p. 61.

Tylor, A. (1872). *Geol. Mag.* **9**, 392 and 485.

Umbgrove, J. H. F. (1930). *Leidse Geol. Meded.* **3**, 261.

Umbgrove, J. H. F. (1947). *Geol. Soc. Am. Bull.* **58**, 729.

Vaughan, T. W. (1914). *Bull. Am. Geogr. Soc.* **46**, 426.

Vaughan, T. W. (1916). *Proc. Natl. Acad. Sci. U.S.A.* **2**, 95.

Vaughan, T. W. (1918). *Smithson. Inst., Annu. Rep.* **17**, 189.

Vaughan, T. W. (1919). *U.S., Natl. Mus., Bull.* **103**, 189.

Vaughan, T. W. (1923). *Proc. Pan-Pac. Sci. Congr., 2nd, 1923* Vol. 2, p. 1128.

Veeh, H. H., and Veevers, J. J. (1970). *Nature (London)* **226**, 536.

von Lendenfeld, R. (1896). *Westermanns Monatsh.* **79**, 499.

Wells, J. W. (1954). *U.S., Geol. Surv., Prof. Pap.,* **260-I**, 385.

Wentworth, C. K. (1928). *Geogr. Rev.* **18**, 332.

Wharton, W. J. L. (1890). *Nature (London)* **42**, 172.

Wharton, W. J. L. (1897). *Nature (London)* **55**, 390.

Wiens, H. J. (1962). "Atoll Environment and Ecology." Yale Univ. Press, New Haven, Connecticut.

Wilson, J. T. (1963). *Nature (London)* **197**, 536.

Wood-Jones, F. (1912). "Coral and Atolls." L. Reeve, London.

Yabe, H., and Tayama, R. (1937). *Proc. Imp. Acad. (Tokyo)* **13**.

THE NATURE AND ORIGIN OF CORAL REEF ISLANDS

D. R. Stoddart and J. A. Steers

There are different opinions amongst ingenious theorists, concerning the formation of such low islands. . . . James Cook, 17 April, 1777

I. Introduction

A fundamental distinction is made in reef areas between "high islands," formed of elevated reef limestones or of nonlimestone rocks, and "low islands," formed by the accumulation of debris derived from the growth and breakdown of reef organisms (Forster, 1778, pp. 148–159; von Chamisso, 1821). Great variety exists within the general class of low islands, in terms of location on reefs, origin and development, topography, and nature of vegetation cover. Studies of sand and shingle cays were first made in the East Indies by Umbgrove (1928, 1929, 1930b, 1947) and Kuenen (1933) and on the Great Barrier Reef by Steers and Spender (Steers, 1929, 1937, 1938; Spender, 1930; Stephenson et al., 1931). David and Sweet (1904) carried out the first detailed study of the islands of an open-ocean atoll (Funafuti, Ellice Islands), and important subsequent studies include those in the northern Marshalls (particularly Bikini), southern Marshalls (Arno), Carolines (Ifaluk, Ulithi, Kapingamarangi), Gilberts (Onotoa), Cooks (Aitutaki), and Tuamotus (Raroia, Mururoa, Rangiroa, and nearby Mopelia). In recent years attention has turned particularly to the islands of the Caribbean (especially to Jamaica, British Honduras, Florida, and the Bahamas) and the Indian Ocean (Addu in the Maldives, Diego Garcia in the Chagos, and several southwest Indian Ocean atolls). The most important of these studies of coral islands are listed in Table I.

These studies permit us to outline a comprehensive classification of islands on reefs; to discuss the major regional differences in reef and island characteristics; and to consider the origin, history and present status of the reef islands of the world.

II. Classification

Spender (1930, pp. 285–286) proposed a fivefold classification of the reefs off the Queensland coast in terms of the nature of sediment accumulation on them:

> Class I. Those reefs where the debris is scattered over the surface without forming a cay or rampart . . .
> Class II. Reefs bearing a sand cay but no rampart . . .
> Class III. Reefs having sand cay . . . and rampart, but no vegetation on the rampart or flat . . .
> Class IV. Reefs having sand cay, and ramparts; the ramparts vegetated, but no distinctive vegetation on the flat . . .
> Class V. Reefs having sand cay, vegetated rampart, and mangrove swamp on the flat.

TABLE I
STUDIES OF REEF ISLANDS

Type of island	Area	Study
Sand cay	Jamaica	Steers, 1940a,b, Steers *et al.*, 1940
	British Honduras	Stoddart, 1962, 1963, 1965a
	Gulf of Mexico	Folk, 1967; Folk and Cotera 1971; Folk and Robles 1964
	Red Sea	Guilcher, 1955
	Madagascar	Guilcher, 1956
	Indonesia	Umbgrove, 1928, 1929, 1930b; Kuenen, 1933; Verwey, 1931
	West Australia	Teichert and Fairbridge, 1948
	Great Barrier Reef	Steers, 1929, 1937, 1938
	New Caledonia	Catala, 1950; Guilcher, 1965
	Solomon Islands	Stoddart, 1969b
Motu	Jamaica	Zans, 1958
	Western Indian Ocean	Stoddart, 1970
	Cocos-Keeling	Guppy, 1889
	Great Barrier Reef	Steers, 1937
	Marshall Islands	Emery *et al.*, 1954; Fosberg and Carroll, 1965; Wells, 1951
	Caroline Islands	McKee, 1956, 1958; Wiens, 1956; Tracey *et al.*, 1961; Schlanger and Brookhart, 1955
	Gilbert Islands	Cloud, 1952
	Ellice Islands	David and Sweet, 1904
	Tuamotu Islands	Newell, 1954a,b, 1956; Stoddart, 1969a; Chevalier 1969
	Cook Islands	Gibbs *et al.*, 1971
	Clipperton	Sachet, 1962
Mangrove	Marquesas (Florida)	Davis, 1942
	Jamaica	Steers, 1940a,b
	British Honduras	Stoddart, 1962
Mangrove-Sand	Marquesas (Florida)	Davis, 1940, 1942
	Jamaica	Steers, 1940a,b
	British Honduras	Vermeer, 1959; Stoddart, 1962, 1963
	Bahamas	Scoffin, 1970
Low wooded island	Great Barrier Reef	Steers, 1929, 1937, 1938; Spender, 1930; Stephenson *et al.*, 1931; Fairbridge and Teichert, 1947, 1948
	Indonesia	Umbgrove, 1928; Verwey, 1931; Kuenen, 1933; Verstappen, 1954
	Jamaica	Steers, 1940a,b
	British Honduras	Stoddart, 1965a

This classification is largely based on the occurrence of a particular type of island, the "low wooded island" or "island reef," recognized by Steers (1929), consisting of a reef patch with windward shingle rampart, leeward sand cay, and intervening area of shallow reef flat and mangrove swamp. Many examples of this type (Class V) have been mapped by Steers (1938). Spender's classes III, IV, and V are geomorphically indistinguishable, the differences depending only on vegetation distribution; Fairbridge (1950, p. 347) has further pointed out that no example of Class IV is known and only two examples of Class III. In essence, therefore, Spender's scheme is threefold:

  I. Reefs without islands
 II. Reefs with a simple sand cay
III. Reefs with a low wooded island in various stages of vegetation development

Fairbridge (1950, pp. 347–349) recognized the incompleteness of this series, and developed a more extensive classification:

Type 1—sand cay, unvegetated.—Commonly unstable and migrating seasonally. Found in all areas [of the Australian reefs] except on the fringing reefs.

Type 2—sand cay, vegetated.—Moderately stabilised, generally with beachrock. Widely distributed, generally missing from the outer more exposed reefs.

Type 3—shingle cay, with or without vegetation.—Moderately stabilised, widely distributed, generally found on smaller, more exposed reefs.

Type 4—sand cay with shingle ramparts (beach ridges).—Unvegetated to completely vegetated islands, including mangrove swamp.

Type 5—island with exposed platform of older, emerged coral-reef material.—With or without a fringe of recent sand or shingle beach ridges or ramparts.

All of these types are represented by islands of the Great Barrier Reefs, and islands of other reef areas can be fitted into the classification. Some of the classes can be further subdivided, as shown by Folk (1967).

Fairbridge's and Spender's schemes, however, were developed for the islands of barrier and patch reefs, and they do not sufficiently emphasize the distinctive character of Indo-Pacific atoll islands. Nor do they consider the various kinds of mangrove islands well described by Davis (1940, 1942) from Florida. An unusually wide range of islands is present on the British Honduras coast (Vermeer, 1959; Stoddart, 1962, 1963), and a fuller island classification has been developed for this area (Stoddart, 1965a). Morphology and sediment character are the main criteria in this classification, with vegetation a secondary criterion; ten categories were first recognized, and others could be added logically to complete the scheme although examples of such additional classes would be nonexistent or unimportant in reality. For present purposes the number of classes is reduced to seven:

A. Sand cay, unvegetated or vegetated
B. Sand cay with shingle ridges (*motu*), generally vegetated
C. Shingle cay, unvegetated or vegetated
D. Mangrove cay, with or without low dry-land areas
E. Mangrove cay with windward sand ridge.
F. Low wooded island *sensu* Steers (1929) (= island reef of Spender (1930), low wooded island-reef of Fairbridge and Teichert (1947), and moat island of Stoddart (1965a)
G. Emerged reef-limestone island

III. Island Types

A. SAND CAYS

The "simple cays" of Steers (1929, p. 247) vary in size, stability and location, and two main types may be recognized. The less permanent are long linear islands forming narrow bars or barriers on reef flats lagoonward of unbroken reef edges. They are usually less than 1 m high at low water, and a few meters wide: they are unvegetated, lack beach-rock, and are often overtopped by swash. They are temporarily emergent sections of more extensive submerged sand bars and ridges on reef flats and platforms (Off, 1963; Ball, 1967), and the term "sandbore" is applied to both the emerged and the submerged sections.

True sand cays are lens-shaped, crescentic, triangular or near-circular islands, constructed of concentric or parallel sand ridges, built by refraction of waves round the underlying reef (Fig. 1). The thickness of sand accumulation varies with tidal range, but many of these islands are only 1 m high at high spring tides. Steers (1929, p. 249) found that the long axis of elongate cays was oriented at 45° to the prevailing wind; Folk (1967, p. 427) found a similar orientation on Alacran in the Gulf of Mexico, but elsewhere the orientation appears to be random or controlled by the orientation of the underlying reef. Steers also found a common difference in alignment between the higher and the lower portions of the cay. Unvegetated sand cays may be 50–100 m in their longest dimension, are awash at high tide, and are clearly ephemeral. Most

Fig. 1. A simple sand cay on a small patch reef: Pompion Cay, British Honduras barrier reef.

workers have recorded variations in the size, location, and even existence of such islands (e.g., Edgell, 1928). Contemporary beachrock is not found on such islands, possibly because the sediments are too mobile for cementation to occur, but relict beachrock may indicate the former existence of a larger island, often confirmed by historical records. Such islands are typically found on barrier reefs and lagoon patch reefs, in both protected and exposed situations.

Vegetated sand cays differ from unvegetated cays in size, morphology, stability, and possible development of beachrock. Smaller islands may be colonized by herbs, grasses, vines, and low shrubs, and in hurricane areas may show evidence in relict beachrock of former larger size. Large cays may support a tall broadleaf woodland, in many places now replaced by coconut plantations. Such islands are usually longer than 100 m and may reach more than 1000 m in length: Great Barrier Reef cays have a mean length of about 300 m and width of 150–200 m (Steers, 1929). They are typically longer than they are broad, and lens-shaped or oval in shape: a few have very complicated shapes as a result of the extension of sand spits to leeward. Few vegetated islands are less than 1 m above high spring tide level, or more than 3 m. On the Great Barrier Reefs, where tides may exceed 3 m, most cays stand between 1 and 2 m above high spring tide level. In areas of seasonally alternating winds such islands are subject to considerable fluctuations of shape over the year, with building and erosion of temporary sand spits (Fairbridge and Teichert, 1948, p. 74). There appears to be no simple relationship between the size of a cay and the size of the underlying reef: cays in Djakarta Bay appear to be the largest relative to their reefs (Kuenen, 1933, p. 84) but elsewhere there is no obvious relationship.

Little is known of sand cays within atoll lagoons, though these are common in some atoll groups, particularly in the Maldives where they have not yet been studied. Fosberg (1956), however, has noted that such islands are often concentrated on patch reefs immediately within lagoon entrances, and has suggested that such islands may have been formed by the exposure of current-deposited sands by a fall in sea level. Alternatively, patch reefs in such situations may be larger and higher than such reefs elsewhere in lagoons.

B. Sand Cay with Shingle Ridges (Motu)

The largest and most stable of inhabitable reef islands, including most of the islands of Indian and Pacific Ocean atolls, fall into this class: Newell (1961) has applied the Polynesian term *motu* to this type, and his usage is adopted here. These islands typically consist of a seaward shingle ridge 3–5 m high, prolonged lagoonwards at each end by lower spits of finer material, and with a lower sand area to leeward (Fig. 2). The seaward shingle complex generally consists of interlocking coral branches (Hedley, 1925) with other coral fragments and mollusc shells, forming an open structure with high permeability. The lagoon beach usually consists of a low sand ridge, in some cases with coral and molluscan gravel. Between the seaward and lagoonward ridges the surface may fall gently or there may be an interior enclosed depression with standing water. Usually the peripheral ridges are single features, though in some cases two, three, or more such ridges have developed.

Motus vary in number, size, and shape on any given reef. Rarely, an atoll may be completely or almost completely rimmed by land (e.g., Diego Garcia); in others a formerly continuous or extensive land rim has been dissected by channels to form a large number of smaller islands.

Islands are particularly numerous on Tuamotu atolls, with 300 on Mururoa, 280 on Raroia and 249 on Rangiroa, covering 30–35% of the atoll rim (10% is common in the Marshall Islands, where Bikini has only 26

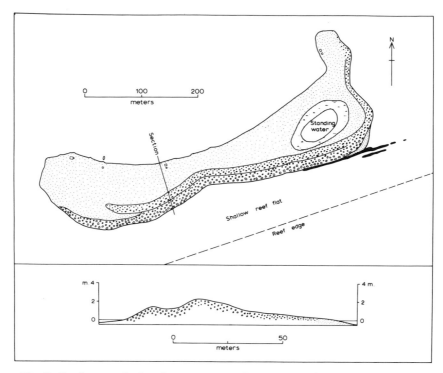

Fig. 2. Sand cay with shingle ramparts on the seaward side: Middle Cay. Glover's Reef, British Honduras. Source: Stoddart (1962).

islands). Larger islands, often with an area of more than 1 sq km, are rectangular and oblong in shape, generally following the plan of the underlying reef; smaller islands are usually convex seaward and concave lagoonward; and the smallest islands may be hammer or T-shaped, with the broader end facing either the lagoon or the sea, depending on their constitution. Figure 3 shows the relationships between length and width of motus in the Marshall Islands: most motus have a length-width ratio of 1.2–5, but in the case of the largest the ratio exceeds 5.5. Most motus are situated on the inner half of the reef flat, with their lagoon beach close to the lagoon reef slope. The seaward beach may be several hundred meters from the seaward reef edge, but frequently overlies a wide exposed area of conglomerate pavement extending seawards. Cases are known, however, of motus lying close to the seaward reef edge, and in some atolls, such as Addu (Stoddart *et al.,* 1966), there is a regular variation in location of the island round the atoll rim, as exposure changes with orientation.

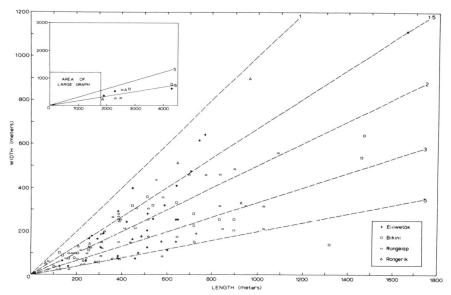

Fig. 3. Relationship between width and length of reef islands on atolls in the Marshall Islands. Data from charts in Emery *et al.* (1954).

C. Shingle Cay

Unvegetated shingle cays are not common on coral reefs, largely because islands built of coarse material are more stable than sand cays and hence are more usually vegetated. Ephemeral shingle cays may be formed during major storms in exposed situations, and are later destroyed by normal wave action. Vegetated shingle islands are also uncommon. They may occur in exposed situations on barrier reefs (e.g., Ragged Cay, British Honduras), but usually they form a shelter in the lee of which fine sediments accumulate, and thus are transformed into motus.

A special case of shingle cay is the hammerhead shingle tongue found on some atoll reefs. These are wider on the seaward side and have a tail directed lagoonward. On Seringapatam Atoll, northwest of Australia, these tongues have a head 50 m wide, with a tail 10 m wide and 200–400 m long (Teichert and Fairbridge, 1948, p. 243).

D. Mangrove Cay

Mangrove islands are common in the Caribbean (Florida, Jamaica, British Honduras), in the East Indies, Melanesia, parts of Micronesia, and the Great Barrier Reef. They are, however, absent over large parts of the reef seas: in the Indian Ocean mangroves are found round continental coasts and on the islands of the southwest; they are present in the northern Maldives but become rarer southwards and are absent in the Chagos and other central Indian Ocean reef areas. In the Pacific there are fewer in the eastern part of the ocean: they are present as far east as Tonga, but are absent in the Cooks, the Society, and the Hawaiian Islands. Guilcher (1967) has drawn attention to contrasts between the Gilberts and the Tuamotus, resulting from the presence of mangroves in the former and absence in the latter.

Mangrove islands are originally formed by the colonization of shoal areas by mangrove species, especially *Rhizophora mangle* and *Avicennia germinans* in the Atlantic and *R. mucronata* and *A. nitida* in the Indo-Pacific. Mangrove islands range from small with individual trees to larger islands hundreds of meters long, intersected by winding tidal channels or "bogues" in Caribbean usage. Though from the sea these larger islands apparently consist of dense mangroves, the interior is often open, with bare mud, dead trees, and standing water. Once established, mangroves promote sedimentation by interfering with water movements; ultimately low-lying areas of silty sand with a strand vegetation or "hammock" woodland may develop. These drier areas may be on the windward or leeward sides of the cay, or in the center, and may be partly or wholly surrounded by mangroves. The ground surface is usually only a few dm above the highest tide levels, and the ground is often waterlogged.

In one case, Turneffe Atoll in the Caribbean, mangroves completely surround the atoll rim and enclose a large lagoon (Stoddart, 1962). However, detailed studies on the ecology and sedimentology of mangrove areas have been carried out mostly near continental coasts, and are well reviewed by Davis (1940), Scholl (1963–1964), and Macnae (1968).

E. MANGROVE CAY WITH WINDWARD SAND RIDGE

This type, first recognized as distinctive by Vermeer (1959) in British Honduras, consists of four main zones: (a) seaward sandy shore with strand vegetation; (b) sand ridge area, 3–5 m high, with littoral woodland and coconuts; (c) mangrove transition zone, with tall mature mangroves being invaded by littoral woodland species; (d) leeward mangrove zone. In some cases the windward beach may be topped by dunes. Outcrops of beachrock are uncommon on the seaward beach, possibly because aggradation is more usual in such cases than erosion, but storm stripping on the beach face may reveal patches of a subaerial cay sandstone. These islands may be several kilometers long and hundreds of meters wide, or they may be quite small. They are usually located some hundreds of meters from the reef edge, in shelf situations. They were first described in detail in the Bogue Islands, Jamaica, by Steers (1940a, pp. 36–37), and in the Marquesas, Florida, by Davis (1942). They are very well developed in British Honduras (Vermeer, 1959; Stoddart, 1962, 1963) and in the Bahamas (Scoffin, 1970).

F. Low Wooded Islands

The term "low wooded island" and similar terms are used for patch reefs bearing on their upper surface: (a) a shingle ridge or rampart, close to the windward reef edge; (b) a sand cay on the lee side of the reef, in some cases joined to the shingle rampart; (c) an area between cay and rampart of open water which may or may not be colonized by mangroves (Fig. 4). One case is known (Hope Islands, Queensland) in which the sand cay is on one reef and the rampart and mangrove swamp on another (Steers, 1929). The area of open water was originally termed

Fig. 4. Moat island (East Snake Cay) in the British Honduras barrier reef lagoon. The three components of shingle rampart (left foreground) shallow lagoon or moat with mangroves, and leeward sand cay (background, with coconut palms) are clearly visible.

the pseudolagoon (Hedley and Taylor, 1907; Steers, 1929), but this term is now rejected (Spender, 1930; Steers, 1937, p. 254).

Islands with these characteristics were first described, and the type recognized, in the Great Barrier Reefs by Steers (1929). Spender (1930), who termed them island-reefs, mapped two in great detail (Low Isles, Three Isles). Steers (1937, 1938) later mapped a further 15, and proposed Bewick Island as the type example. Low Isles itself (Fig. 5) has been studied in great detail ecologically over a period of years (T. A. Stephenson *et al.*, 1931; W. Stephenson *et al.*, 1958), and geomorphic changes have also been observed (Fairbridge and Teichert, 1947, 1948). Umbgrove (1928) described rather similar islands in Djakarta Bay, Java, and Steers (1940a, pp. 32–35) recognized the basic similarity of the

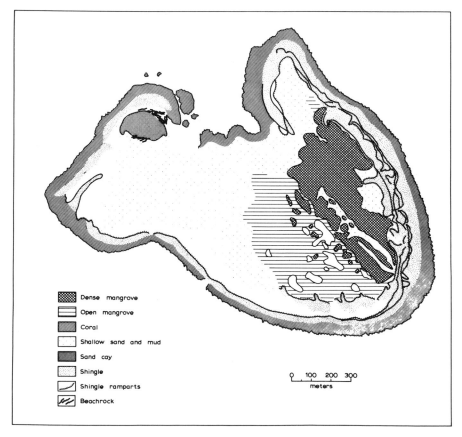

Key:
- Dense mangrove
- Open mangrove
- Coral
- Shallow sand and mud
- Sand cay
- Shingle
- Shingle ramparts
- Beachrock

0   100   200   300
meters

Fig. 5. Low wooded island (Low Isles) in the Great Barrier Reef lagoon, Queensland. From Stephenson *et al.* (1931) with additions from Fairbridge and Teichert (1948).

Pigeon and Salt Cays, Jamaica, to the low wooded island type. Finally, analogous forms have been found in the British Honduras barrier reef lagoon, in situations similar to those of the Australian low wooded islands (Stoddart, 1965a).

G. Emerged Limestone Island

Emerged limestone islands range from very slightly emerged coral reefs to completely raised atolls (such as Aldabra), barrier reefs (such as those around New Georgia), and fringing reefs (well developed in the New Hebrides); some may be tens or hundreds of meters high in the mobile belts of Melanesia and the East Indies. We deal here only with slightly emergent reefs which in height and size resemble detrital reef islands, and which, because they are referred to as keys or cays, may be confused with some of the islands already discussed. Such islands are best developed in Florida, the Bahamas, and Bermuda.

The Florida keys consist of reef limestones generally raised up to 1 m above high tide level, with the lower Key Largo Limestone overlapped in the Lower Keys by the Miami Oolite (Stanley, 1966; Hoffmeister and Multer, 1968). The Upper (northern) Keys, formed of raised reef limestone, are linear islands arranged in an arcuate manner parallel to the seaward shelf edge 8 km distant; mangroves grow in their lee. The Lower Keys are larger and wider. It is believed that the channels between them are inherited from tidal channels cut in the original submarine oolite bar (Hoffmeister and Multer, 1968), and similar contemporary sedimentary features have been found in the Bahamas (Hoffmeister *et al.*, 1967). The size of the Lower Keys results largely from mangrove growth on both their seaward and lagoonward sides. Because of their raised limestone nature, beaches are rare throughout the keys.

The Bahamas and Bermuda are also formed by elevated limestones, including oolites, but in addition large areas are covered with cemented aeolianites. There is a large literature on the complex Pleistocene history revealed by these deposits (Land *et al.*, 1967).

IV. Geomorphic Features

A. Uncemented Topographic Features

1. Beaches and Beach Ridges

Beaches of reef islands vary widely in height, texture, and composition and are correspondingly difficult to classify. A simple distinction between windward and leeward beaches does not account for the further distinction between seaward and lagoonward beaches. They may be grouped in terms of sediment caliber (boulder beach, cobble beach, gravel beach, sand beach), but these terms often do not convey the peculiar characteristics of reef-derived sediments. They may be classed in terms of dominant process in terms of normal (tidal) ridge, storm ridge, and dune ridge (Fosberg, 1956), though these categories are not normally exclusive. The characters of island beaches may also differ markedly if the island is close to or distant from the reef edge (Stoddart et al., 1966, p. 31). In general, however, there is a fairly direct relationship between beach characteristics (height, width, angle, sediment caliber) and exposure (dependent on wind speed, direction, frequency, and fetch, and on local offshore topography) (Newell, 1954b, p. 24).

Windward ridges on seaward coasts are usually high and steep, with relatively narrow beaches rising from a shallow or intertidal reef flat. Such ridges have been widely described in reefs in the trade-wind belts (Steers, 1937; Cloud, 1952; Newell, 1956; Stoddart, 1962), where they may consist of gravel, cobbles, and in some cases boulders up to 1 m in diameter (Emery et al., 1954). Such exposed beaches may have angles of 15–20° (e.g., Wells, 1951, p. 51, on Arno). Such steep angles can be easily maintained on beaches built of highly permeable, interlocking coral shingle (Hedley, 1925; Kuenen, 1933; Folk and Robles, 1964). It is possible that the highest ridges are constructed during exceptional storms, when vegetation on the ridge crest may act as a sediment trap for overtopping waves (Stoddart, 1965b). Frequently, however, storm effects are most evident on coasts other than normal windward coasts, e.g., on northwest coasts in the central Great Barrier Reefs, on west shores in the Tuamotus, and on south shores in the Marshalls. Moreover, not all windward seaward ridges are high and steep. In the equatorial belt of less constant winds coarse material may be largely absent and the ridges sandy and lower, as on Diego Garcia, Indian Ocean (Stoddart, 1971b). Alternatively, windward ridges may be altogether absent where the windward shore of an island is formed by a wide, low, conglomerate platform thinly veneered by sediments. In such cases, the only ridge is the lagoonward beach ridge and the normal island asymmetry is reversed [e.g., the seaward sides of Kapingamarangi (McKee, 1956, p. 8) and Aitutaki (Gibbs et al., 1971)]. When present, high ridges may be double or even treble features. Windward beaches throughout the world are undergoing erosion and retreat, and they are characterized by outcropping beachrock and other lithified sediments (Section IV,B).

Windward beaches on lagoon shores lack most of the features of seaward shore beaches. They are dominantly sandy, with patches of gravel, and there is usually no pronounced ridge. Vegetation approaches close to the water level. Dunes may form in such situations, especially on shores facing wide, shallow lagoon flats (e.g., on Rangiroa: Stoddart, 1969a). On channel shores between islands, where sediments are in continuous transit from the sea toward the lagoon, flat cobble and gravel beaches alternate with steep-sided, flat-topped sand lobes and spits. Sand spits commonly form at the inner ends of channels, and prolong lagoon beaches.

Seaward beaches on the leeward sides of reefs are usually wide and may be high; they are generally sandy except where coarse sediments are deposited from time to time by major storms. The reef flats immediately offshore are deeper than on windward reefs, and coral may grow close inshore. These beaches are generally prograding, and beachrock exposures are much less common than on windward shores. On leeward lagoon shores beaches may be low and narrow, or they may be absent, with the shore formed by a low cliff of eroding sediments or by a low calcarenite platform. Both types of shore are found on the Clipperton Atoll, for example (Sachet, 1962, pp. 39–41). Coarse sediments are uncommon except in the lower intertidal and subtidal zones. Where the beaches are prograding there is a characteristic overlap of well-sorted fine sediments over a pavement of poorly sorted coarse sediments.

Gan Island, Addu Atoll, in the Indian Ocean, is perhaps exceptional in that its seaward coast is relatively protected by a wide reef flat and its lagoon coast is rather more exposed to lagoon wave action, but its beaches have been studied in detail. Here and at Hitaddu Island the width of the beaches is remarkably constant, at 14–16 m, and there is a direct correlation between beach height and beach angle (Fig. 6). At Hitaddu 19 seaward profiles gave a mean angle of 9.5°, and 11 lagoon profiles a mean angle of 5.2° (Stoddart *et al.*, 1966). Minor sedimentary structures also differ between seaward and lagoon beaches, with beach cusps more common on the former and ripple marks on the latter, especially on flatter beaches.

2. Ramparts and Shingle Tongues

Shingle ramparts (meaning ridges distinct from a sand or shingle cay) were classically described at Low Isles, Great Barrier Reefs. Here there are two main ramparts on the windward side of the reef flat; they are asymmetric, with a gentle outer slope and a steeper inner scarp, with mean angles 2–5° and 45° respectively, the scarp being about 1 m high. The ramparts extend inwards across the surface of the reef for about 215 m, and their crests are generally between mean sea level and extreme

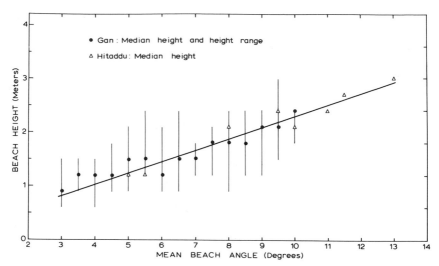

Fig. 6. Relationship between beach height and beach angle for beaches on two islands of the Addu Atoll, Maldive Islands. The median heights and height range are shown for each class of angles. From Stoddart *et al.* (1966).

high water springs. Tongues of shingle extend inwards from these ramparts from place to place, especially at each end of the series. In addition to the two main ramparts, the inner of which consists of old, blackened, partially-cemented shingle and the outer of fresh, white, loose shingle, later work revealed portions of an even older consolidated rampart lagoonward of the inner rampart, while a fourth, the outermost ridge has been formed since the 1928–1929 survey (Steers, 1929; Spender, 1930; Stephenson *et al.*, 1931; Moorhouse, 1933, 1936; Fairbridge and Teichert, 1947, 1948). Steers (1937) has found similar features on other islands, except that in general the ramparts are high enough on other islands to support terrestrial vegetation.

Elsewhere, where similar features have been mapped, they are much less complex. In both Djakarta Bay and on the British Honduras coast, the ramparts are simple shingle ridges 1–2 m high and a few meters wide, with patches of pioneer strand vegetation and mangroves (Fig. 7). The development and significance of these features is considered in Section VIII,D.

3. Central Flat

Between the seaward and lagoonward ridges the island surface may be horizontal, with a slight fall from sea to lagoon, or with a steeper fall to a central depression. Immediately in the lee of the seaward ridge, the

Fig. 7. Shingle rampart and moat with mangroves, West Snake Cay, British Honduras barrier reef lagoon.

surface may be raised by the deposition of washover sediments during storms. Such "island gravel" spreads have been described at Raroia (Newell, 1956) and similar "sand wash" at Clipperton (Sachet, 1962, p. 35). In the northern Marshalls Fosberg (1956) has described storm-deposited blockfields up to 1.5 m thick, with large boulders deposited up to 200 m inland. Except where dunes or depressions are present, however, inland surfaces are relatively featureless. Fosberg and Carroll (1965) have described "blowdown mounds" as minor features; these are 0.6–1.5 m high and several meters in diameter, formed of small sharp coral fragments with no fine material present. They suggest that these fragments result from the washing of sediments from the roots of fallen trees, though the absence of an adjacent hollow presents a major problem.

4. Central Depression

Depressions in island surfaces are usually closer to lagoon than to seaward coasts; they may be dry, marshy, or with standing water. The latter are known as *kuli* in the Maldives (Willis and Gardiner, 1901, p. 78) and as *tairua pape* in the Tuamotus (Stoddart, 1969a). The size

of pools in interior depressions may fluctuate widely with seasons, with
individual rainstorms, and even with tides. Such depressions were described in Funafuti by David and Sweet (1904), who thought they
resulted from solution. Steers (1937, p. 7) argued that the surrounding
rim was higher because it was newer, while Wiens (1962) believed that
sea-level changes were important causes of differences in height. In many
cases, however, the presence of an interior depression probably results
simply from the spatial separation of seaward and lagoonward beach
ridges and an inadequate supply of sediment to fill the intervening gap.
At Maraki Atoll in the Gilberts the interior depression is so extensive that
it gives the impression that the island rim consists of two narrow parallel
linear islands (Agassiz, 1903, pp. 255–257, plates 149–150).

A second type of depression, especially common in the West Pacific, is
anthropogenic though many may have originated as natural depressions.
These are taro or puraka pits, often narrow and winding, with angular
outline, 1–2 m deep, with steep sides, and floored with organic soil. On
Kapingamarangi these pits occupy 9% of the total land area, and on one
island, Warua, the proportion rises to 25% (Wiens, 1956, p. 36).

Other kinds of depressions, generally transverse, may be formed by
hurricane scour which fails to divide an island completely, or, alternatively, by the growth of beach ridges across the seaward and lagoonward
ends of a former interisland channel (see Section IV,A,7).

5. Barachois

The term *barachois* was used by Gardiner (1903) to refer to tidal
depressions opening onto the leeward shores of islands, particularly in
the Maldive and Chagos archipelagoes. Barachois may have very irregular
outlines on their landward sides, as a result of the development of spits
and washover features, but their lagoon beaches are generally straight
and smooth. Barachois are flooded at high tide, and large areas of silty
sand may be exposed at low tide; at lower levels meandering creek systems may develop. These features are best developed in areas such as
the central Indian Ocean where mangroves, which would normally
colonize such a habitat, are absent. Gardiner believed the Indian Ocean
barachois to be actively enlarging but although outcropping lithified
sediments form low cliffs around the margins of many, there is very little
evidence of significant enlargement since his surveys (Stoddart, 1971b).

6. Dunes

Dunes may form on either windward seaward shores (e.g., Cosmoledo
and Farquhar Atolls: Stoddart, 1970) or windward lagoon shores (e.g.,

the southern Marshalls: Wells, 1951; northern Marshalls: Fosberg, 1956; Tuamotus: Stoddart, 1969a), or more rarely inland. They are best developed in arid or semi arid trade-wind areas, and are usually absent or weakly developed in humid equatorial areas. They are more common in areas of high tidal range, where wide areas are exposed at low water. Thus high dunes are found on many southwest Indian Ocean atolls but are absent in the Chagos and Maldives. Dune deposits in humid areas, e.g., in the Caribbean, are often less than 2 m thick (Stoddart, 1962; Folk, 1967). Very high dunes reported at Christmas Island (Pacific) in the past are now known to be only 10.7 m high (Jenkin and Foale, 1968). Dunes on several southwest Indian Ocean atolls reach 15 m (Stoddart, 1970). Massive dune development in the past is indicated by the presence of aeolian ridges 30–45 m thick forming a windward fringe on Andros and covering most of Cat, Eleuthera, Great Abaco, and Great Exuma Islands in the Bahamas (Ball, 1967).

7. *Hoa*

The Tuamotuan term *hoa* is applied to erosional channels cut through formerly continuous atoll rims; they are particularly well developed on atolls in the Tuamotus, but are found elsewhere in the Indo-Pacific. They may be open at both ends, permitting free flow of water through them at high tide, or they may be closed by intertidal conglomerate ledges at their seaward end or by temporary sand bars and spits at their lagoonward end. Lithified sediments may outcrop in the walls of the hoa toward the seaward side, but towards the lagoon the sediments are sandy and the walls lower. Hoa have been described at Rangiroa (Stoddart, 1969a) and other Tuamotu atolls (Chevalier, 1969; Chevalier *et al.*, 1969). Newell (1956) counted 260 hoa on Raroia. Chevalier made a distinction between functional (i.e., open) and nonfunctional (closed) hoa, the latter being properly termed *tairua*, and suggested that transverse faulting combined with sea-level change has resulted in the initiation of hoa-cutting. Where hoa are common there has certainly been a change from the period when the continuous land rim was built to the present time when dissection is taking place. Hoa features are so uniform and widespread that it is difficult to ascribe them to storm action, and it is possible that they are partly the result of sea-level changes altering the balance of sedimentation and erosion.

B. CEMENTED TOPOGRAPHIC FEATURES

Sediments of reef islands may be lithified to form a variety of rocks with different morphologic expressions. These processes of lithification,

which play an important part in stabilizing loose sediments and protecting
them from wave attack, are not well understood and are probably diverse.
It is important and often difficult to distinguish these rocks from true elevated reef-rock and other lithified submarine sediments (Teichert, 1947,
p. 152). Four types of topographic feature resulting from such lithification are distinguished here.

1. Beachrock Ridges

Beachrock is normally exposed intertidally, or within the range of
swash or spray on more exposed coasts, and has long been considered
characteristic of tropical carbonate beaches (Jukes, 1847, I, pp. 7–9). It
forms seaward-dipping plates extending along the foot of the beach (Fig.
8). Primary cementation takes place within the beach, and the rock
retains the dip of the original bedding planes, generally 5–30° seawards.
The width (generally 1 to a few meters) and thickness (generally 1 m or
less) of each plate are functions of tidal range and beach dimensions.
Following initial cementation beachrock is exposed by beach retreat.
Once exposed, the surface may be beveled by erosion to form a surface

Fig. 8. Single line of beachrock showing marked seaward dip: Laurie Island, East
Guadalcanal, Solomon Islands.

with lower dip than that of the bedding planes, plates may be under-mined and fractured by wave action, and a secondary cementation commonly takes place to form a very strongly bonded rock. Since it is formed beneath the surface of the beach, beachrock may be more exten-sive than actual exposures indicate, but on most islands it has a very discontinuous distribution. On atolls it is usually more extensive on eroding windward shores and more rare on aggrading leeward shores, but this is not invariable. On Kapingamarangi and Ifaluk it is apparently not forming on lagoon shores (McKee, 1956, p. 7; Tracey *et al.*, 1961, p. 33), while on Onotoa it is absent from seaward beaches and common on lagoon shores (Cloud, 1952, p. 28).

Once effectively formed, beachrock records a beach position: lines of beachrock record successive beach positions. Complex overlapping pat-terns of beachrock are common near shores of smaller reef islands and around variable sand spits, recording minor fluctuations around mean positions. On many islands throughout the world, however, there are suc-cessive lines of beachrock extending up to 50 m or more offshore from the seaward beach (e.g., Newell, 1956, on Raroia; Emery *et al.*, 1954, p. 151, on Bikini; McKee, 1958, on Kapingamarangi; Tracey *et al.*, 1961, on Ifaluk; Sewell, 1935, 1936a,b, on the Maldives; Stoddart, 1962, on British Honduras). The lagoonward migration thus indicated may be a universal feature of reef islands at the present time (Fig. 9). In addition to leaving a trail of beachrock to seaward, continued lagoonward migra-tion of an island may ultimately expose lagoonward-dipping beachrock from the lagoon shore on the seaward beach. In spite of this migration, beachrock affords at least partial protection to beaches once it is formed; waves may overdeepen the sea floor on the outer edge of the beachrock and expend energy there rather than on the beach. Undermined older beachrock may sink from intertidal to subtidal levels, and be colonized by corals and algae.

Russell (1959, 1962, 1963; Russell and McIntire, 1965) has suggested that the cement of beachrock is calcite derived from ground water, and that beachrock is thus marine only in the sense that it is exposed at the sea. It has, however, been shown in other cases that the cement is aragonite derived from seawater, though the mechanism of precipitation is still uncertain (Daly, 1924; Stoddart and Cann, 1965).

2. Cay Sandstone Ledges

Kuenen (1933, pp. 86–88) and Sewell (1935, p. 520) applied the term "cay sandstone" to lithified island sediments that form above high water level and that lack the seaward dip and linear form of typical beachrock.

Fig. 9. Extensive beachrock on a retreating island shore: Heron Island, Great Barrier Reef.

Such cay sandstone is exposed on retreating shores as massive, homogeneous, unstratified layers or ledges (Fig. 10), and similar layers may also be found by digging in the cay surface. These layers are more common on lagoon shores than on seaward shores (e.g., Newell, 1954a, p. 32, on Raroia; Sachet, 1962, on Clipperton). There is little doubt that cay sandstone is formed at the water table, with a calcite cement derived from ground water, but whether there is any continuity between this rock and beachrock is not clear. It is, however, useful to maintain the distinction on morphologic grounds. The phosphate rock formed under *Pisonia* woodland on some Pacific islands (Fosberg, 1957) may be regarded as a variant of cay sandstone.

3. *Conglomerate Platforms*

Horizontal platforms of coarse coral rubble are found at the foot of seaward beaches on many Indo-Pacific atoll islands. They were classically described at Cocos-Keeling by Darwin (1842) and Guppy (1889, p. 462). The upper surface of these platforms is usually slightly above high-water level. Where they are exceptionally wide the island beach may be weakly developed. They are often continuous between islands, across the sea-

Fig. 10. Cay sandstone forming a massive outcrop on East Island, Diego Garcia Atoll, Chagos Archipelago.

ward mouths of hoa, and they may also form isolated sheets on the reef flat with no island sediments now associated with them. Fragments of such sheets form the "coral horses" described by Sewell (1936b) in the Maldive Islands. These platforms on atolls resemble the "promenades" of coral conglomerate on the low wooded islands of the Queensland coast, where they probably originated as subaerial shingle ramparts or ridges (Steers, 1937, p. 119), and on the Morant Cays, Jamaica, where, however, their composition resembles that of beachrock (Steers *et al.*, 1940, p. 309).

There is considerable controversy over whether any of these platforms represent former reefs exposed by a Recent fall in sea level. In some cases, corals in the position of growth have been reported and Holocene radiocarbon dates obtained (e.g., by Guilcher *et al.*, 1969, on Mopelia), while on other nearby islands similar platforms are detrital and probably of storm origin (e.g., Gibbs, *et al.*, 1971, on Aitutaki). On Funafuti, David and Sweet (1904, p. 70) thought the "breccia sheet" to be of submarine origin and to have formerly been continuous around the atoll rim. Fosberg and Carroll (1965, p. 9) and other workers (notably Gardiner,

1931, pp. 34–35; Sewell, 1935, pp. 493–498) have argued strongly that many reef islands exist only because of the presence of these conglomerate platforms, which they consider relics of a former higher sea-level stand.

Most windward islands in the northern Marshalls possess such conglomerate platforms, often extending for one third to one half the width of the island and standing up to 2 m above low tide level. The surface of such platforms is pitted and rough, and the edges abrupt or sloping (Fosberg and Carroll, 1965). Wells (1951, p. 4), however, has described identical platforms at Arno Atoll in the southern Marshalls which he considers to be entirely detrital and to have been formed at their present elevation with respect to sea level. Platforms are also well exposed in the Tuamotu Islands, where they are termed *pakakota*. Newell (1954a, p. 15) describes cases where platforms outcrop all around the shores of an island but cannot be found by digging in the center of the island; he therefore believes that the cementation is recent, and that the islands cannot be said to rest on a preexisting older platform. A further complication is suggested by Tracey *et al.* (1961, p. 32) at Ifaluk: that the constituent sediments of the platforms were formed and accumulated in their present positions during a former higher sea level, but that the lithification postdates the formation of the island at present sea level. The implications of these views will be discussed in Sections VIII and IX.

4. Erosion Ramps

Among the most common lithified exposures on motus of atolls are smooth, well-cemented surfaces of coral fragments and sand-size sediments, inclined towards the sea at angle of up to 15°, in places apparently continuous with the horizontal reef flat surface, and exposed on the lower sections of seaward beaches. Often the surfaces are highly polished by wave action, and coral fragments in the rock are smoothly beveled. These erosion ramps (Fosberg, 1956) occupy a position similar to that of ordinary beachrock, but the morphologic features of beachrock are usually lacking. In some cases "morphologic beachrock" overlies the ramps, and sometimes the rock of the ramps appears to pass laterally into beachrock. Rocks underlying erosion ramps are of controversial origin. Sections are rare, but they may be either beachrocks or conglomerates similar to those under the conglomerate platforms; they may be of contemporary origin or may indicate a former high stand of the sea. Well-developed and extensive erosion ramps are present around much of the seaward coast of Diego Garcia, Indian Ocean, and are discussed by Stoddart (1971b).

V. Sediments

A. CHARACTERISTICS AND CONSTITUENTS

1. *Characteristics*

Reef island sediments are variable in texture and origin; they differ markedly from sediments of adjacent depositional environments such as reef flat, reef slopes, and lagoon floor (Maxwell *et al.*, 1961, 1964). This is illustrated in general terms by over 1100 samples from Addu Atoll, Maldive Islands, where median diameter (ϕ_{Md}) and sorting (ϕ_σ) of sediments from various environments were as shown in Table II. Island sediments include boulders, cobbles, gravel (together generally termed rubble or shingle, depending on degree of weathering and rounding), and sand. Other sediment sizes, especially silt and clay sizes, are rare except in mangrove areas, although they are common in other reef environments such as lagoon floors. Some indication of the relative importance of the main sediment types on islands is given by the record of 1700 beach observations at Bikini Atoll, 66% of which were sand, 19% gravel, and 15% beachrock (Emery *et al.*, 1954).

Sedimentary units are usually areally extensive and sharply delimited, but show abrupt vertical discontinuities. Most writers recognize units such as boulder beds, gravel beds, sand flats, and dunes (Emery *et al.*, 1954), qualified as older or younger (Tracey *et al.*, 1961), and further subdivided in terms of caliber and composition (Cloud, 1952). Geological maps using such units have been prepared for islands on Funafuti, Kapingamarangi, Ifaluk, Ulithi, and Mururoa; these show considerable differences in complexity, ranging from the simplicity of Ifaluk to the more complex build of islands on Funafuti. In general, coarse sediments show poor or inconspicuous bedding, while sand deposits are well

TABLE II

MEDIAN SIZE AND SORTING OF SEDIMENTS
AT ADDU ATOLL[a]

	ϕ_{Md}	ϕ_σ
Beach	0.85	0.5–0.6
Reef flat	0.5	0.7–2.1
Lagoon	1.0–4.0	poor

[a] Stoddart *et al.* (1966).

bedded, with a primary lagoonward dip in island interiors and primary
seaward dip on seaward beaches (McKee, 1958, 1959). Bedding in
mangrove areas is originally horizontal but is rapidly destroyed by bur-
rowing organisms (Shinn *et al.*, 1968).

2. Composition

Reef sediments are derived almost entirely from reef organisms (pumice forming the only widespread exception: Sachet, 1955), and the
mechanical characteristics of the sediments are largely derived (a) from
the sizes of the skeletal components of constituent organisms, and (b)
from the breakdown patterns of particular particles, which are largely
determined by the structural organization of the skeletons themselves,
and which differ between organisms. These principles have been largely
established by Folk and his co-workers in the Gulf of Mexico (Folk,
1962; Folk and Robles, 1964; Folk and Cotera, 1971), and have been
confirmed and extended in other reef areas (e.g., Fosberg and Carroll,
1965; Stoddart, 1964). These principles do not directly apply to secondary sediments formed by the fragmentation of cemented deposits such
as reefrock, conglomerate, and beachrock; and different principles apply
to sediments not subject to sorting by wave action or in which mobility
is inhibited by vegetation (Scoffin, 1970).

Sedimentary particles of organic origin differ in many ways from
terrigenous particles of similar sizes. Apart from differences in chemical
composition, hardness, and specific gravity, their shapes are directly inherited from organic forms: corals form sediments of irregular spheres
or sticks, and calcareous algae such as *Halimeda* form sediments of
flattened plates. These sediments behave differently from quartz particles of equivalent diameter (Folk and Robles, 1964; Maiklem, 1968).
They also break down differently under mechanical stress: whole *Acropora* colonies break into sticks and then directly into grit, while *Halimeda* plates break first directly into two or three fragments and then into
dust. Chave (1960) has simulated such breakdown experimentally. Only
a few organisms, mainly molluscs such as the Caribbean *Strombus
gigas* with massive shells, undergo gradual reduction in size. Because
many carbonate particles are softer than terrigenous particles they
abrade more rapidly, although this is unlikely to be of geomorphic
importance except where the two types of sediment are found together
(Moberly, 1968). It has been shown that equidimensional carbonate
particles such as oolites do not differ significantly from terrigenous
quartz particles in beach-forming properties (Monroe, 1969), but such
particles, while common in some reef environments, are rarely found

on beaches and islands. Where breakdown of particles is by chemical weathering rather than mechanical stress, very different sediment characteristics result (Swinchatt, 1965).

Folk has shown that the characteristics of sediments, particularly size and sorting, are largely dependent on organic composition and that, because island sediments of a given caliber usually consist of a single type of organic material, size and sorting are closely linked. Most of the limestone-secreting organisms of reefs are represented in island sediments, but the relative proportions of each depend on the caliber of the sediments present. Thus on the largely sandy Isla Perez, Yucatan, Folk and Robles (1964) found up to 60% *Halimeda,* 25% coral, 15% Foraminifera, and 4–8% Mollusc: with coarser sediments the percentage coral would be much higher. There are also considerable local and regional differences in the abundance of organisms in island sediments. Beaches in the Marshalls, Carolines, Tuamotus, and Tonga, for example, generally have abundant tests of Foraminifera (e.g., *Amphistegina, Marginopora, Calcarina*), giving atoll beaches a pink color, but these are unimportant in many other reef areas (e.g., the Solomons, Cooks, many Indian Ocean and Caribbean islands). *Halimeda* is extremely important in the Caribbean where it may be an exclusive beach constituent, but in the Indo-Pacific it is always a minor component.

3. Types

The main types of sediment present on reef islands are as follows:

a. Boulders. When rounded these are usually whole coral colonies, storm-deposited; when angular they may be sections of beachrock or other lithified bedded material. In British Honduras, such sediments up to -8ϕ to -10ϕ (25.6–102.4 cm) in diameter are well sorted (Stoddart, 1964). The largest known boulder deposits on reef islands are storm blocks on the lee sides of Tuamotu atolls, some of which may be more than -12.5ϕ (6 m) in longest diameter (Stoddart, 1969a; Newell, 1956).

b. Shingle or Coral Gravel. Where rounded, this is directly derived from small hemispherical coral colonies, but it is more often composed of sticks derived from the breakdown of branching coral colonies. In the case of *Acropora cervicornis* these sticks are 2.5–15 cm in length and up to 2.5 cm in diameter; they form very well sorted sediments.

c. Coarse Sand. Mainly composed of *Halimeda* flakes, ranging in size from $+1$ to -1ϕ (0.5–2 mm), mean 0ϕ (1 mm), and usually very well sorted. These flakes break down directly into 10ϕ (0.001 mm) dust (Folk and Robles, 1964).

d. Coral Grit. *Acropora* sticks and other coral colonies break down directly into sand-sized particles 1.5–2.5ϕ (0.18–0.35 mm) in diameter, again forming well-sorted sediments.

e. Foraminiferal Sand. Peneroplid Foraminifera form a coarser sand of discoid grains in the size range $+1$ to -0.5ϕ (0.5–1.4 mm). A finer sand of smaller Foraminifera concentrates in the range $+1$ to $+2.5\phi$ (0.5–0.2 mm), and consists of mainly globular or irregular grains. Foraminiferal tests, being light, often concentrate on swash marks on beaches.

Other organisms also contribute in varying amounts to sediment on reef islands. Mollusc shells range from cobble size to sand size, and though most are initially irregular and convoluted in shape, the more massive shells abrade to form cobbles more similar to terrigenous cobbles in shape and size than most other reef sediments.

Sediments which accumulate away from the sorting agency of waves and currents lack the good sorting and restricted organic composition of beach sediments. Sediments of enclosed depressions and mangrove swamps, like those of reef flats, consist of poorly sorted mixtures of transported material and of particles derived from the *in situ* growth and death of organisms. Mechanical abrasion is of small importance in such environments, especially where sediments are bound by mangroves, benthic algae, and marine phanerogams (Scoffin, 1970), and chemical breakdown processes become more important (Swinchatt, 1965). In such circumstances the breakdown characteristics of particular types of organic grain may no longer be the same as on beaches under wave action (Jindrich, 1970). Fine sediments, including algal and chemically derived silt- and clay-sized particles, may be quantitatively important in such environments, they are rare on beaches and in island sediments generally, although silt-sized particles may reach 5% of the total in island interiors.

B. STRATIGRAPHY

Little is known of the vertical distribution of sediments in islands, except for the frequent observation of buried organic soil horizons in areas of hurricane activity (e.g., Newell, 1956, on Raroia). At Eniwetok Atoll, however, 17 shallow holes have been drilled on islands, 9 of them on the single island of Engebi (Ladd and Schlanger, 1960, pp. 872–884). These show great lateral variation in the island sediments, which are mostly unconsolidated. A cemented horizon up to 2 m thick generally occurs at the water-table, but the next lithified horizon in the holes is found at depths greater than 6 m (the island surface varying from 2–3.35 m above sea level). Most of the shallower sediments in the cores appear to be of shallow lagoonal origin. In particular, and surprisingly, there is no continuous rock substratum or reef flat beneath these islands.

The lack of such a reef flat or of cemented reef limestones beneath reef islands has also been demonstrated on islands in Djakarta Bay, where sand extends from the island surface to the bottom of the bay, suggesting considerable infilling behind a reef "breakwater" (Umbgrove, 1928), and at Low Isles, Great Barrier Reef, where less than 2 m of sand of the sand cay rest on a mud stratum which also extends beneath the mangrove swamp and shingle ramparts (Marshall and Orr, 1931). Much more information is needed on the stratigraphy of islands to supplement these rather enigmatic records before general conclusions can be drawn and, in particular, it may be noted that there are almost no island sediments, other than conglomerate platforms, which have been dated by radiometric methods so that the recent history of sedimentation can be determined.

VI. Distribution

In any discussion of the origins of reef islands, distribution patterns on both world and local scales must be considered.

A. WORLD SCALE

Guilcher (1971) has called attention to wide differences over the reef seas in the abundance of coral islands. He points out that on some barrier reefs, such as those of New Caledonia, Mayotte, and New Guinea, and also on the outer reefs of the Great Barrier system itself, reef islands are very rare. On others, such as that of the British Honduras, they are common but small; while on others, including those of Bora-Bora, Maupiti, and Mangareva, islands cover a large proportion of the reef flat. Similarly, on open-ocean atolls, some, such as Diego Garcia, have a continuous or almost continuous land rim, while others may have no islands at all. The contrast in island frequency between the atolls of the Tuamotus and those of the Marshalls has already been noted. There is no simple pattern in these distributions, which are often complicated by local conditions, e.g., lagoon reefs in the lee of outer reefs devoid of islands may support many cays, as on the Great Barrier Reefs.

B. Regional Scale

The first detailed discussions of the distribution of islands over a reef province were those of Steers (1929, 1937) and Spender (1930) on the Queensland coast. Spender believed that the distribution of reefs without islands, reefs with sand or shingle cays, and low wooded islands is the result of a systematic difference in the height of reef surfaces, caused by tilting at right angles to the mainland coast. Raised wave-cut platforms on the mainland coast indicate uplift in this scheme; low wooded islands are formed on unusually high reefs near the coast; sand and shingle cays are formed on lower reefs further seawards; and no islands at all are formed on the lowest, most seaward reefs. Spender argued rather unconvincingly that sand cays could form on deeper reefs than those on which shingle ramparts would form, and stated that it was "inconceivable" that ramparts could be formed on any reef on which no sand cay existed. (Such cases are, however, known in reef areas). Steers (1937) doubted Spender's general explanation, and argued that differences in exposure of reefs to wave action accounted for the main pattern of island distribution, although this was complicated in the Australian case by probable eustatic changes of sea level (Steers, 1931). Data collected by Umbgrove (1928, 1929, 1930a,b, and 1939) in Djakarta Bay, the Duizend-Eilanden, the Spermonde Archipelago, and the Togian Islands, and by Guilcher (1965) in New Caledonia supported the exposure hypothesis, but the range of islands found in these areas was much smaller than that on the Great Barrier Reefs.

The reefs of the British Honduras coast parallel in miniature those of the Great Barrier Reefs (Stoddart, 1965a). Mapping of the distribution of island types defined in Section III demonstrates a zonation over the reef province very similar to that on the Queensland coast. On very exposed reefs with high wave action (or, elsewhere, extreme tidal range) sediments are transported across the reef flats and no islands form. On slightly less exposed reefs shingle cays are formed. Sand cays are built in more protected situations, and in areas of low wave energy mangrove and mangrove-sand cays are found. On patch reefs in deeper lagoons, wave action is sufficient to build windward shingle ridges at the same time as leeward cays are formed by wave refraction, and conditions are sufficiently protected for mangroves to grow between them, thus forming an analogue of the Queensland low wooded islands.

Similar zonations involving fewer types of islands have been described elsewhere. Thus at the southern extremity of the Great Barrier Reefs, islands of the Bunker Group, such as One Tree, are shingle cays, while in the Capricorn Group to leeward the islands, such as Heron, are sand cays (Steers, 1937, pp. 5–12). Sand and shingle cays are also spatially segregated in the Port Royal cays, Jamaica (Steers, 1940a,b; Steers *et al.*, 1940).

C. LOCAL SCALE

Most individual atoll reefs show variations in the distribution of islands, though these are less instructive than the main barrier reef provinces because of the small number of island types involved. In a survey of 125 Pacific atolls, Wiens (1962) found 6 with a complete land rim; 22 with between one half and two thirds of the rim occupied by land; 34 with between one third and one half; and 55 with less than one third. In this sample the frequency of occurrence of land in the west, northwest, and southwest sectors of the atoll rim was one half of that in the north, northeast, east, southeast, and south sectors. This distribution, reflecting the importance of the Trade Winds in building motus, holds for most of the northern Pacific atoll groups; in the Carolines and Gilberts land tends to occur more frequently in the southeast, in the Marshalls in the east, and in the Tuamotus in the northeast. Among the well-studied atolls with islands concentrated on the windward side are Kapingamarangi, Onotoa, Funafuti, Tarawa, and Farquhar. In the Indian Ocean this relationship is less evident, and it has been stated that on some reefs land is concentrated on the leeward rather than the windward sides (e.g., in the Laccadives: Gardiner, 1903, p. 149). The situation is complicated in the Indian Ocean, however, by seasonal reversal in monsoonal winds, and in the southern Trade Wind belt windward concentration of land is again the rule. In the Maldives and Chagos, which have considerable latitudinal range—from monsoonal circulation to equatorial calms—there are marked changes in the orientation of islands on the atoll rims (Sewell, 1935).

Variations in frequency of atoll islands have already been noted. Many Tuamotu atolls have 100–300 islands each, largely as the result of dissection of a formerly more continuous atoll land rim. Wiens (1962) finds that most atolls in the Marshalls, the Gilberts, and the Ellice groups have more than 20 islands on each atoll, while atolls in the Carolines generally have less than 10 islands.

VII. Processes

Processes involved in constructing reef islands include those involved in moving sediments, both accretionary and erosional, and those involved in stabilizing sediments that have been accumulated. These are primarily lithification and the growth of vegetation. The characteristics of these processes are briefly considered in this section.

A. SEDIMENT-MOVING PROCESSES

1. Waves

Wind-driven waves form the primary process in island construction. Few direct studies have been made of the effects of wave characteristics on reefs, and most workers substitute measures of wind force, frequency, and direction (Umbgrove, 1947; Verstappen, 1954). Kuenen (1933, p. 84) also emphasized fetch. The effects of wind waves are seen not only on many windward reefs, but also in more protected situations where wind effects become less apparent. Umbgrove referred to cases in the East Indies, an area of alternating monsoonal winds, where, in the Spermonde Archipelago, a high island limited wind-wave action from the normally dominant direction, and in another case, in the Togian Island reefs are surrounded by high land and consequently no cays can form (Umbgrove, 1930a,b, 1939).

2. Swell

Long ocean swell of distant origin is important in sediment accumulation on many atolls, and is particularly significant on reefs in the zone of equatorial calms where normal wave action is reduced. Swell traversing the Pacific from the Roaring Forties is responsible for boulder ramparts on the south sides of many central and northern Pacific atolls, for example, in the Marshalls (Emery et al., 1954) and in the Tuamotus (Newell, 1956; Stoddart, 1969a).

3. Storms

Ridges formed by major storms have been noted by many workers. Hedley (1898) noted that individual storms could form fresh shingle ridges on islands, and prominent ridges can in many cases be attributed to particular storms, e.g., the 1878 storm ridge on Funafuti (David and Sweet, 1904, p. 87) and the 1907 ridges on Ifaluk (Tracey et al., 1961, p. 32). Constructional effects of such storms have been studied in detail at Jaluit Atoll in the Marshalls (Blumenstock, 1961; McKee, 1959) and in the Caribbean (Stoddart, 1963) (Figs 11 and 12). Severe erosion can also occur, resulting in beach retreat, surface scouring, and channel cutting. It has been suggested that the reason why many storms now result in erosion rather than aggradation is because island stability has been disturbed by man during the replacement of natural vegetation by coconut plantations (Stoddart, 1965b). Generalization is difficult, however, since successive hurricanes in the same area may have differing results (Ball *et al.*, 1967; Perkins and Enos, 1968); the subject is reviewed by Stoddart (1971a).

4. Tides

Tidal conditions vary widely in the reef seas, from the almost tideless conditions in parts of the Caribbean to the tides of more than 3 m in the southwest Indian Ocean and in parts of the Great Barrier Reefs. Since reefs cannot grow above the level of low water neaps, it is clear that the greater the tidal range the greater the submergence at high tide and the greater the effective exposure. Hence reef islands are more easily formed in areas of low tidal range than of high.

Fig. 11. Ridge of fine sand deposited over an older land surface by a minor hurricane: Placencia, British Honduras, July 1961.

Fig. 12. Ridge of shingle and rubble deposited over an older land surface by a major hurricane in October 1961: Colson Cay, British Honduras.

5. Currents

Tides, together with winds, are also important in generating surface currents in reef areas. Tidal currents set into lagoons at high tide, generally across reef flats and between islands, and set out of lagoons at low tide, generally through large gaps; constant unidirectional winds, especially in the Trades, also set up a head of water on the seaward edge of windward reefs which generates rapid currents flowing across the reef flat into the lower lagoon. Tidal and other currents are capable of transporting sand-sized sediments, maintaining and perhaps eroding submarine channels, and depositing submarine sediments (Jindrich, 1970). McKee (1956, p. 28) has also shown at Kapingamarangi how currents flowing across the reef flat from sea to lagoon carry sediments from the seaward flat along the channel shores of islands, helping to build bars, spits, and deltas at the ends of lagoon beaches, and how sediments from the seaward flat are finally deposited on the lagoon beaches themselves.

6. Wind

Wind is important in building dunes, especially in dry areas, on shores facing wide drying reef flats with an abundant supply of sediment. The Trades blow with great constancy over large parts of the reef seas with velocities of more than 5 m/sec. Minor physiographic effects may also result from vegetation damage caused by occasional hurricane-force winds with velocities of 40–50 m/sec.

B. SEDIMENT-STABILIZING PROCESSES

1. Lithification

The contemporary formation of beachrock and cay sandstone, and the possible present-day formation of other kinds of lithified materials on reef islands, help to stabilize otherwise loose accumulations of sediment. Beachrock can form very rapidly, as shown by the presence of human artifacts in it and also by direct observation over a period of years (Emery *et al.*, 1954; Stoddart, 1963); when first formed it is very friable and subject to destruction, but once secondary cementation has occurred it forms a very tough rock. Well-cemented beachrock can only be removed (a) by undermining, fracturing, and transport during storms; (b) by slow biological and other erosion (McLean, 1967a,b); or (c) by mining for constructional and other purposes (Bowen, 1951).

2. Vegetation Growth

Colonization by vegetation takes place on islands even before subaerial lithification begins. In spite of the high insolation, salinity, and salt spray, and the lack of fresh water and soil, seeds of herbs, vines, and grasses brought by the sea, wind, and birds begin a vegetational succession which culminates in high woodland on humid islands and in scrub on dry islands. Vegetation helps stabilize sediments by several means: (a) binding loose sediments with roots; (b) protecting the surface from wind scour; (c) acting as a baffle for the entrapment of wind-carried sand and particularly of particles carried in overtopping storm waves. The effects of vegetation can often best be seen when it is removed by man: destruction of the littoral vegetation hedge almost invariably results in continued marginal erosion, which may reach catastrophic proportions during storms. The effects of vegetation on island sediments may be compared with the effects of mangroves, algae, and marine grasses on submarine vegetation in reef areas.

VIII. Origins

The distribution patterns outlined in Section VI,B strongly suggest that each of the island types distinguished are equilibrium forms in adjustment with the controlling processes of the reef environment at any given location in space and time, and that they do not form stages in any evolutionary sequence, nor are they inherited from environmental conditions of the past. Each main type of island will be considered in turn.

A. SAND CAYS

Sand cays of Type A (page 63) are typically developed at gaps in linear reefs, on small arcuate or oval reef segments, or at prominent elbows or bends in reefs. The presence of gaps or bends in reefs causes refraction of surface waves, and convergence of wave trains leads to sediment accumulation and cay formation. This was recognized in general terms by Darwin (1842, p. 16) and Davis (1928, p. 13), and has been confirmed in the field (Steers, 1929, p. 250; 1931, p. 12; 1937, p. 13; Stoddart, 1962) and in wave tank experiments (Stoddart, 1962). Spender (1937, 141) believed the cay to be a "perfect equilibrium structure," the stability of which depends on the equilibrium of various opposing forces brought into play at the leeward extremity of a reef (Spender, 1930, p. 200). Such equilibrium can only, of course, be instantaneous, for environmental conditions are constantly changing and, with low inertia, cays are constantly responding to changes on time scales of years, seasonal and tidal cycles, and even individual storms.

Patterns of refraction in reef areas are well shown on a larger scale at Cocos-Keeling by Wood-Jones (1910) and at Low Isles, the Great Barrier Reefs, by Stephenson et al. (1931). The degree of refraction and hence the locus of accumulation of the cay on the reef flat will depend on reef geometry, strength and direction of wave action, and tidal range. On more exposed reefs and in areas of greater tidal range the cay will form further from the seaward reef edge than in less exposed areas or those with smaller tidal range. Most cays form on the leeward sides of reefs (Steers, 1929; Stoddart, 1962; Guilcher, 1965), but some may form centrally, perhaps under the influence of reversing winds, e.g., on the Sahul Shelf (Teichert and Fairbridge, 1948) and in eastern Guadalcanal (Stoddart, 1969b). Sand cays are common on the British Honduras reefs, with a tidal range of less than 0.5 m, and are relatively less so on the Great Barrier Reefs with a tidal range in places greater than 3 m. In more exposed areas the width of the reef flat may be a critical control of cay formation, but even where this is constant, cays are often not found where they would be expected. Steers (1929, p. 250) has observed that the simplest unvegetated cays are often found in the most exposed locations, suggesting that constant mobility prevents vegetation colonization or lithification, and on the British Honduras reefs many small islands are known to have been destroyed by storm action during the last century (Stoddart, 1965b).

B. SAND-SHINGLE CAYS (MOTUS)

Some sand-shingle cays are also located at reef gaps or elbows, where refraction is a factor in their formation, but others lie on unbroken reefs on many atoll rims, though they are more common on convex than on concave reef sectors and they are wider at reef angles. Motus are formed primarily by the deposition of their windward shingle ridge: (a) as a storm boulder ridge lodged on the reef flat; (b) by the deposition of sediment from waves losing energy as they cross the reef flat from sea to lagoon; and (c) by the neutralization of sea waves by lagoon waves on the reef flat and consequent deposition of sediment in transportation. Coarse sediments thus accumulate to form a ridge which serves as a further barrier to water and sediment movement. Motus thus originate by the deposition of coarse material near the seaward reef edge, with infilling of sand (much of it derived from the seaward reef and reef flat) on the leeward side of the coarse material (Emery *et al.*, 1954). In general, therefore, the shingle ridge forms before the sand area, for fine sediments could not normally accumulate on such seaward flats (Wells, 1951, p. 3 on Arno; Cloud, 1952, p. 48 on Onotoa; Zans, 1958, p. 21 on Pedro Bank; Stoddart, 1962, on British Honduras atolls). However, cases are known where shingle ridges can be added to ordinary sand cays during storms to form, in effect, a sand-shingle cay (Folk, 1967, p. 433–434).

This model, while applying to some smaller motus, is undoubtedly oversimplified. It implies, as stated by Emery *et al.* (1954, p. 35), that motus form "on top of the reef flat," but we know from borings (Section V,B) that reef flats apparently do not exist beneath many islands. The reasons for the present dissection and erosion of many motus are still largely unknown. On Raroia, for example, the largest and least dissected islands are on the lee side, while on the windward side a formerly almost continuous land rim has been dissected by large numbers of hoa. Finally, if it can be demonstrated that the conglomerate platform dates from a former high stand of the sea, and that motus have formed in the lee of these platforms, a historical element will have to be included in our explanation.

C. MANGROVE AND MANGROVE-SAND CAYS

Problems of mangrove islands are primarily ecological rather than physiographic (Davis, 1940), though in most cases a preexisting shoal, perhaps of pre-Recent limestones, forms the foundation of the mangrove cay. Once such a shoal is colonized, the root systems of the mangroves may encourage sedimentation, raise the floor, and initiate a succession from pioneer mangroves to dry-land littoral woodland.

Mangrove-sand cays are morphologically more interesting. They are often located close to reef edges, separated from them by shallow moats up to a few meters deep. They are usually linear islands, aligned parallel to the reef edge, and not associated with reef gaps. Vigorous reef growth is rare near such islands, and coarse sediments on them are unusual. Vermeer (1959, p. 116) believed that the sand ridge formed first, with subsequent colonization of mangroves to leeward. No examples of such an initial stage in the formation of such islands is known. It has since been shown that large "shelf islands" of this type owe their location to underlying topographic eminences on a karst-eroded Pleistocene reef surface (Ebanks, 1967; Scoffin, 1970, pp. 271–272). These eminences control the location of the original sediment accumulation, which rapidly becomes adjusted to prevailing exposure conditions, and the vegetation of the mangrove area, once developed, acts as a further sediment trap.

D. LOW WOODED ISLANDS

Low wooded islands and analogous forms of the Great Barrier Reefs, the British Honduras barrier reef lagoon, and Djakarta Bay are all formed in deep lagoon conditions, in which short steep waves are capable of fragmenting light-skeletoned lagoon corals, lifting fragments on to the reef edge to form a windward shingle rampart. Wave refraction round the reef patch leads to the formation of a sand cay on the leeward side, and the proximity and even continuity of ridge and cay depends largely on the size of the patch reef. The mangrove woodland on the intervening flat is thus not geomorphically significant; it simply colonizes a suitably protected site. Presence or absence and state of development of the mangroves depends on such factors as size and depth of the moat, proximity of cay and rampart, and frequency of storms, rather than simply on age. The ramparts themselves are clearly not ephemeral storm constructions which disappear within a few years of formation: in many cases they are partly lithified and vegetated.

The Queensland low wooded islands are larger and more complex structures than this brief summary suggests: the ramparts in particular are more complex features than in the Caribbean and East Indian cases. Spender (1930, pp. 208–209) suggested a cyclic development of the ramparts. In his view, a sheet of shingle moves inwards from the reef edge until it reaches a limiting width, when a second sheet of shingle begins to form outside it and then this also moves inward. He pointed out that lithified remnants of inner ramparts at Low Isles were at the same high intertidal level as outer newer ramparts, and thus could not agree with Steers (1931, 1937) that eustatic changes of sea level may have caused repeated rampart formation. The concept of cyclic formation is, however, difficult to understand, and irregular storm action has since been shown to be sufficient to initiate rampart formation (Fairbridge and Teichert, 1948).

The leeward sand cay of the low wooded island complex may be as well defined as any solitary sand cay (Low Isles, Two Isles, Three Isles), it may be an indeterminate sand patch (Low Wooded Island, Turtle Island), or it may adjoin and merge into the mangrove area (Enn or Nymph Island). The mangrove area itself may be sharply defined, as at Three Isles, rather diffuse, as at Low Isles, or it may spread over the whole flat, as at Houghton and Enn Islands. Fairbridge and Teichert (1948, p. 851) have recorded considerable changes in mangrove distribution on low wooded islands over the period 1928–1947, and it is clear that the extent of mangrove has no general significance in any scheme of island classification. Steers (1929, p. 257) believed that the sand cay formed before the rampart, since the latter could limit sediment supply to the former (also Stephenson *et al.*, 1931, p. 97), or that both could form at once (Steers, 1937, p. 132). If the sand cay does usually form before the rampart, the situation is different from that common on motus (Folk, 1967, p. 433).

Kuenen (1933, p. 85) believed that the shingle tongues within ramparts formed before the ramparts themselves, but this seems an unnecessary complication.

IX. Equilibrium and Evolution

A. EQUILIBRIUM

We thus conclude that certain types of islands are restricted to certain reef areas, defined in terms of exposure to wave action (a function of strength, constancy and direction of winds, effective fetch and depth of water offshore, and tidal range). Since these factors may vary independently and are not constant over time, and since reefs themselves are very variable features, the limits under which cays of different types develop can only be stated in general terms.

While reef islands may thus be arranged together in an apparent series, into which all islands may at some point be fitted, the series is one of location and not of stage: it represents a series of equilibrium positions, not an evolutionary succession (Fairbridge, 1950). Within the scheme of island types proposed, there are of course some exceptions to this generalization. Thus Folk (1967) has documented the growth and changes in sand cays on the Alacran reef over the period 1842–1960, showing that most have increased in area from 1.5 to 7 times since 1896. Most such developmental sequences are, however, ecological rather than geomorphic (Steers, 1937, p. 14; Fairbridge and Teichert, 1948). The main types of islands treated in Section VIII—sand cays, motus, mangrove cays, mangrove-sand cays, and low wooded islands—are distinct entities which will remain in the same category unless external conditions change.

However, some workers have taken a rather different view. For example, Umbgrove (1928, p. 64) discerned an evolutionary trend in the islands of Djakarta Bay. His "younger" islands have a small sand cay separated by a wide reef flat from a low shingle ridge, whereas on "older" islands the sand cay is larger, the ridge higher, and the intervening depression smaller, so that ultimately in the oldest islands the cay adjoins the rampart. This might suggest that the only difference between low wooded islands and motus is one of age. Verwey (1931) and Kuenen (1933, p. 71) both produced evidence against any such evolutionary trend, and the latter pointed out the mobility of those features which to Umbgrove indicated age. It would, however, be unfair to extend Umbgrove's generalization from the specific local area in which it was made and to apply it to islands of different sizes and characteristics and of very different reef areas.

B. Catastrophic Change

The present equilibrium is not static, however. Major changes are taking place in islands over a time scale of years and decades, primarily as a result of catastrophic storms. These changes include: (a) the building of storm ridges and the deposition of boulder, gravel, and sand sheets; (b) the erosion of shores, stripping of surface sands; and (c) the cutting of channels (Figs. 13 and 14). Some of these changes are temporary while others, notably the cutting of hoa in Pacific atoll land rims, completely change the character of reef islands. The long-term effects of such storms have been studied in recent years by Blumenstock *et al.*

(1961) and Stoddart (1969d), and catastrophic storm effects are reviewed by Stoddart (1971a). More work is needed on contrasts between reef islands within and outside of the hurricane belts, for it is likely that there are systematic differences in topography and stratigraphy which have not yet been fully recognized.

C. Secular Change

A number of long-term trends are visible in reef island geomorphology. First, erosion and retreat of seaward shores appears to be worldwide. Most of the evidence in offshore beachrock trains and shore cliffing is geomorphic (McKee, 1958, on Kapingamarangi; Tracey *et al.*, 1961, at Ifaluk; Guilcher, 1965, on New Caledonia; Stoddart, 1962, in the Caribbean), but Nicring (1962) has also described truncation of vegetation patterns by such erosion. Such retreat is not invariable, for cases of accretion on seawardshores are known (e.g., Guppy, 1889, pp. 463–464, on Cocos-Keeling), but these are usually caused by specific major storms. Second, lagoonward accretion is concurrent with seaward erosion. Accretion is recorded in the historical record (e.g., Folk, 1967, on Alacran), in vegetation patterns (e.g., Fosberg, 1962, also on Alacran), and especially by lines of stranded pumice which can be ascribed to known volcanic eruptions (e.g., Sewell, 1936a, p. 77, on Addu; McKee, 1956, p. 8, on Kapingamarangi). The net result of these two trends is the bodily migration of reef islands across reef flats from the sea towards the lagoon. In some cases this process may have gone so far that islands are being pushed off the reef flat into the lagoon. Measured rates of migration reach 1 m per year for periods of up to 75 years. The evidence of beachrock suggests that such migration is slightly episodic but generally steady: it appears too regular to be ascribed to storm action alone.

Further evidence of long-term change comes from Verstappen's (1954) work on changes in the orientation and size of shingle ridges on the island of Njamuk Besar in Djakarta Bay over the period 1875–1950. He was able to explain substantial changes in the topography of the island in terms of meteorological records showing a marked increase in the frequency of southwesterly winds over this 75-year period. It is likely that other such effects are operative in the coral seas but have not yet been recognized.

D. Sea-Level and Reef Islands

Throughout this treatment of the nature and origin of reef islands we have tacitly implied that one major boundary condition in the island

Fig. 13. English Cay, British Honduras barrier reef, before Hurricane Hattie in October 1961.

environment—sea level—has been stable. In effect, we have argued that sea level has been essentially stable since the sea reached its present level about 3000 years ago, and that most reef islands have formed during this period of stability. This is a large assumption, and one on which it is difficult to reach agreement, and it is fair to say that it is not one which would be accepted by many past students of coral islands.

Gardiner (1931), Sewell (1935, pp. 512–518), and Kuenen (1933, pp. 70–73, and Fig. 85) all believed that reef islands were formed (though not exclusively so) as a result of the exposure of abnormally high reef flats following a slight negative shift of sea level. They have argued that the sea is at present reducing these high reef flats to a new lower position, and that this down-cutting results in the seaward erosion of cays and their migration lagoonward over reef flats. Kuenen (1933, p. 70) states that:

> very many and perhaps most islands have been formed as a result of the emergence of their flats. Without the negative movements the number of islands on reefs would be quite small and if no further movements occur their number and extent will in the course of time undergo considerable reduction.

Fig. 14. English Cay following Hurricane Hattie.

This hypothesis is attractive; it has the support of Tayama (1952) after many years of work in the west Pacific; and it has been adopted as a general principle of reef morphology by Cloud (1954). Nevertheless it is largely inferential, and the history of Holocene sea level which it implies is at variance with the now generally accepted picture. It is quite clear in many areas, particularly the Great Barrier Reefs, that fluctuations of sea level have taken place, but we believe that these are not necessary in explaining the features of present reef islands. First, there is almost complete agreement that there have been no Holocene high stands of the sea in the Caribbean area; for a contrary view, see Lind (1969). Yet there are no systematic differences in the character of islands in this area compared with islands in areas where such stands are said to have occurred. Second, if we study the islands of a reef province which is known to have been tectonically active and to have abnormally high reef flats as a result, we again find no systematic differences between these islands and those of other, more stable reefs (see Stoddart, 1969b, on the reefs of Guadalcanal).

Furthermore, it is possible to argue that the widespread retreat of cays results from a gradual rise of sea level rather than a recent fall.

Such a rise has been recorded instrumentally over the last century, and would have the effect of increasing the depth of reef flats and hence of increasing effective exposure (Stoddart, 1962). It is possible that the low conglomerate platforms of Pacific atoll islands record a former higher sea level, but there is complete lack of agreement on their interpretation among those who have studied these and related features on other reefs, and a recent catalogue of radiometric dates from reef areas reveals no unequivocal evidence for sea levels of the Holocene higher than those of the present (Stoddart, 1971c). In addition, with the development of uranium series dating methods, which have a much greater range than radiocarbon dating, many of the features used by Cloud and others as evidence for Holocene high stands have been shown to be of much greater age, mostly last Interglacial.

While stressing the need for further work, we prefer to rest our analysis on the geomorphology of the islands themselves rather than on theoretical sea-level curves. The morphology of most types of islands is now reasonably well known, though much work needs to be done on the stratigraphy and foundations of islands, which may itself throw light on the sea level question.

References

Agassiz, A. (1903). *Mem. Mus. Comp. Zool. Harvard* **28**, 1.

Ball, M. M. (1967). *J. Sediment. Petrol.* **37**, 556.

Ball, M. M., Shinn, E. A., and Stockman, K. W. (1967). *J. Geol.* **75**, 583.

Blumenstock, D. I. (1961). *Atoll Res. Bull.* **75**, 1.

Blumenstock, D. I., Fosberg, F. R., and Johnston, C. G. (1961). *Nature* (Lond.) **189**, 618.

Bowen, R. LeB., Jr. (1951). *Geogr. Rev.* **41**, 384.

Catala, R. (1950). *Bull. Biol. Fr. Belg.* **84**, 234.

Chave, K. E. (1960). *Trans. N.Y. Acad. Sci.* [2] **23**, 14.

Chevalier, J.-P. (1969). *Abstr. Pap., Symp. Corals Coral Reefs Mar. Biol. Assoc. India, 1969* p. 27.

Chevalier, J.-P., Denizot, M., Mougin, J.-L., Plessis, Y., and Salvat, B. (1969). *Cah. Pac.* **12**, 1.

Cloud, P. E., Jr. (1952). *Atoll Res. Bull.* **12**, 1.

Cloud, P. E., Jr. (1954). *Sci. Mon.* **79**, 195.

Daly, R. A. (1924). *Carnegie Inst. Washington, Pap. Dep. Mar. Biol.* **19**, 95.

Darwin, C. R. (1842). "The Structure and Distribution of Coral Reefs." Smith, Elder & Co., London.

David, T. W. E., and Sweet, G. (1904). *In* "The Atoll of Funafuti" (Coral Reef Committee of the Royal Society), pp. 61–124. Royal Society, London.

Davis, J. H. (1940). *Pap. Tortugas Lab.* **32**, 302.

Davis, J. H. (1942). *Pap. Tortugas Lab.* **33**, 113.

Davis, W. M. (1928). "The Coral Reef Problem." Am. Geogr. Soc., New York.
Ebanks, W. J., Jr. (1967). Ph.D. Thesis, Rice University, Houston, Texas.
Edgell, J. A. (1928). *Rep. Great Barrier Reef Comm.* **2**, 57.
Emery, K. O., Tracey, J. I., Jr., and Ladd, H. S. (1954). *U.S., Geol. Surv., Prof. Pap.* **260-A**, 1.
Fairbridge, R. W. (1950). *J. Geol.* **58**, 330.
Fairbridge, R. W., and Teichert, C. (1947). *Rep. Great Barrier Reef Comm.* **6**, 1.
Fairbridge, R. W. and Teichert, C. (1948). *Geogr. J.* **111**, 67.
Folk, R. L. (1962). *Trans. N.Y. Acad. Sci.* [2] **25**, 222.
Folk, R. L. (1967). *J. Geol.* **75**, 412.
Folk, R. L., and Cotera, A. S. (1971). *Atoll Res. Bull.* **131**, 1.
Folk, R. L., and Robles, R. (1964). *J. Geol.* **72**, 255.
Forster, J. R. (1778). "Observations made during a voyage round the world, on physical geography, natural history, and ethic philosophy." G. Robinson, London.
Fosberg, F. R. (1956). "Military Geography of the Northern Marshalls." Intelligence Division, Office of the Engineer, H.Q. U.S. Army Forces Far East.
Fosberg, F. R. (1957). *Am. J. Sci.* **255**, 584.
Fosberg, F. R. (1962). *Atoll Res. Bull.* **93**, 1.
Fosberg, F. R., and Carroll, D. (1965). *Atoll Res. Bull.* **115**, 1.
Gardiner, J. S. (1903). *In* "The Fauna and Geography of the Maldive and Laccadive Archipelagoes" (J. S. Gardiner, ed.), Vol. I, pp. 12–50, 146–183, 313–346, and 376–423. Cambridge Univ. Press, London and New York.
Gardiner, J. S. (1931). "Coral Reefs and Atolls." Macmillan, New York.
Gibbs, P. E., Stoddart, D. R., and Vevers, H. G. (1971). *R. Soc. N.Z., Bull.* **8**, 91.
Guilcher, A. (1955). *Ann. Inst. Oceanogr. Monaco* **30**, 55.
Guilcher, A. (1956). *Ann. Inst. Oceanogr. Monaco* **33**, 64.
Guilcher, A. (1965). *Exped. Fr. Recifs Coral. Nouv.-Caledonie* **1**, 113.
Guilcher, A. (1967). *J. Soc. Oceanist.* **23**, 101.
Guilcher, A. (1971). *In* "Regional Variation in Indian Ocean Coral Reefs" (D. R. Stoddart and C. M. Yonge, eds.), pp. 65–86. Academic Press, New York.
Guilcher, A., Berthois, L., Doumenge, F., Michel, A., Saint-Requier, A., and Arnold, R. (1969). *Mem. ORSTOM* **38**, 1.
Guppy, H. B. (1889). *Scott. Geogr. Mag.* **5**, 281, 457, and 569.
Hedley, C. (1898). *Nat. Sci.* **12**, 174.
Hedley, C. (1925). *Rep. Great Barrier Reef Comm.* **1**, 66.
Hedley, C., and Taylor, T. G. (1907). *Aust. Assoc. Adv. Sci.* **2**, 397.
Hoffmeister, J. E., and Multer, H. G. (1968). *Geol. Soc. Am. Bull.* **79**, 1487.
Hoffmeister, J. E., Stockman, K. W., and Multer, H. G. (1967). *Geol. Soc. Am. Bull.* **78**, 175.
Jenkin, R. N., and Foale, M. A. (1968). "An Investigation of the Coconut-growing Potential of Christmas Island." Directorate of Overseas Surveys, Land Resources Division, Tolworth.
Jindrich, V. (1970). *J. Sediment. Petrol.* **39**, 531.
Jukes, J. B. (1847). "Narrative of the Surveying Voyage of H. M. S. 'Fly'." Boone, London.
Kuenen, P. H. (1933). *Snellius-Exped. Rep.* **5** (1), 1.
Ladd, H. S., and Schlanger, S. O. (1960). *U.S., Geol. Surv., Prof. Pap.* **260-Y**, 863.
Land, L. F., Mackenzie, F. T., and Gould, S. J. (1967). *Geol. Soc. Am. Bull.* **78**, 893.

Lind, A. O. (1969). *Res. Pap. Dep. Geogr. Univ. Chicago* **122**, 1.
McKee, E. D. (1956). *Atoll Res. Bull.* **50**, 1.
McKee, E. D. (1958). *Geol. Soc. Am. Bull.* **69**, 241.
McKee, E. D. (1959). *J. Sediment. Petrol.* **29**, 354.
McLean, R. F. (1967a). *Mar. Geol.* **5**, 181.
McLean, R. F. (1967b). *Bull. Mar. Sci.* **17**, 551.
Macnae, W. (1968). *Adv. Mar. Biol.* **6**, 73.
Maiklem, W. R. (1968). *Sedimentology* **10**, 101.
Marshall, S. M., and Orr, A. P. (1931). *Sci. Rep. Great Barrier Reef Exped.* **1**, 93.
Maxwell, W. G. H., Day, R. W., and Fleming, P. J. G. (1961). *J. Sediment. Petrol.* **31**, 215.
Maxwell, W. G. H., Jell, J. S., and McKellar, R. G. (1964). *J. Sediment. Petrol.* **34**, 294.
Moberly, R., Jr. (1968). *J. Sediment. Petrol.* **38**, 17.
Monroe, F. F. (1969). *Misc. Pap., Coastal Eng. Res. Cent.* **1–69**, 1.
Moorhouse, F. W. (1933). *Rep. Great Barrier Reef Comm.* **4**, 35.
Moorhouse, F. W. (1936). *Rep. Great Barrier Reef Comm.* **4**, 37.
Newell, N. D. (1954a). *Atoll Res. Bull.* **31**, 1.
Newell, N. D. (1954b). *Atoll Res. Bull.* **36**, 1.
Newell, N. D. (1956). *Bull. Am. Mus. Nat. Hist.* **109**, 311.
Newell, N. D. (1961). *Z. Geomorphol.* [N.S.] Suppl. **3**, 87.
Niering, W. A. (1962). *Ecol. Monogr.* **33**, 131.
Off, T. (1963). *Am. Assoc. Pet. Geol. Bull.* **47**, 324.
Perkins, R. D., and Enos, P. (1968). *J. Geol.* **76**, 710.
Russell, R. J. (1959). *Z. Geomorphol.* [N.S.] **3**, 227.
Russell, R. J. (1962). *Z. Geomorphol.* [N.S.] **6**, 1.
Russell, R. J. (1963). *J. Trop. Geogr.* **17**, 24.
Russell, R. J., and McIntire, W. G. (1965). *Geogr. Rev.* **55**, 17.
Sachet, M.-H. (1955). *Atoll Res. Bull.* **37**, 1.
Sachet, M.-H. (1962). *Atoll Res. Bull.* **86**, 1.
Schlanger, S. O., and Brookhart, J. W. (1955). *Am. J. Sci.* **253**, 553.
Scholl, D. W. (1963). *Mar. Geol.* **1**, 344.
Scholl, D. W. (1964). *Mar. Geol.* **2**, 343.
Scoffin, T. P. (1970). *J. Sediment. Petrol.* **40**, 249.
Sewell, R. B. S. (1935). *Mem. Asia. Soc. Bengal* **9**, 461.
Sewell, R. B. S. (1936a). *Sci. Rep. Murray Exped.* **1**, 63.
Sewell, R. B. S. (1936b). *Sci. Rep. Murray Exped.* **1**, 109.
Shinn, E. A., Ginsburg, R. N., and Lloyd, R. M. (1968). *J. Paleontol.* **42**, 879.
Spender, M. A. (1930). *Geogr. J.* **76**, 194 and 273.
Spender, M. A. (1937). *Geogr. J.* **89**, 141.
Stanley, S. M. (1966). *Am. Assoc. Pet. Geol. Bull.* **50**, 1927.
Steers, J. A. (1929). *Geogr. J.* **74**, 232 and 341.
Steers, J. A. (1931). *C. R. Congr. Int. Geogr. Paris* **2**, 164.
Steers, J. A. (1937). *Geogr. J.* **89**, 119.
Steers, J. A. (1938). *Rep. Great Barrier Reef Comm.* **3**, 51.
Steers, J. A. (1940a). *Geogr. J.* **95**, 30.
Steers, J. A. (1940b). *Geogr. Rev.* **38**, 279.
Steers, J. A., Chapman, V. J., Colman, J., and Lofthouse, J. A. (1940). *Geogr. J.* **96**, 305.

Stephenson, T. A., Stephenson, A., Tandy, G., and Spender, M. A. (1931). *Sci. Rep. Great Barrier Reef Exped.* 3, 17.

Stephenson, W., Endean, R., and Bennett, I. (1958). *Aust. J. Mar. Freshwater Res.* 9, 261.

Stoddart, D. R. (1962). *Atoll Res. Bull.* 87, 1.

Stoddart, D. R. (1963). *Atoll Res. Bull.* 95, 1.

Stoddart, D. R. (1964). *Atoll Res. Bull.* 104, 1.

Stoddart, D. R. (1965a). *Trans. Inst. Br. Geogr.* 36, 131.

Stoddart, D. R. (1965b). *Proc. Conf. Coastal Eng., 9th, 1964* p. 893.

Stoddart, D. R. (1969a). *Atoll Res. Bull.* 125, 1.

Stoddart, D. R. (1969b). *Philos. Trans. R. Soc. London, Ser. B* 255, 403.

Stoddart, D. R. (1969c). *Biol. Rev. Cambridge Philos. Soc.* 44, 433.

Stoddart, D. R. (1969d). *Atoll Res. Bull.* 131, 1.

Stoddart, D. R. (1970). *Atoll Res. Bull.* 136, 1.

Stoddart, D. R. (1971a). *In* "Applied Coastal Geomorphology" (J. A. Steers, ed.), pp. 155–197. Macmillan, New York.

Stoddart, D. R. (1971b). *Atoll Res. Bull.* 149, 1.

Stoddart, D. R. (1971c). *J. Mar. Biol. Assoc. India* 11, 44.

Stoddart, D. R., and Cann, J. R. (1965). *J. Sediment. Petrol.* 35, 243.

Stoddart, D. R., Davies, P. S., and Keith, A. C. (1966). *Atoll Res. Bull.* 116, 13.

Swinchatt, J. (1965). *J. Sediment. Petrol.* 35, 71.

Tayama, R. (1952). *Bull. Hydrogr. Off. Tokyo* 11, 1.

Teichert, C. (1947). *Proc. Linn. Soc. N. S. W.* 71, 145.

Teichert, C., and Fairbridge, R. W. (1948). *Geogr. Rev.* 38, 222.

Tracey, J. I., Jr., Abbott, D. P., and Arnow, T. (1961). *Bull. Bishop Mus., Honolulu* 222, 1.

Umbgrove, J. H. F. (1928). *Wet. Meded. Dienst Mijnb. Ned.-O.-Ind.* 7, 1.

Umbgrove, J. H. F. (1929). *Wet. Meded. Dienst Mijnb. Ned.-O.-Ind.* 12, 1.

Umbgrove, J. H. F. (1930a). *Proc. Pac. Sci. Congr., 4th, 1929* Vol. 2, p. 49.

Umbgrove, J. H. F. (1930b). *Leidse Geol. Meded.* 3, 227.

Umbgrove, J. H. F. (1939). *Leidse Geol. Meded.* 11, 132.

Umbgrove, J. H. F. (1947). *Geol. Soc. Am. Bull.* 58, 729.

Vermeer, D. E. (1959). "The Cays of British Honduras." Department of Geography, University of California, Berkeley.

Verstappen, H. T. (1954). *Am. J. Sci.* 252, 428.

Verwey, J. (1931). *Treubia* 13, 199.

von Chamisso, A. (1821). *In* "A Voyage of Discovery into the South Sea and Beering's Straits in the Years 1815–19" (O. von Kotzebue), Vol. II, pp. 349–433; Vol. III, pp. 1–318, 331–336, and 436–442. London.

Wells, J. W. (1951). *Atoll Res. Bull.* 9, 1.

Wiens, H. J. (1956). *Atoll Res. Bull.* 48, 1.

Wiens, H. J. (1962). "Atoll Environment and Ecology." Yale Univ. Press, New Haven, Connecticut.

Willis, J. C., and Gardiner, J. S. (1901). *Ann. R. Bot. Gardens Peradeniya* 1, 1.

Wood-Jones, F. (1910). "Coral and Atolls." Reeve, London.

Zans, V. A. (1958). *Jamaica, Geol. Surv., Bull.* 3, 1.

4

ECONOMIC GEOLOGY AND FOSSIL CORAL REEFS

Richard E. Chapman

I. Introduction

The stimulus that economic geology has provided for the investigation of coral reefs, old and new, has been as great as that for other geological and geographical phenomena. As is commonly the case with natural resources, exploitation has to some extent preceded study and research; but detailed study subsequently contributed significantly to development. This interaction has continued, as one would expect in an empirical science, to the benefit of both.

While there may be some doubt as to whether the Romans exploited mineralized Carboniferous (Mississippian) reefs in Britain, Belgian ornamental stone has been quarried since the twelfth century. It is only in the last century that "les marbres rouges" have had their picturesque trade names, such as "rouge royal" and "rouge imperial," qualified by such terms as "middle Frasnian." The quarries of Frasnes and Couvin have contributed more than just building stone and "marbres."

Important as the studies of European Palaeozoic reefs were, and are, for our understanding of the structure, palaeoecology, and stratigraphy of old reef environments, they suffer from the limitations of surface geology. In particular, reconstruction of the palaeogeography of these reefs is hindered by distortions imposed by subsequent geologic events.

Today it is no exaggeration to say that the bulk of our knowledge of fossil reefs is derived from drilling for petroleum; and much of the surface work on fossil reefs in outcrop, and on present-day reefs, is explicitly or implicitly directed towards a better understanding of coral reefs as an aid to petroleum exploration (e.g., Playford, 1969; Logan *et al.*, 1969).

Of critical importance in the study of fossil reefs was the publication of Lloyd's paper in 1929 with the thesis that the oil that had been found in 1926 in the Hendrick Field in north Texas, United States, came from a Permian coral reef.

Lloyd's paper was not the first to postulate petroleum production from a reef reservoir. Ellisor (1926) had already recognized that petroleum occurred in Oligocene coral reefs around salt domes in the coastal plains of Texas, and it is apparent that other workers were thinking along these lines at that time. However, it was Lloyd's paper that indicated to petroleum geologists that fossil coral reefs could be important petroleum reservoirs—and this in spite of the fact that there is a dearth of corals in the Permian reef faunas of West Texas.

Girty's early work on the Guadalupe fauna (1908) had not revealed a coral fauna, and this may have delayed the discovery that the Permian Basin of Texas contained reefs.

Lloyd (1929) accepted the absence of identifiable corals, but nevertheless insisted that the reservoir of the Hendrick Field was a coral reef. This interpretation was to be the basis of the enormous development of West Texas and New Mexico during the 1930's.

If corals are still regarded as unimportant frame builders in these Permian reefs, it has not prevented the use of such names as "Horseshoe Atoll." Nor did the paucity of corals detract from the area and the thesis as a source of encouragement to explore for reefs elsewhere.

Petroleum production from reef reservoirs was not confined to West Texas during the 1920's and 1930's. The Golden Lane of Mexico had reef production, and reef production had been obtained in Canada (Stewart, 1948) and in Netherlands New Guinea (now West Irian) at Klamono in 1936 (Visser and Hermes, 1962). The great oilfield of Kirkuk in Iraq, discovered in 1927, perhaps did for the Middle East what the Hendrick Field did for Texas, and its recognition as a Tertiary reef complex no doubt stimulated intensive studies in that part of the world (Henson, 1950).

The investigation of coral reefs received a new stimulus with the discovery of the Leduc Field in Alberta, Canada, in 1947; most major oil companies engaged in intensive research into fossil and modern coral reefs during the 1950s.

Although the development of the Devonian reefs of Canada was the most dramatic reef activity of the 1950s, receiving wide publicity even outside professional circles, reef reservoirs continued to be discovered in the United States, New Mexico, the Middle East, and Southeast Asia (New Guinea).

By the 1960s, fossil reefs had acquired the reputation of being harder to find than conventional structural reservoirs, but many were much more prolific. This view was reconfirmed in 1968 when Occidental's well D1-103, 15 miles southeast of the Intisar[1] Field, Libya, yielded 74,867 barrels a day of clean 37° API. oil from 731 feet of fossil reef on its production test (World Oil, January 1968), thus eclipsing individual well productions of about 50,000 bbl/day from over 900 feet of Paleocene reef limestones in the Intisar Field itself. (Even these production rates are dwarfed by the estimated 260,000 bbl/day obtained from Cerro Azul No. 4 in Mexico's Golden Lane—also from a reef complex.)

The search for fossil reefs as potential petroleum reservoirs continues in many parts of the world today, notably in Saskatchewan, the north-central United States, and the New Golden Lane of offshore Mexico.

The discovery of fossil coral reefs in the subsurface is a difficult and uncertain operation. Indeed, it is probably true that most new discoveries have been made largely by chance, either by drilling to an anticlinal structure that was subsequently found to be a drape structure over a buried reef, or by drilling to a seismic structure that had been induced by the "pull-up" of subreef reflections by the higher velocity within the reef limestones. The subsequent discoveries in a reef trend often result from recognizing that these features may indicate a buried reef.

The intensive research that has been devoted to modern and fossil coral reefs and their environments, in outcrop and in the subsurface, has been directed towards the understanding of the biofacies and lithofacies relationships so that the presence and position of reef structures can be inferred from associated facies.

One important general conclusion that can be made from this wealth of data is that those reef provinces that have been revealed to us by drilling have had a strong tendency to form on relatively stable shelf areas that have received very little contemporaneous or subsequent tectonic deformation. Another is that the reefs grew during dominantly transgressive phases of sedimentary basin development, when the sea tended to encroach upon the land, sediment accumulation did not keep

---

[1] Known as "Idris" in the literature from discovery in 1967 until 1969.

pace with subsidence, and the depth of the sea tended to increase with time.

As present-day reefs, viewed at this moment of geologic time, present very varied characteristics, it is hardly surprising that the study of fossil reefs has revealed so many points of difference and so few points of similarity with present-day reefs.

Present-day reefs have been strongly affected by sea-level changes since the Pleistocene (also broadly transgressive), and most are either oceanic or situated on eastern continental margins.

The fossil record is predominantly of reef provinces in shallow shelf seas, often nearly enclosed. And, as most writers point out, the influence of winds, tides, currents, salinities, temperatures, and other factors controlling reef ecology and morphology, are not easily studied.

Though "coral reefs" of economic importance have been found extensively in space and time, and many of these areas have been studied closely, it is evident that economic geology has tended towards a morphological, rather than an ecological, definition of reefs. The proliferation of definitions is indicative only of the wide variety of parameters, and of their local importance. This is particularly true of the Permian Basin of west Texas and New Mexico where, as has been noted, corals are rare.

Lowenstam (1950) set high standards for himself and others in stipulating that a reef must properly contain organisms that were frame builders, and that these organisms grew to produce a wave-resistant structure, with sediment retention and sediment binding playing an important part.[2]

Defining reefs in terms of their biologic potential to build wave-resistant structures, it must be concluded that many fossil "reefs" are not true reefs (e.g., the Capitan Formation in the Guadalupe Mountains of New Mexico—Achauer, 1969) and many others have yet to be demonstrated to be such.

On the other hand, delineation of a subsurface "reef" structure by drilling is unlikely, by the very nature of the investigative process, to yield all the information needed to satisfy a definition of reefs in biological, ecological, and morphological terms. Well spacing is usually too coarse to reveal the degree of detail required, even if all wells are cored through the entire reef complex.

---

[2] This definition has been retained not only for reasons of priority. Some subsequent definitions in use include thickness as a criterion. Thickness is merely a measure of the response of the reef-building organisms to subsidence relative to sea level. Any definition must be as applicable at the beginning of reef growth as at the end.

The chances of penetrating and coring the reef-rock itself is small. For example, only two wells (one onshore, one offshore) are thought to have penetrated the reef core of the Cretaceous reef complex of Mexico (Viniegra O. and Castillo-Tejero, 1970). If the gas or oil/water contact is well above the base of the complex, knowledge of the water-bearing part will be fragmentary or nonexistent.

A further complication in the classification of subsurface reefs is that the density of drilling is related to the occurrence of petroleum, and hence to porosity and permeability characteristics, so that the weight of evidence tends to be unevenly distributed. Once the petroleum-bearing part of the reef complex has been roughly delineated, information in adjacent areas will be obtained only from wildcats.

On the other hand, if surface exposure is required to demonstrate the essential reef characters, surface geology will rarely demonstrate the gross morphology.

It is for these reasons that the (mainly) Upper Devonian reef complexes of the Canning Basin, Western Australia, are of such importance, for here the accidents or designs of nature have exposed a reef complex at the surface, almost undisturbed, in such a way that three-dimensional studies in great detail are possible (Playford and Lowry, 1966; Playford, 1967, 1969).

The definition of *coral* reefs must, by analogy, be reefs in which corals played an important part as a frame builder. Problems arise, then, of the degree of importance of corals in this role if the structure is to be called a coral reef. The mere presence of corals is not sufficient.

Again, the information from the subsurface will often be insufficient to assess these roles. Playford (1969) assessed corals as third in importance as frame builders in the Devonian reefs of Alberta and those of Western Australia, following calcareous algae and stromatoporoids.

Finally, diagenesis and subsequent alteration of the carbonate rocks may obscure the structure of the rock to varying degrees.

These difficulties account in the main for the rather loose usage of the term "reef" in subsurface geological work, and have tended to place emphasis on morphology, which is also important on a wider scale when exploring for or extrapolating from reef trends. The common term "reef complex" is necessary to allow for ignorance of detail.

The foundations of fossil coral reefs show the same general ranges as those of the present day. They grew on carbonate banks or shelves, and as fringing reefs round granitic islands in Alberta. Silurian reefs in Illinois grew on bioclastic calcarenites and limy muds into which the reef mass settled (Shrock, 1939; Lowenstam, 1950). And in places the reefs themselves grew to modify their own environment, sometimes to the point of self-extermination by creating hypersaline conditions.

Of considerable importance to the petroleum potential of reef areas is the nature and extent of permeable beds under the foundations of the reefs, the nature and extent of impermeable sediments deposited intermittently during the development of the reef complex, and the manner in which the environment that was once favorable for reef growth changed to terminate it.

If the reef structure has formed substantial relief on the sea floor, compaction of surrounding sediments results not only in drape structures but also in the expulsion of the interstitial fluids. These fluids will normally be expelled with an important vertical component.

The presence of extensive, relatively impermeable beds has the effect not only of inducing greater horizontal displacement of the fluids in associated permeable beds, but also of acting thereafter as permeability barriers to the vertical migration of petroleum, leading to its accumulation within the porous and permeable zones of the reef complex.

Evaporites have evidently been important in this respect in the Permian basin of west Texas and New Mexico, and in the prolific reef complexes of the Middle East. Both evaporites and shales have been important in the Devonian reefs of western Canada.

Before reviewing some specific reef provinces it is necessary to discuss briefly the terminology of reefs and reef environments.

The proliferation of descriptive terms has resulted not, I believe, from scientific individuality, but rather, as has been said above, from the natural differences of ecological history and geographical environment, and from the more fundamental geological role of time. Time has blurred some relationships to such an extent that one wonders whether some biostromes would not have been called reefs at some moments during their formation. Time has distorted most petroleum-bearing reefs in the vertical dimension, so that their thickness far exceeds the presumed depth tolerance of the reef-building organisms. In other cases, time has hardly blurred the record at all. Between these extremes is a series of more or less known and preserved reef complexes.

Reef complexes are divided into three major zones—fore reef, reef, and back reef—and these three zones are also broadly depositional and ecological zones, and may therefore be recognized by their lithology and fossil content. In particular cases it may be possible, and useful, to erect a more detailed classification. Here we shall confine ourselves to the simplest classification of general applicability.

The *reef* is the wave-resistant organic structure. The *back reef,* on the sheltered side of the reef, is contiguous with the reef, and consists of organic, bioclastic, and sometimes terrigenous sediments. The *fore reef* is the apron of detritus around the reef, characterized by an original dip often in excess of 30°. The fore reef may grade into interreef deposits.

Geomorphologically, reefs on the continental shelf fall into two main divisions—*barrier reefs* and *random* or *patch* reefs. *Fringing reefs* also occur in the fossil record. Confusion occurs only when "random" reefs become ordered to some extent by structural control, which, while developing reef trends, falls short of forming a barrier reef.

II. Petroleum in Reef Reservoirs

While petroleum is not the only resource of economic importance obtained from fossil reefs, it is undoubtedly the most significant.

The net result of the reef-building processes is the localization of large volumes of clastic material of various origins. This material is subject to wide variations in composition, because the growth of the reef itself affects its own environment; and because of its topographical features, it is subject to marked lateral facies changes, not all of which are necessarily strictly contemporaneous.

A *reef complex* is therefore a structure of variable porosity and permeability, often enveloped by sediments of an entirely different character. These enveloping sediments are almost always relatively impermeable shales or evaporites.

Reef complexes are potential reservoirs due to their porosity and permeability, and they are potential traps if they are enclosed by impermeable sediments before petroleum migration takes place. Not all reefs in an area are found to contain petroleum.

The organic matter from which the petroleum in reef reservoirs was generated is clearly not the organic matter of the reefs themselves. Areas of ecologically similar reefs may contain some reefs with gas, some with oil, and some with water. The environment around a living reef is oxidizing, with high water energy. This is inimicable to the preservation of organic matter for subsequent conversion to petroleum in the sediments. Small quantities of petroleum may be generated *in situ,* as in the Silurian reefs near Chicago mentioned below. But deepening water between reefs may lead to euxinic conditions at the sea floor (as indicated, for example, by the Keg River Formation in the Rainbow area of western Canada—Barss *et al.,* 1970) in which sufficient quantities of organic matter may be preserved in the sediments, to be subsequently converted to petroleum when favorable temperature and pressure conditions have been created by burial. Secondary migration paths may lead to some reefs but not others.

A. SILURIAN REEFS OF THE GREAT LAKES AREA

Lowenstam (1950) demonstrated that the Silurian (Niagaran) reefs of the Great Lakes area could properly be called reefs because the stromatoporoids not only had the potential to build a wave-resistant structure, but clearly did so.

Silurian reef production was first obtained in Illinois in 1943 in the Marine Pool, Madison County (Lowenstam, 1948). Silurian reefs are still regarded as attractive prospects in the Great Lakes area.

Silurian rocks outcrop along the western and northern shores of Lake Michigan and around the north shore of Lake Huron to the Bruce Peninsula, from which the outcrop extends to the southeast.

The main reef trend in the subsurface is around the eastern and southern shores of Lake Huron, in Ontario and Michigan. From south of Lake Huron, this trend probably extends westward towards Lake Michigan.

Reefs are observed at the surface along the west coast of Lake Michigan, in northeastern Illinois, and in eastern Wisconsin. It was from Wisconsin that the first account of fossil reefs in America came (Chamberlin, 1877).

In this area, during Niagaran and earliest Salinian times, there was a general carbonate deposition with associated evaporites. Silurian reefs along the margin of the Michigan basin appear to have developed periodically into effective barrier reefs, leading to the deposition of evaporites in the backreef areas. Jodry (1969) has pointed out that these lateral facies changes are probably not strictly contemporaneous, differential compaction having obscured their relationships.

The reefs and associated carbonates have been extensively dolomitized, and they are quarried for commercial dolomite at the surface. In the subsurface they contain important petroleum reserves. Gas reserves in reefs in the Michigan basin are given by Fenstermaker (1969) as 50×10^9 cu ft.

These Niagaran reefs grew in a shallow shelf sea, apparently with unconsolidated sediments as their foundations. Reef heights range from 10 ft to over 300 ft, with a tendency for height to increase going north to south (Lowenstam, 1950; Jodry, 1969).

Their morphology is similar to that of the Devonian reefs of Alberta (see below) in the occurrence of barrier reefs, patch reefs, and pinnacle reefs. *Pinnacle reefs* (normally defined as patch reefs without an inter-reef facies) are the only type found to be petroleum bearing in Ontario; but further south petroleum is also found in patch reefs.

The Niagaran reefs of the Great Lakes area have no present-day analogues. The presence and importance of evaporites in the former is a point of difference that is shared with most other reef areas of economic importance.

A well-documented Niagaran reef of Illinois is the Thornton reef complex (Ingels, 1963) which is exposed in a commercial dolomite quarry that is situated about 10 miles southwest of the southern tip of Lake Michigan.

This is a roughly circular, dolomitized reef that is about 575 ft thick at its maximum development, occupying about 2 sq miles at the surface.

Albertite, a variety of solid bitumen, occurs in vugs and pores, and on a hot day, hydrocarbons ooze from a fresh quarry face. Ingels mentions that attempts were made in 1864 to produce this type of oil within the city of Chicago.

The foundation for the reef is a bioclastic calcarenite on the southern flank, and Ingels suggests that coral colonies coalesced to form the original growth. Subsequently, rising sea level led to upward and outward growth of the reef. Water depth is inferred to have been about 200 ft outside the reef, implying about 400 ft of subsidence relative to sea level during the growth of the reef.

The prevailing winds were west-southwesterly (by present geography), and this led to ecological zoning, with biotopes and lithotopes trending southeast to northwest.

To windward there was a stromatoporoid ridge that was wave resistant. This is followed to leeward by a coral rampart and back reef lagoonal sediments with crinoids, and then another stromatoporoid ridge to leeward. Peripherally to the main reef some satellite reefs are found. The fore reef detrital apron dips radially away from the reef at 34°–45°.

Ingels recognized nine biosomes, of which three were flank biosomes to leeward of the reef. The main biosomes on the reef are, from windward, stromatoporoid–coral–trilobite–cephalopod–crinoid–stromatoporoid. The three flank biosomes to leeward are brachiopod–gastropod–sponge. These biosomes are not, of course, exclusive. Trilobites, for example, occur in both adjacent biosomes, and in the sponge biosome to leeward.

The main frame builders were stromatoporoids, *Pycnostylus, Halysites,* and *Favosites*. Crinoids, as elsewhere in the area, were important sources of bioclastic material.

There are numerous examples of subsurface Niagaran reefs in the Great Lakes area, but these cannot be known in the same detail as those that are exposed at the surface, particularly those (like Thornton) that are artificially and progressively exposed by quarrying.

Marine Pool was not only the first reef to produce oil in Illinois, but it was also studied closely by Lowenstam (1948, 1950). Drilling has revealed it as a horseshoe-shaped structure, open to the north, that is regarded as a reef with detrital bars. Only one well has penetrated the reef itself. The structure is surrounded by terrigenous clastics of the Moccasin Springs deposits.

The clays in these surrounding sediments are normally red, but in the vicinity of the reef there is a change of color to green. From this it is inferred that the Moccasin Springs deposits are strictly contemporaneous with the reef and that, whereas the sea floor away from the reef was below the euphotic ceiling, near the reef it was in the zone of turbulence above the euphotic ceiling, allowing the activity of reducing bacteria.

In addition to the color change, the Moccasin Springs deposits are thicker, in the form of a ridge, to the north of the reef. This suggests that water energy redistributed the sediment, which accumulated to a greater thickness in the lee of the reef. This, and the shape of the reef, suggests prevailing southerly winds.

B. DEVONIAN REEFS OF WESTERN CANADA

The history of petroleum exploration is punctuated by significant discoveries that have provided special impetus, and stirred the petroleum industry well beyond the area affected most by the discovery. One such recent event was the discovery of oil on the North Slope in Alaska, with potentially far-reaching consequences on North American domestic supplies, and on petroleum technology, transport, and marketing patterns.

In 1947 the discovery of oil at Leduc was such an event. News of the success went far beyond Canada, and received such publicity that it is still hardly necessary to state that Leduc is in Alberta.

The discovery that the oil at Leduc came from an Upper Devonian reef put reefs once again at the forefront of petroleum exploration prospects.

The Western Canada basin (Fig. 1) extends from the Arctic islands in the north, through the Northwest Territories and Alberta, east of the Rockies, to the northcentral United States.

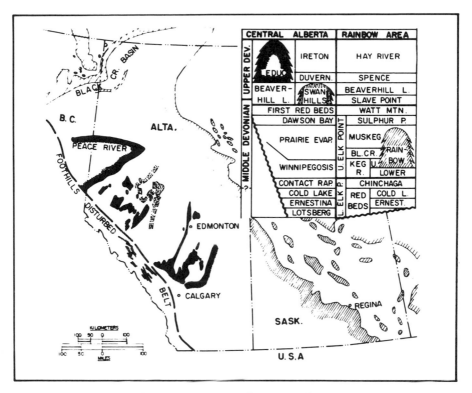

Fig. 1. Devonian reefs of Western Canada. (Reproduced with permission from Barss *et al.*, 1970, p. 20.)

Over much of its area, it developed with very little tectonic deformation. Dips are generally very low (commonly less than 1°), and structures are simple.

The dominant feature of the Western Canada Basin in British Columbia and Alberta is the Peace River Arch, which remained an influence on sedimentation throughout most of the Palaeozoic, and eventually formed an island of Precambrian granodiorites, schists, and gneisses in late Devonian seas. Around this island fringing reefs grew, depositing in cyclic succession reef rock, detritus, and clastic, often dolomitic, limestones (de Mille, 1958).

North of this island, important reef complexes grew in middle Devonian times in the Rainbow-Zama area. South of this island, younger reefs grew in late Devonian times to form the Swan Hills and Leduc complexes. All three areas have reefs with important petroleum reserves.

In the absence of clear structural control, geological research concentrated on the stratigraphical aspects of reef associations.

The development of petroleum in the Rainbow-Zama area since its discovery in 1965 affords an example of how detailed geological and geophysical work has revealed a pattern of reefs unlike any of the present day, but with points of similarity with other fossil reef areas (particularly the Niagaran reefs of the Great Lakes area). As elsewhere, not all reefs found contain petroleum (see Barss *et al.*, 1970, for a recent description).

In the Rainbow area a barrier reef separates two areas of smaller reefs. This barrier reef, running northeast to southwest, was a dominant influence on the environment, separating an area of terrigenous sediments to the west and north from an area of evaporites to the south and east.

These reefs were built on limestones and argillaceous limestones (with crinoids) of the Lower Keg River Formation, and there was apparently basement structural control for the barrier reef (Langton and Chin, 1968). Crinoids may have formed the nucleus for reef growth.

No petroleum has been found in the barrier reef so far, but it extends northwards into the Northwest Territories where, on the south shore of the Great Slave Lake, it is host to the lead-zinc deposits of Pine Point (Jackson and Folinsbee, 1969).

Langton and Chin (1968) have shown how detailed lithofacies studies have revealed two distinct types of reef to the south and east of the Presqu'ile barrier reef. These Middle Devonian (Givetian) reefs are either pinnacles (without lagoonal or back reef facies) or atolls. While corals are present, stromatoporoids appear to be the main builder. Structures are up to about 800 ft thick, and the reef-rim development suggests that the winds were from the northeast.

Porosity in these reefs is mainly attributable to diagenetic causes, but original porosity is also present. Variations in reservoir characteristics are mainly due to local differences in diagenetic history.

These reefs were terminated by environmental changes, doubtless largely contributed to by the reefs themselves, that resulted in the deposition of evaporites and carbonates.

To the north and west of the Presqu'ile barrier reef, isolated reefs occur within marine shales. These reefs range in size up to 15 or 20 miles in diameter, but they have been economically disappointing as regards petroleum (Hriskevich, 1967).

It is tempting to speculate that the contrast in petroleum potential across the Presqu'ile barrier reef is related to the presence and absence of evaporites, and that the evaporites here, as in other important areas of reef production, can contribute directly to the petroleum potential. It is also likely that a permeable base enhances the petroleum potential of areas of reefs, providing a migration path from areas of petroleum generation to some of the reefs.

To the south of the Peace River Arch a second, younger phase of reef building occurred in the Swan Hills area in late Middle and early Late Devonian times (Murray, 1966; Hemphill *et al.*, 1970).

The reefs of the Swan Hills area[3] grew on a carbonate shelf in normal marine waters. They are constructed almost entirely of stromatoporoids, with solenoporid algae and corals locally. The dendritic stromatoporoids *Amphipora* and *Stachyodes* were important as sediment retainers. Murray (1966) describes the Judy Creek complex in this area as a reef-fringed bank.

These banks with their reefs (up to 400 ft thick) show a vague northerly trend and are separated by channels. No channeling has been revealed in the Judy Creek complex itself.

Deeper-water sediments of the Waterways Formation lie to the east and north of the Swan Hills Member. Dolomitic argillaceous limestones and calcareous shales envelop the Swan Hills Member.

Reef growth was apparently terminated by drowning, and Murray suggests that this took place earlier in the north than in the south. The Swan Hills Member was terminated by the deposition of calcarenite-stromatoporoid beds. Evaporites are not important in this area.

Still further south, the prolific reef area that was discovered with Leduc in 1947 represents a third, and still younger, period of reef growth in the Alberta basin (Western Canada basin).

These reefs are of Late Devonian age (Frasnian), younger than those to the north, and many are prolific oil fields. Again, not all reef structures contain petroleum.

Three major trends have been revealed. The most prominent is the Rimbey-Meadowbrook trend (Downing and Cooke, 1955), which includes the Leduc field itself. This runs close to the west of Edmonton in a north-northeasterly direction. To the southeast there is a parallel trend through Bashaw that includes the Duhamel reef, and to the south of this there is a north-south trend of reefs in the Stettler area.

[3] Swan Hills Member of Beaverhill Lake Formation. Stratigraphic nomenclature of reefs is rather inconsistent. Hemphill *et al.* (1970) follow an unpublished work and elevate the Swan Hills Member to Formation rank. The writer is of the opinion that the spirit of the Code of Stratigraphic Nomenclature is better served by ranking reefs as Members within a Formation. He retains Murray's terminology for this area, but does not seek to alter the accepted terminology of the Leduc reefs.

The stratigraphic relationships are shown in Fig. 2. The lower part of the Ireton Formation is probably contemporaneous with the upper part of the Leduc Formation due to the presence of large fans of detrital limestone that are thought to have originated from the Rimbey-Meadowbrook reef trend (Newland, 1954).

The seas in which the Leduc Formation reefs grew were shallow and appear to have transgressed from the northwest over a carbonate platform that lay on the eastern side of an open basin. The prevailing winds have been from the northeast.

The conditions that localized reef growth are not clear, but the Cooking Lake Formation of the Woodbend Group—the carbonate platform underlying the Leduc Formation reefs—shows two features of interest. In the Leduc and other areas, the facies of the generally biostromal Cooking Lake Formation changes to become indistinguishable from the overlying Leduc Formation, and the boundary there is artificial. Furthermore, in areas adjacent to the main reef development, the Cooking Lake Formation thickens with a compensating thinning of the overlying Duvernay Formation, consisting of brown and black marine shales with argillaceous limestones (Andrichuk, 1958).

These features suggest that some form of shoal or wave-resistant structure existed in Cooking Lake times, before the main period of reef growth in Duvernay times, and they are reminiscent of the relationship of the Moccasin deposits to the Marine Pool in Illinois.

Reef growth was terminated by the deposition of grey marine shales

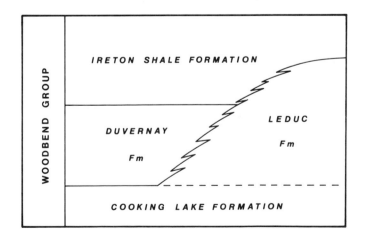

Fig. 2. Stratigraphic relationships in the Leduc reef complexes.

and argillaceous limestones of the Ireton Formation, which in turn was followed by carbonates and evaporites. Evaporites increase in importance in the Duvernay Formation and equivalents to the southeast of the Stettler area, suggesting that the waters became restricted by the reefs themselves.

As in the Devonian reef provinces to the north, the area suffered little deformation, and no obvious structural features control the reef trends.

The morphology of the reefs is variable. In general they are flat-topped "table" reefs (Andrichuk, 1958) ranging in size to about 20 miles in maximum dimension. Thicknesses of the reef complexes are 600–1000 ft.

Most of these reefs are completely dolomitized, with the destruction of the original diagnostic features of texture and organic contents. The Leduc reef itself is dolomitized, and the type section of the Leduc in British-American's Pyrcz No. 1 consists of 603 ft of crystalline dolomite (Imperial Oil Ltd., Geological Staff, 1950). Undolomitized reefs include Redwater and Golden Spike (west of Leduc); Duhamel is partly dolomitized.

The Redwater field, discovered in 1948 by seismic reflection survey, produces oil from the most important reef that has not been dolomitized. It lies slightly east of the main Rimbey-Meadowbrook trend, at depths below 3000 ft. The petroleum reservoir is confined to the upper 150 ft of the reef, so that little information is available on this reef as a whole.

The Redwater reef complex grew on a carbonate bank, and it is enveloped in marine shales and argillaceous limestones of the Duvernay and Ireton Formations. Reef growth, as elsewhere in the area, appears to have been terminated by the deposition of shale (Klovan, 1964).

The main frame builders in the upper, better-known part of the reef were stromatoporoids, and there is a general parallelism between biosomes, lithosomes, and the northwest-southeast trending reef (Andrichuk, 1958; Klovan, 1964).

Klovan distinguishes seven facies: a *Megalodon* (pelecypod) facies along the outer edge, tabular stromatoporoid reef, massive stromatoporoid detritus, skeletal calcarenite, and a back-reef facies that encloses an *Amphipora* limestone facies.

Corals (*Phillipsastrea* and *Alveolites*) appear to have been important locally as frame builders, and they occur in most areas of the reef. *Alveolites* was apparently able to adjust to various reef environments, for it appears commonly in all but the back reef facies. *Syringopora* and *Syringoporella* are found in the fore-reef, in which original dips of 20° have been recorded (Klovan 1964).

Such details cannot be known from the dolomitized reefs; and if dolomitization is in any way related to reef environment, it cannot be assumed that the details known of Redwater are applicable to the other reef complexes.

Playford (1969) has drawn interesting comparisons between the Devonian reef complexes of Alberta and those of the Canning Basin, Western Australia.

The economic importance of Canada's Devonian reef complexes is immense. In 1968, 88% of Canada's oil reserves and 80% of her marketable gas reserves were in Alberta. Most of these were in Devonian reef complexes.

Sulphur, a common contaminant of "sour" petroleum, is now an important by-product of Canadian refining. The presence of sulphur and sulphur compounds in petroleum is commonly associated with limestone and dolomite reservoirs. This association is probably intimately related to the environment of deposition of the source and host rocks.

C. Other "Reef" Provinces

Space does not permit more than a very cursory mention of other "reef" provinces that are important for their petroleum content. If it seems unreasonable to include three major petroleum provinces under this heading, it is because for one reason or another they are not considered to be as instructive as those already described.

The Permian basin of west Texas and New Mexico is not a major coral reef province. (It should be noted that the "Permian" basin area includes rocks, mainly carbonates, of all the Palaeozoic Systems, and all these systems contain petroleum to a greater or lesser extent.)

During the Permian Period, this basin was bordered by carbonate banks with deeper, euxinic subbasins of marine clastic sedimentation (Galley, 1958). Corals, frame-building organisms, reefs, and fore-reefs (as defined in this chapter) are rare.

These carbonate banks often resemble atolls and barrier reefs, but the resemblance is morphological only. Apparently they accumulated to form sufficient relief on the sea floor to modify their environment, and the period closed with the spread of evaporites and red beds over the basin area from the north.

The well-known Scurry Reef in Scurry County, Texas, is a carbonate bank of Upper Carboniferous (upper Pennsylvanian) age that is over 20 miles long and about 5 miles wide. It is composed of organic debris and calcareous sediments with calcite cement.

The Golden Lane of Mexico, discovered in 1908, is part of a large carbonate complex of Early Cretaceous age. Since 1963, offshore exploration has indicated a marine Golden Lane that, with the Golden Lane, is morphologically a huge atoll measuring 85 miles by 40 miles, partly underlying the Gulf of Mexico (Viniegra and Castillo-Tejero, 1970).

Corals are rare. Mollusca (Rudistids) and foraminifera are the commonest organic components; and the internal "lagoonal" facies is variable, including carbonates and evaporites. Apparently the "reef" formed a topographic elevation sufficient to modify the internal environment, but there is little other evidence that this was a true reef of frame-building organisms.

The Middle East is a carbonate province with important petroleum production and reserves. Reef complexes occur extensively in the Cretaceous and Tertiary (Henson, 1950; Kent *et al.*, 1951) and some of these may be coral reefs. Many are carbonate banks with Rudistids and foraminifera.

The difficulty of synthesizing the reef complexes of the Middle East may be related to its tectonic deformation. Unlike the North American reef provinces, the Middle East has been folded into long anticlines, some of which contain important petroleum reservoirs. Drilling has naturally been concentrated on these structures.

Henson (1950) described the Kirkuk Field, Iraq, as a *regressive* Tertiary reef, with fore reef, reef, and back reef facies—but stated that reefs had not been developed as a research project. Van Bellen (1956) discussed the stratigraphy in some detail.

Kirkuk Field is on a reef belt that is over 250 miles long, ranging in age from Middle Eocene to Miocene. The reefs are composed of algae with foraminifera and some corals; they apparently formed wave-resistant structures.

The lower part of the Asmari Limestone (the main petroleum reservoir in southwest Iran) was interpreted as comprising Oligocene to lower Miocene reef complexes in which calcareous algae, foraminifera, and corals occur (Kent *et al.*, 1951). These complexes form a belt trending northwest-southeast across the northern end of the Persian Gulf. Recently, however, Hull and Warman (1970) insisted that the Asmari Limestone shows no reef features.

Undoubtedly other areas will become important for the understanding of fossil coral reefs. In particular we may look to North Africa, where important petroleum reserves have recently been found in Paleocene reef complexes in Libya.

There are at least six Upper Paleocene (Landenian) reefs in the central part of the Sirte basin, Libya, of which three contain commercial quantities of oil (Terry and Williams, 1969). These are in part true coral reefs in which coral grew to form a wave-resistant structure, and played an important part in retaining and binding sediment.

These reefs grew contemporaneously in an epicontinental sea, on thick mounds of foraminifera and algae. They formed subcircular reefs about 3 miles in diameter and up to 500 ft thick, consisting largely of coral, coralline algae, bryozoa, and foraminifera.

Their location was apparently controlled by preexisting structures over which shoaling of the sea took place. Wind and current influences do not seem to have been marked, for these reefs do not have a well-developed reef-rim enclosing a back reef or lagoonal facies. They may have been protected to the southwest and northeast by a barrier reef.

Reef growth was terminated by drowning during the late Paleocene transgression, and the deposition of argillaceous sediments.

III. Metalliferous Ores in Fossil Coral Reefs

The economic geology of fossil coral reefs is not confined to petroleum, for there are important occurrences of base metals in carbonate rocks, some of which are associated with fossil coral reefs.

The principles governing the emplacement of ore bodies in fossil coral reefs may not be very different from those leading to the accumulation of petroleum. Hodgson and Baker (1959) noted an association between nickel, asphaltenes, resins, vanadium, sulphur, and the density of the associated oil. There is a growing opinion that both base metals and petroleum migrate with the interstitial fluids expelled by the compaction of sediment. The chemistry of these interstitial fluids depends largely on the sediments from which they were expelled, and their subsequent migration depends on porosity and the permeability paths available, and on the fluid potential gradients in them.

A fresh mud or clay may have a porosity of about 80% at the surface; but on burial, the load of overlying sediment begins a process of compaction that increases the bulk density of the clay, reduces its porosity, and expels a commensurate amount of its interstitial liquids (see Chapman, 1972b, for a discussion of these processes). For example, compaction of 1 km^3 of clay from 40% to 20% porosity, which could occur during burial from about 300 to 2500 m (1000–8000 ft), requires the expulsion of about 250×10^6 m^3 of water. This is the *total* pore volume of 1 km^3 of sandstone with 25% porosity. Trace quantities of base metals in such volumes of displaced fluids have a potential for economic accumulation by physicochemical processes. Fossil coral reefs have the potential to alter the physicochemical environment of migrating fluids.

There is little doubt that an important proportion of the world's petroleum originated in clays and marls, migrated through permeable rocks, and accumulated when further migration was prevented by a trap. Many accumulations occur in the porosity of fossil coral reefs, as we have seen, from which further migration has been prevented by cap rocks of relatively impermeable shale or evaporites. The same confidence cannot be shown for the processes of emplacement of ore bodies in fossil coral reefs. This may be due in part to the small number of known mineralized reefs, and of mineralized carbonates in which the role of volcanic activity can be fully assessed.

Current interest in mineralized fossil coral reefs is focused on Pine Point, on the southern shore of the Great Slave Lake, Northwest Territories, Canada. This important lead-zinc ore occurrence lies on the northern edge of the Presqu'ile Barrier Reef, which is associated with petroleum-bearing fossil coral reefs to the southwest (Jackson and Beales, 1967; Jackson and Folinsbee, 1969). The ore occurrence is near the surface: the petroleum, at depths below 5000 feet.

As noted earlier, this barrier reef divides a shale province in the north and west from an evaporite province in the south and east, and thus forms a clear porosity and permeability path for the interstitial fluids expelled by compaction of sediments in the adjacent basins, particularly the Mackenzie Shale basin. (There are difficulties in accepting important lateral migration of fluids through shales because the principal fluid potential gradients are vertical; but the paths could be through a permeable underlying formation that is in hydraulic continuity with the barrier reef.)

The host rocks form part of a barrier complex that has been interpreted as an organic bank consisting mainly of stromatoporoids, crinoids, and corals, that was periodically exposed above sea level (McCamis and Griffith, 1967). The complex faced the north-northwest, with a fore reef facies, an indistinct organic reef facies, and a back reef facies to the south-southeast.

The ore bodies occur in the back reef and organic reef facies, but they do not correspond geometrically with bedding or facies units. They occur as "pods" in breccia zones that are probably due more to solution than tectonism. The host rocks were dolomitized before ore emplacement (*op. cit.*).

The development of the barrier reef complex was terminated by drowning in a late Middle or Late Devonian marine transgression.

In the *eastern Alps*, from Switzerland to Austria, there has been a long history of lead-zinc mining from upper Middle Triassic reefs (Schneider, 1964). As with Pine Point, mineralization is confined to the reef and back-reef facies, but ore bodies are both discrete and bedded. These reefs were extensive, perhaps tending to form barriers with hypersaline lagoons.

Research into mineralized sediments has tended to emphasize mineralogy rather than sedimentology. This tendency, with relatively small areas occupied by ore bodies, has restricted geological synthesis in studies of mineralized areas.

IV. Conclusions

The geology of fossil coral reefs revealed by drilling allows us to make a number of important generalizations, the significance of which can not yet be fully assessed.

Fossil coral reefs that contain petroleum (and the associated reefs that do not) are characteristically unfolded. The petroleum accumulations are due to the morphology of the reef complex, sealed by a fine-grained, relatively impermeable rock. While this may give the appearance of a fold, and overlying sediments may drape over the reef complex, the platform or base on which the reefs grew has not been folded significantly. This is generally true of the Silurian reefs of northern North America, the Devonian reefs of west Canada, the Cretaceous reefs of Mexico (see Viniegra, and Castillo-Tejero, 1970, pp. 320–323, Figs. 6 and 7, for a clear illustration, noting the exaggeration of the vertical scale) and, judging by the little information available, also for the Paleocene reefs of Libya. Thus, those reef areas in which significant petroleum accumulations of widely different ages have been found were relatively stable continental areas in which subsidence and some tilting was the dominant contemporary influence. Little deformation has been suffered since.

The subsidence is also revealed by the vertical dimensions of the reefs, commonly 500–1000 ft thick. It seems unreasonable to postulate a depth tolerance of past reef-building organisms much greater than about 300 ft, so the observed thicknesses imply an upward growth with time compensating for the subsidence of the reef foundation (relative to sea level). This is a transgressive situation in which the sea tends to deepen with time and to encroach upon the land. With clastic sedimentation, transgressions lead to the migration of facies landward with time, but organic reefs that can build upwards faster than the rate of subsidence respond by developing reef structures that grow vertically with time. This growth has the economic result that large potential traps for petroleum are created.

Sedimentary basins characteristically begin with a transgressive phase, when the volume vacated by subsidence is greater than the volume of sediment supplied to the area. Subsequently, the volume of sediment may increase and lead to a regressive phase, in which the volume of sediment supplied to the area exceeds the volume vacated by subsidence. The basin then tends to fill up and may become land—part of the continent to which the reefs were peripheral. Some reef growth is terminated by drowning during the transgressive phase—that is, the rate of subsidence exceeded the rate of growth of the reef. Others are terminated by smothering under clastic sediments (usually fine grained) during the regressive phase (or a regressive interlude). A few reefs have been terminated, apparently, by an unfavorable, hypersaline environment created by their own growth.

The significance of these generalizations is not clear (but broad conclusions from the occurrences of petroleum are outlined in Chapman, 1972a). Not all fossil coral reefs are unfolded. The Silurian reefs of northeastern Queensland, the Devonian reefs of Belgium, and indeed, the Devonian reefs in the foothills of western Canada (see Fig. 1) have been folded and faulted. The association is between petroleum-bearing reefs and relative stability and the significance of this association may be simply that deformation of the reefs leads to the escape of the petroleum. However, geological associations can be complex and the causal relationships obscure. A reef complex 1000 ft thick is easier to find in the subsurface than one 100 ft thick, so our information may be biased.

As for the extension of these observations to the present day, we cannot judge the stability of continental reef areas except in so far as they are generally devoid of seismic activity. But this is not evidence that these areas are not subsiding. Present-day reefs share with their ancestors a transgressive tendency, for the sea level has risen about 400 ft above the lowest point reached during the Pleistocene Ice Age.

The broad field of economic geology of fossil coral reefs must lean on concurrent studies of modern reefs. If this has been more true of petroleum geology than of base-metal geology, it is only because the need for this understanding has been greater in petroleum geology. Fossil coral reefs have the dimension of time, but not that of life: modern reefs, life but not time.

Studies of coral reefs, old and new, may have economic motivation; but the interplay of research and exploration has contributed much to both "pure" and applied geology. Better progress will no doubt be achieved when biologists and geologists view their problems jointly.

References

Achauer, C. W. (1969). *Am. Assoc. Pet. Geol. Bull.* **53**, 2314.
Andrichuk, J. M. (1958). *Am. Assoc. Pet. Geol. Bull.* **42**, 1.
Barss, D. L., Copland, A. B., and Ritchie, W. D. (1970). *Mem., Am. Assoc. Pet. Geol.* **14**, 25.
Chamberlin, T. C. (1877). *Geol. Wis.* **2**, 360.

Chapman, R. E. (1972a). *J. Aust. Petrol. Explor. Ass.* **12**, 36
Chapman, R. E. (1972b). *Am. Assoc. Pet. Geol. Bull.* **56**, 2185.
de Mille, G. (1958). *Alberta Soc. Pet. Geol. J.* **6**, 61.
Downing, J. A., and Cooke, D. Y. (1955). *Am. Assoc. Pet Geol. Bull.* **39**, 189.
Ellisor, A. C. (1926). *Am. Assoc. Pet. Geol. Bull.* **10**, 976.
Fenstermaker, C. D. (1969). *Ky. Geol. Surv., Ser. 10, Spec. Publ.* **17**, 50–77.
Galley, J. E. (1958). *In* "Habitat of Oil" (L. G. Weeks, ed.), pp. 395–446. Am.
    Assoc. Pet. Geol., Tulsa, Oklahoma.
Girty, G. H. (1908). *U.S., Geol. Surv., Prof. Pap.* **58**, 651 pp.
Hemphill, C. R., Smith, R. I., and Szabo, F. (1970). *Mem., Am. Assoc. Pet. Geol.*
    14, 50–90.
Henson, F. R. S. (1950). *Am. Assoc. Petrol. Geol. Bull.* **34**, 215.
Hodgson, G. W., and Baker, B. L. (1959). *Am. Assoc. Petrol. Geol. Bull.* **43**, 311.
Hriskevich, M. E. (1967). *World Pet. Congr., Proc., 7th, 1967* Vol. 3, p. 733.
Hull, C. E., and Warman, H. R. (1970). *Mem., Am. Assoc. Pet. Geol.* **14**, 428–437.
Imperial Oil Ltd., Geological Staff. (1950). *Am. Assoc. Pet. Geol. Bull.* **34**, 1807.
Ingels, J. J. C. (1963). *Am. Assoc. Pet. Geol. Bull.* **47**, 405.
Jackson, S. A., and Beales, F. W. (1967). *Bull. Can. Pet. Geol.* **15**, 383.
Jackson, S. A., and Folinsbee, R. E. (1969). *Econ. Geol.* **64**, 711.
Jodry, R. L. (1969). *Am. Assoc. Pet. Geol. Bull.* **53**, 957.
Kent, P. E., Slinger, F. C., and Thomas, A. N. (1951). *World Pet. Congr., Proc., 3rd,*
    1951 Sect. 1, p. 141.
Klovan, J. E. (1964). *Bull. Can. Pet. Geol.* **12**, 1.
Langton, J. R., and Chin, G. E. (1968). *Bull. Can. Pet. Geol.* **16**, 104; *Am. Assoc.*
    Pet. Geol. Bull. **52**, 1925.
Lloyd, E. R. (1929). *Am. Assoc. Pet. Geol.* **13**, 645.
Logan, B. W., Harding, J. C., Ahr, W. M., Williams, J. D., and Snead, R. G. (1969).
    Mem., Am. Assoc. Pet. Geol. **11**, 1–198.
Lowenstam, H. A. (1948). *In* "Structure of Typical American Oil Fields" (J. V.
    Howell, ed.), Vol. 3, pp. 153–188. Am. Assoc. Pet. Geol., Tulsa, Oklahoma.
Lowenstam, H. A. (1950). *J. Geol.* **58**, 430.
McCamis, J. G., and Griffith, L. S. (1967). *Bull. Can. Pet. Geol.* **15**, 434.
Murray, J. W. (1966). *Bull. Can. Pet. Geol.* **14**, 1.
Newland, J. B. (1954). *Alberta Soc. Pet. Geol., News Bull.* **2**(4), 1.
Playford, P. E. (1967). *In* "International Symposium on the Devonian System"
    D. H. Oswald, ed.), Vol. 2, pp. 351–364. Alberta Soc. Pet. Geol., Calgary.
Playford, P. E. (1969). *West. Anst., Geol. Surv., Rep.* No. 1.
Playford, P. E., and Lowry, D. C. (1966). *Geol. Surv. West. Aust., Bull.* **118**.
Schneider, H.-J. (1964). *Dev. Sedimentol.* **2**, 29–45.
Shrock, R. R. (1939). *Geol. Soc. Am. Bull.* **50**, 529.
Stewart, J. S. (1948). *In* "Structure of Typical American Oil Fields" (J. V. Howell,
    ed.), Vol. 3, pp. 86–109. Am. Assoc. Pet. Geol., Tulsa, Oklahoma.
Terry, C. E., and Williams, J. J. (1969). *In* "The Exploration for Petroleum in Europe
    and North Africa" (P. Hepple, ed.), pp. 31–48. Inst. Pet. London.
van Bellen, R. C. (1956). *J. Inst. Pet., London* **42**, 233.
Viniegra O., F., and Castillo-Tejero, C. (1970). *Mem., Am. Assoc. Pet. Geol.* **14**,
    309–325.
Visser, W. A., and Hermes, J. J. (1962). *Verh. K. Ned. Geol. Mijnbouwkd. Genoot.,*
    Geol. Ser. **20**, 1.

5

ASPECTS OF SEDIMENTATION IN THE CORAL REEF ENVIRONMENT

G. R. Orme

I. Introduction

The Great Barrier Reef Expedition of 1928–1929, during which the relationship between environment and reef growth was studied, marked a significant departure from the early preoccupation with reef morphology and geological history. The many and diverse investigations that followed, concerned with the biological, ecological, geomorphological, sedimentological, geochemical, and systematic aspects of coral reefs demonstrate not only the diversity of approach, but also the interdisciplinary nature of coral reef studies. A good illustration of this is provided by the complex interrelationship between sedimentation and ecology which prevails in the coral reef environment.

The total sedimentary characteristics of a reef complex are the result of the development of an ecological system, and the wide range of biological, physical, and chemical processes involved in sedimentation and diagenesis. An urgent desire to formulate principles for the interpretation of such characteristics in terms of depositional environment and sedimentary process was stimulated by the recognition of the importance of ancient reefs and bioherms in many oil fields. Properly interpreted facies patterns, sedimentary structures, textures, and fabrics, provide evidence of a range of sedimentary processes and ecological situations.

In recent years there have been several excellent reviews, e.g., Stoddart (1969a), Newell (1971), and Milliman (1974, Chapter 6) dealing with Recent or ancient reef complexes and it is not my aim to try to duplicate such masterly contributions which are of immense value to laymen and serious students alike. Rather, the present purpose is to focus attention on the sediments and sedimentary processes (excluding those involved in diagenesis) characteristic of reef complexes.

Descriptive accounts of sediments and the distribution of sedimentary facies are incorporated in many studies of reef complexes, but due largely to inherent practical difficulties, few include investigations of the processes of sedimentation. In some accounts, basic petrography has been elevated to the status of petrology through the application of informed or intuitive reasoning, but there are limitations to this deductive approach, and relatively few quantitative studies of sedimentary processes in the coral reef environment have so far been made. Consequently the answers to a number of questions implicit in such studies still appear to be either totally inadequate or, at best, only superficially adequate.

Questions that arise are fundamentally concerned with the overall sediment budget of a reef complex, i.e., factors which under normal conditions control it, or under abnormal conditions interfere with it. How are sedimentary particles formed and at what rate? How are they transported and broken down? What are the true implications of sediment textures? What is the significance of sedimentary structures? What will result if the normal hydrological regime is interfered with by man's engineering constructions and mining operations? What are the sedimentological consequences of sea level changes or the effects of cataclysmic events such as the passage of a hurricane or the occurrence of an earthquake? What factors disturb and modify sediment after deposition, how important are they, and how can they be recognized in the sedimentary record? The answers to some questions are already emerging and perhaps the present impetus in coral reef studies foreshadows solutions to many others involved in coral reef sedimentology. A more complete understanding of sedimentary facies and texture in turn will enhance the interpretation of ancient carbonate sequences.

II. The Coral Reef and the Coral Reef Complex

> Coral reefs consist essentially of calcium carbonate secreting organisms with the capacity to build skeletons capable of withstanding wave action, and as a result they form topographic features which rise from the sea floor to sea-level (Cloud, 1952; Lowenstam, 1950)—Stoddart (1969a, p. 473).

This valuable, succinct definition has a wide application to both Recent and ancient reefs but does not express the vitality which is characteristic of a coral reef ecosystem. The viable coral reef is a dynamic, constantly changing structure, sensitive to fundamental properties of the marine environment such as salinity, light intensity, temperature, oxygen supply, etc.; it is a vital force affecting and effecting ecological, bathymetric, hydrological, and sedimentological aspects of the marine environment in its vicinity, and the geographical extent of its influence may be considerable, extending well beyond its active growth limits. Under normal conditions the reef constantly renews itself against decay and erosion, but adverse conditions which interfere with its viability will lead to its decline and partial removal by the unhindered process of mechanical erosion.

The reef complex (Henson, 1950, pp. 215–216; see Gary *et al.*, 1972) is "The solid reef proper (reef core) and the heterogeneous and contiguous fragmentary material derived from it by abrasion; the aggregate of reef, fore-reef, back-reef and interreef deposits, bounded on the seaward side by the basin sediments and on the landward side by the lagoonal sediments" (Nelson *et al.*, 1962, p. 249). The term is sometimes loosely applied by some authors to embrace a wider range of facies. However, it is not proposed to extend the scope of this chapter to a discussion of the situations that prevail in some extensive barrier reef lagoons or the particular sediments and conditions which exist on the Bahama Banks, but to consider the sedimentary characteristics of the reef and peripheral areas, together with some facets of adjacent lagoonal sedimentation which are relevant to the general understanding of the development of sediment textures in and near a reef complex.

III. Morphological, Structural, and Ecological Characteristics of Coral Reefs

A coral reef is an expression of an interdependence between environment and organic activity. Temperature, salinity, depth, water circulation, nutrient supply, turbulence, and turbidity are among the determining factors for coral reef formation and form, and relatively small variations in certain of these parameters will strongly influence the viability of a coral reef. On the other hand, a reef strongly affects the environment; it brings about changes in bathymetry, influences water circulation and nutrient supply, and affects sedimentary conditions. Under normal circumstances of stable sea level and stable sea floor, it reflects a balance between the forces of construction (reef growth) and the forces of destruction (erosion and solution). Further, being the provenance of much carbonate sediment, it is a major factor influencing the distribution of sedimentary facies.

The concept of a reef being essentially composed of frame-building and frame-binding organisms with sediment fill (Ginsburg and Lowenstam, 1958) has become well established. Organisms such as corals, or in some instances crustose coralline algae and hydrozoans (Ginsburg and Schroeder, 1973), may be described as primary frame builders. Secondary frame builders are those organisms which have no potential to erect a frame or which are in this capacity subordinate to an associated organism. Thus, where hermatypic coral is the primary frame former, frame binders such as crustose coralline algae and all other sessile organisms which live on the reef surfaces would fall into this category.

The nature of living reefs and the sequences of unconsolidated sediment encountered in deep bores on atolls suggest that a wave-resistant skin of massive corals and encrusting algae on the seaward sides encloses and protects a less rigid structure of skeletons and sediments. It is a product of *in situ* biogenic activity and sedimentation which, despite its rigid outer surface, remains a remarkably porous complex (Ladd, 1961; Ladd and Schlanger, 1960; Stoddart, 1969a, p. 473).

Recent studies, e.g., that of Schroeder and Zankl (1974), and of Scoffin and Garrett (1974). have clearly demonstrated that reef formation and maintenance is a very dynamic process, every surface undergoing modification while the reef remains in its original environment.

Morphological and ecological zonation parallel to the reef edge is characteristic of most fringing reefs, barrier reefs, and atolls, and is exemplified by the classic studies of Indo-Pacific reefs by Emery *et al.* (1954) and by Wells (1954, 1957). The main features encountered in a transect across

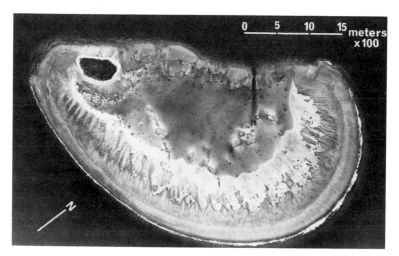

Plate I. Vertical aerial photograph of Lady Musgrave Reef, Great Barrier Reef Province. (Reproduced by permission of the Great Barrier Reef Committee, Brisbane.) See Fig. 3, p. 136.

a windward reef from open sea to lagoon may be summarized as follows (see Figs. 1, 2, and 3, and Plate I).

A. MORPHOLOGICAL ZONES

1. Seaward or Outer Slope

The seaward slope is the steeply inclined outer reef surface below approximately 18 m. In some works (e.g., Goreau and Goreau, 1973; Lang, 1974), a number of morphological zones and environments have been recognized, viz., fore reef slope, and deep fore reef.

2. The Reef Front and Terrace

This is the upper seaward face of the reef extending from the seaward edge of the reef flat to the dwindle point of abundant living coral. Groove-and-spur development is characteristic of the reef front. A terrace (14.5–18 m) of eustatic significance, "the ten fathom terrace," which may have a veneer of organic growth, is normally a feature of the reef front, and if broad and well developed, it may be considered as a separate zone.

3. Algal Ridge and Boulder Zone (Seaward Reef Margin) (Plate II,A)

The seaward reef margin is cut by surge channels which connect with grooves of the reef front. An algal ridge parallel to the reef front may be present in places and is characteristic of many Indo-Pacific reefs, although

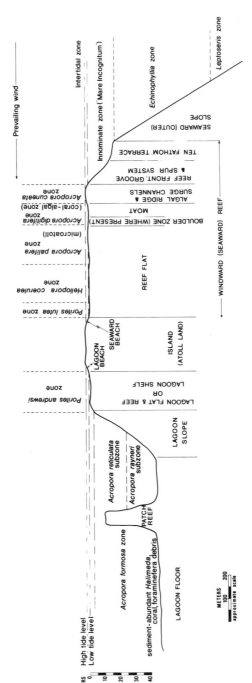

Fig. 1. Schematic cross section of an Indo-Pacific atoll showing the main morphological features and ecological zones (after Emery *et al.*, 1954; Tracey *et al.*, 1955; and Wells, 1957).

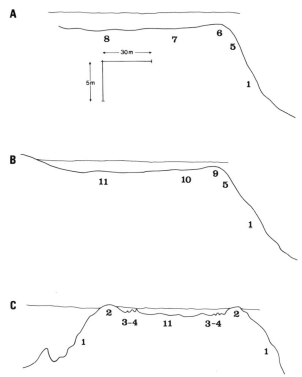

Fig. 2. Three windward reef flats. A: typical Bahamian-Floridan reef flat; B: southern Caribbean reef flat; C: platform reef, Great Barrier Reef Province. 1. Fore reef; 2. algal ridge; 3. *Acropora digitifera* (coral-algal) zone; 4. *Acropora palifera* and *Heliopora* zone; 5. *Acropora palmata-Montastrea* zone; 6. *Millepora* zone; 7. hemispheric coral zone; 8. inner reef flat (alcyonaceans and gorgonians); 9. *Millepor-red* algae-*Playthoa* zone (ecological equivalent of 2 and 6); 10. *Diploria* zone (ecological equivalent of 7 and 4); 11. inner reef flat (sand, gravel, some grasses and corals, few gorgonians. After Milliman (1974). Compare with Fig. 1.

it may not be developed in more protected situations. A ridge, very similar to those occurring on Indo-Pacific reefs, has been described by Glynn (1973) in coral reefs off Panama. The algal ridge has been called the "Lithothamnion Ridge" but is really dominated by *Porolithon*, and may grow to a height of a meter above low tide level (Milliman, 1974, p. 164).

Boulders of reef rubble may form a well defined zone at the outer edge of the reef flat.

4. Reef Flat

The reef flat is commonly awash or exposed at low tide, being the upper surface of the reef (Plate II,B,C). It is mantled with sand and

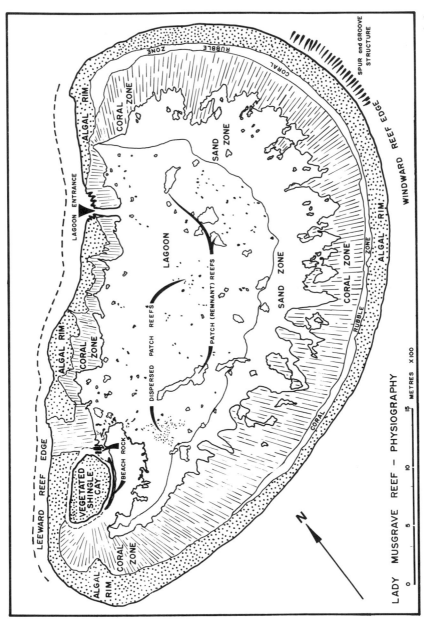

Fig. 3. Plan of physiographic zones of Lady Musgrave Reef (Orme *et al.*, 1974). Reproduced by permission of the Great Barrier Reef Committee, Brisbane. Compare Plate I.

Plate II. A: Heron Island Reef, northern edge, showing a low *Algal* ridge just awash at low tide, to the northern side of Heron Island, Great Barrier Reef Province. B: The reef flat, just awash at low tide. Biogenic sediment occupies the depression between coral heads. C: *Acropora* and encrusting coralline algae (light grey), devoid of sediment, completely exposed at low tide. Outer reef flat, Heron Island, Great Barrier Reef Province. D: *Montipora* (microatoll) approximately 3 ft in width, with a veneer of encrusting algae and some sediment. The adjacent light grey patches occurring between the corals are areas of "coralgal" sediment lying in shallow protected pools. Reef flat, Heron Reef, Great Barrier Reef Province.

gravel (Plate II,D). Corals may be sparse on windward reef flats, although a great variety of stunted branching corals and microatolls may be present in a moat landward of the boulder zone, and the small-scale relief may be complex. Subdivision into inner and outer reef flats is commonly possible, but the presence of cays and/or islands on the reef flat, with their effects on the ecology of the reef, will necessitate a modification of this simple subdivision, and sediments will show the influence of subaerial as well as submarine processes.

5. Island and Cays (Atoll Land) with Seaward- and Lagoon-Facing Beaches

6. Lagoon Reef Margin

A lagoon reef margin may consist of a narrow lagoon flat and lagoon reef. A lagoon terrace may be present if the lagoon reef margin is well defined. However, the lagoonward margin of the reef may not be defined by reef growth (Tracey et al., 1955). The lagoon reef may be poorly defined, especially where islands lie close to the lagoon, and the reef flat or lagoon beach may be separated from the lagoon slope by a lagoon shelf.

7. Lagoon slope

This term is given to the slope which descends from the lagoon reef margin or lagoon beach to the lagoon floor.

8. Lagoon Floor with Reef Knolls

The morphology of the lagoon floor is mainly the result of sediment accumulation and in situ organic growth. The organic growth structures may be merely small algal mounds or narrow coral pinnacles, or they may be extensive patch reefs or widespread Acropora thickets (Plate III,A.).

Similar features have been recognized in the reefs of the Great Barrier Reef Province, e.g., Fairbridge (1950) and Maxwell (1968a), but Maxwell (1968a, p. 98) emphasizes the distinction between oceanic and shelf reefs, and classifies the shelf reefs of the Queensland continental shelf on the basis of form, the patterns being a gradational sequence in reef development which is primarily controlled by the "hydrologic-bathymetric-biological balance."

However, there is a growing weight of evidence in support of the Antecedent Platform Theory of Hoffmeister and Ladd (1944), and Purdy (1974a) has suggested that the shapes of modern reefs are basically karst induced rather than growth induced. Indeed, the distribution pattern of Holocene nonreef sediment facies on the British Honduras Shelf is a result of water circulation and turbulence imposed by bathymetric features which have been inherited from an underlying pre-Holocene limestone surface (Purdy, 1974b).

B. Ecological Zones

The distribution of reef communities and zonation in terms of indicator coral species has been described from many Indo-Pacific and Caribbean reefs, e.g., by Wells (1954, 1957), Logan (1969), and Barnes *et al.* (1971). Thus, as would be expected, there is a change in the composition of reef communities with depth on the reef front (Barnes *et al.*, 1971; Lang, 1974). At Bikini the seaward slope below 145 m is characterized by a zone of the ahermatypic corals *Sclerhelia* and *Dendrophyllia;* from 90 to 145 m there is a zone of *Leptoseris* (elsewhere in the Indo/Pacific region the *Leptoseris* zone has been recognized between 25 and 40 fm); and a zone of *Echinophyllia* at 18–90 m. Above this, the reef front (fore reef) is dominated by crustose coralline algae with few corals (see Fig. 2).

Across the reef flat on windward reefs of the Marshall Islands several important zones have been defined. At the seaward margin there is the *Acropora cuneata* zone, a zone of encrusting corals and coralline algae including *Porites, Montipora, Pocillopora* and *Millepora*. This is followed on the outer reef flat by the *Acropora digitifera* zone, with microatolls. The *Acropora palifera* zone, which contains scattered microatolls, comes next, followed by a zone with enormous microatolls of *Heliopora coerulea* and finally the inner flat with *Porites lutea* adjacent to the beach is reached. Stoddart (1969a) discusses some deviations from this general pattern which occur in the Indo-Pacific region.

Some striking contrasts between the general ecological zonation of reef flats on Indo-Pacific and Caribbean reefs are discussed by Milliman (1974, see Figs. 1 and 2). Basically coral growth is more important on Caribbean reef flats which are deeper, and encrusting coralline algae are of lesser importance than on Indo-Pacific reef flats. Green algae (including *Halimeda*) are also more abundant on the outer reef flats of the Caribbean, and patches of green algae and marine grasses (including *Thalassia*) are common on the sandy inner reef flats.

Observations on Indo-Pacific reefs also suggest that gastropods are more common on hard reef substrates, and bivalves are more common on unconsolidated sediments (Odum and Odum, 1955; Taylor, 1968). In protected areas of the inner reef flat where algae are available for grazing, benthonic Foraminifera are plentiful (e.g., Jell *et al.*, 1965). The sea grasses (Humm, 1964) and mangrove communities (Macnae, 1968) which are associated with the coral reef environment, are important in the trapping and stabilization of sediment (see Section VI).

C. CATACLYSMIC EVENTS

A number of factors may cause some modification of the normal morphological and ecological characteristics of coral reefs. Pronounced changes in sedimentation and ecology result from the raising or lowering of base level (Barrell, 1917) which accompanies changes in the relative positions of sea level and sea bed. Such changes may be local, being induced by sediment accumulation, tectonic movements, or isostatic adjustments, or they may be of a eustatic nature. Thus, reefs may be drowned or exposed to subaerial erosion, and the stratigraphic consequences may be graphically recorded.

However, on a smaller scale, interference with the viability of a reef and its normal sedimentological regime may be brought about in several ways, some causes such as earthquakes (Stoddart, 1972), and hurricanes being abrupt, others extending over a short period of time, but all being instantaneous when considered in terms of the geologic time scale.

Cataclysmic events such as the passage of hurricanes show some periodicity and the effects of a major disturbance such as hurricane Hattie, which devastated the British Honduras reefs in 1961 may last for more than 25 years (Stoddart, 1974). In British Honduras the evidence suggests that the frequency of severe hurricanes such as Hattie coincides with the period of reef recovery in the zone of maximum damage (Stoddart, 1974, p. 479). In recent years the effects of such atmospheric disturbances on reefs have been studied by a number of workers, including Blumenstock (1958, 1961), Blumenstock et al. (1961), Stoddart (1963, 1965, 1969b, 1974), and Baines et al. (1974), and there is a clear implication that the sedimentological result of such phenomena may have been preserved in the features of older reef complexes. Some minor disconformities in the geologic record of a reef complex may reflect such events.

Predation of reef-building corals by *Acanthaster planci* on an abnormal scale through its infestation of reefs in plaguelike proportions has caused particular concern in recent years (see Endean, 1973). Apart from ecological effects, the wholesale mortality of hermatypic corals would render large sections of the reef more susceptible to mechanical erosion, locally increasing the rate of sediment supply. Opinion varies regarding the time span needed for a reef to recover following the devastation caused by a population explosion of *Acanthaster planci,* and a divergence of opinion prevails regarding the frequency or the cyclic nature of such events. Investigations of subsurface reef sediments for aggregations of *Acanthaster planci* skeletal debris (see Frankel, 1975) may provide evidence of the presence of this predator in abnormally large numbers during past phases of reef development.

Man's erratic, deliberate, or accidental interference with the coral reef environment through pollution (Sorokin, 1973, pp. 37–39; Loya, 1975), tourism, or engineering constructions (Stoddart, 1968; Flood, 1974) may locally change sedimentation patterns, instigating or intensifying marine erosion, and sediment production, transport and deposition.

IV. Sedimentary Environments, Sedimentary Facies, and Constituents

With the exception of terrigenous particles, which may be significant on coastal fringing reefs, and pumice fragments, which are occasionally washed onto reef flats and beaches, sediments accumulating on and near coral reefs are entirely of skeletal origin, but in the reef environment *sensu lato* profound differences may exist within the depositional basin resulting in contrasting sedimentary facies defined by such properties as mineralogy, grain type, texture, and depositional structures. The pure carbonate facies of the reef complex *sensu stricto* may pass laterally into neritic and/or bathyal and abyssal deposits of barrier reef lagoons or into deep fore reef deposits and deep basin deposits respectively. Considered on such a broad scale, the range of sedimentary facies extends from the more or less autochthonous part of the complex, the reef itself, through the dominantly coarse bioclastic sediments associated with it (Plate III,B,C, and D), to the fine grained "chemical deposits" of restricted bays and lagoons and the heterogeneous sediments of bathyal slopes. In situations such as that which exists in the Great Barrier Reef Province, a change across the shelf from the pure carbonate sediment of the outer barrier to the terrigenous sediment of the near shore and inner shelf regions may occur. However, sediment provenance may be complex owing to the effects of Pleistocene and Holocene sea level changes (Maxwell 1968a, p. 183), and relict sediment may become an important facies component (Maxwell, 1968b).

Although "lagoons" are distinct depositional environments within which sediments may vary considerably in petrography according to local circumstances, and may under unusual environmental conditions give rise to a special range of sedimentary facies such as that which characterizes the Bahama Banks (Purdy, 1964a), several authors have demonstrated a basic similarity in the range of sediments and environments associated with the reef complex *sensu stricto*.

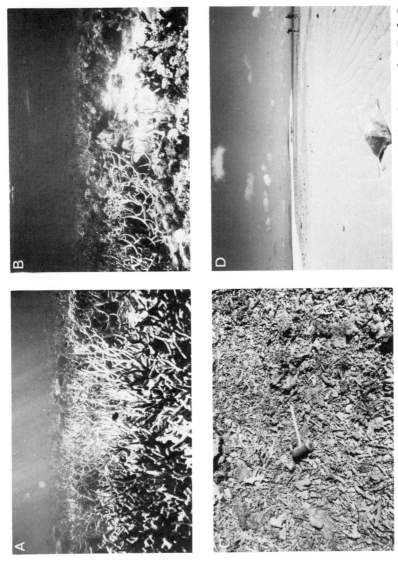

Plate III. A: *Acropora* thicket beneath approximately 12 ft of water on the leeward side of Myrmidon Reef, Great Barrier Reef Province. B: Biogenic carbonate sediment occupying channel floor, leeward side of Carter Reef, Great Barrier Reef Province. C: "Coral Stick Gravel" in a rubble zone on Heron Island Reef, Southern Region, Great Barrier Reer Province. D: Current ripple marks in biogenic carbonate sand on the reef flat near Ingram Island, Northern Region of the Great Barrier Reef Province.

In the coral reef ecosystem a particular environment is characterized by such factors as topography, oxygen supply, wave energy, diurnal temperature range, light intensity, and sedimentation rate. The basic parameters, being those required for hermatypic coral growth, may vary only within narrow limits and, although significant differences may exist between individual reefs even within the same province (Maxwell, 1968a), there is a similarity of elements in the contemporary coral reef communities in broad terms. Since natural skeletal populations supply the basic constituents, a corresponding characteristic petrography of reef sediments prevails. Furthermore, the similarity of ecological and morphological features and associated hydrological factors across a reef establishes a general facies arrangement, which defines the fore reef, reef, and back reef in a reef complex (Ginsburg, 1956). Such distinctive facies characteristics as grain types and texture are reflected in particular ecological zonations, differing energy levels, and water circulation patterns. The energy level reflected by the sediment has been used as an index for classification of carbonate deposits (e.g., Plumley *et al.*, 1962). This general facies arrangement is recognized in widely separated reef provinces; analogous situations and facies patterns are well known in many ancient reef complexes (e.g., Newell *et al.*, 1953; Wolfenden, 1958; Eden *et al.*, 1964; Orme, 1971).

Complex local patterns may occur within these broad facies divisions due to several local factors among which irregular reef topography may be important. Thus Kukal (1971) tabulates reef sediment grain size and sorting characteristics "which in places correspond to clear cut sedimentary environments," e.g., the leeward shelter afforded by coral heads or boulders cemented to the reef flat will allow finer grains of sediment to accumulate locally. Such deposits may subsequently become incorporated with the reef structure, if they are not finally removed to become part of back reef deposits. The regular zonation of the sedimentary facies of atoll lagoons, which is considered to correspond to the normal habitats of the contributing organisms (McKee, 1956; Stoddart, 1969a, p. 474), is complicated by patch reefs each of which forms a locus for local facies development, although the geographical extent of patch reef influence might be limited (Garrett *et al.*, 1971).

The actual reef structures developed *in situ*, and their associated clastic deposits in, on, and adjacent to, the reef represent the decay and disintegration of carbonate secreting organisms on a vast scale. Although the reef itself may constitute only a small part of the entire complex—Stoddart (1969a, p. 465) estimates that four or five times more loose sediment is produced than is incorporated as reef proper—it is the focus of carbonate production, the primary sources being the organisms growing on the seaward sides of the reef (Emery *et al.*, 1954), and is directly or indirectly the provenance of much peripheral sediment in many carbonate provinces.

The diverse assemblages of reefs and reef flats, typically including corals, bryozoans, coralline algae, foraminiferans, molluscs, worms, and alcyonarians, give rise to the distinctive sedimentary facies-coralgal facies of Purdy (1963). The reef-derived sediment of the outer slopes changes in composition and decreases in grain size with increase in depth. However, sediment slumping, and the release of huge blocks of reef rock which slide down the outer slope, may complicate this general arrangement (Lang, 1974; Orme, this volume, Chapter 10).

A. SEDIMENTARY COMPONENTS

In addition to the debris of organisms which dwell in and on the reef, and the contribution made by organisms living on or in the peripheral sea bed, the fabric of the reef itself is a source of sediment (Kornicker and Boyd, 1962) and, where cays and/or islets have developed on a reef flat, additions may be made from the supratidal and subaerial environments. Contributions to the sediment budget of a reef complex may therefore be grouped as follows:

1. Reef rock, solid biogenic material, a complex of framework elements hermatypic corals, coralline algae, etc., often altered by diagenesis, modified by bioerosion and cavity filling processes, which is conspicuous particularly on the open sea side of the reef and on the reef flat where it is exposed by constant wave and surf action. Enormous blocks of this material dislodged and thrown up from the reef front during storms may become quickly cemented to the reef on which they come to rest. Boulders of reef-rock may become lodged in channels and depressions on the reef flat. Large blocks may separate from the reef front and slide down the seaward slope to become part of slumped deposits in deep fore reef zones.

2. Clastic sediment derived from the disintegration of solid biogenic material. In the reef environment the disintegration of solid biogenic structures occurs on a large scale. Disintegration is accomplished jointly by mechanical and biological means. Gravels and sands, particularly from the fragmentation of corals and coralline algae, may accumulate in pockets in the reef flat or may form prominent rubble zones (see Plate II, B and D, and

Plate III, B and C). At times of storms very coarse sediment of this type may be carried far from the reef to contribute to the finer sediments of lagoons.

3. Primary clastic debris, i.e., particulate organic remains (foraminiferans, molluscs, etc.) which are not firmly attached to the reef framework. Peripheral reef sediment of the lagoons and fore reef, and temporarily accumulated finer-grained sediment of the reef flat, are included here and give rise to distinctive deposits such as *Halimeda* sand, foraminiferan sand, etc.

4. Beach rock and cay rock fragments may be contributed locally to reef sediments.

5. Fecal material produced by sediment-ingesting and lithophagic organisms.

The presence or absence of these constituents in a deposit depends on their availability through transportation or *in situ* postmortem disintegration, on durability, and on chemical stability (resistance to solution).

B. DISTRIBUTION OF COMPONENTS

The high wave energies characteristic of the reef tract are reflected in the general coarseness of the sediment found there—silt and clay-grade sediment constitutes less than 1–2% of the total carbonate (Milliman, 1974, p. 168). Many reef flats contain little loose sediment except where it falls into depressions or occurs leeward of large coral heads. Coralline algae, corals, *Halimeda*, foraminiferans, and molluscs are usually the main contributors to reef flat sediments, although bryozoans (Cuffey, 1972) and echinoids may also be important. The relative abundances of such components is controlled largely by population and productivity factors. Depth may determine the distribution of components in deep lagoons, and in deep atoll lagoons a concentric arrangement of sedimentary facies may be the result (McKee *et al.*, 1959).

Although marked changes occur in mineralogy, geochemistry, and grain types related to changes of biota across a reef complex, and may indeed show variations from reef to reef even within the same province (Maxwell, 1968a), the overall constitution of the sediments of reef complexes shows basic similarities over a particular region. However, some contrasts in sediment composition between regions are evident (see Table I, which shows the general composition of lagoon and reef sediments from the Indo-Pacific and Atlantic regions).

Coral fragments constitute between 15–40% of coral reef sediments (Maxwell, 1968a; Milliman, 1974; Stoddart, 1969a). Mollusc fragments

TABLE I

SOME EXAMPLES OF THE COMPOSITION OF PERIPHERAL REEF SEDIMENT[a]

	Coral	Coral-line Algae	Hali-meda	Forami-nifera	Mol-luscs	Miscellaneous skeletal (Echinoids, Bryozoans, Crustaceans)
Caribbean Reefs						
Florida Reefs	20[b]	10	30	6	12	7
Alacran Reef	26	11	40	8	7	1
Andros Bahamas	24	33	17	12	6	6
Hogsty Reef	27	19	2	5	22	1
Courtown Cays	35	21	28	3	10	—
Albuquerque Cays	30	21	32	2	9	1
Indo-Pacific Atolls						
Bikini lagoon	1	—	43	8	8	35
Eniwetok lagoon	12	—	28	30	3	26
Cocos lagoon, Guam	45	18	11	3	15	—
Midway	25	35	6	14	12	8
Great Barrier Reef						
Reef	20–40	17–40	10–30	8–20	4–15	Less than 5
Inter-reef	5–10	0–15	5–65	15–40	20–35	5–30

[a] Sources, Maxwell (1968a), Stoddart (1969a), Milliman (1974).

[b] Figures refer to percentage of skeletal components in total sediment.

make up from 4–22% of reef flat sediments, and are important constituents of the interreef sediments of the Great Barrier Reef Province. *Halimeda* plates are the most important biogenic component of most Caribbean reef flat sediments (Milliman, 1973, 1974) and although this component might not be important on Pacific reef flats, it forms the major constituent of deep Pacific atoll lagoons and of many back reef and interreef sediments of the Great Barrier Reef Province. Coralline algae are important in Caribbean reefs, but not as abundant as in Indo-Pacific reef sediments. A major contribution to Indo-Pacific reef flat sediment is made by benthonic foraminiferans; they are particularly conspicuous in reef flat and interreef sediments of the Great Barrier Reef Province.

C. CARBONATE PRODUCTION RATE

Carbonate production is not uniform over the reef surface and there is a difference of opinion as to where it occurs most rapidly. Using ^{14}C dating, estimates of the vertical accretion rate of the frame of the reef off Discovery Bay, Jamaica, range from 1 m/1000 years (fore reef slope at —36 m) to 0.55 m/1000 years ("Pinnacle Reef extending above the fore-reef slope" —36 m); a horizontal seaward accretion rate of 0.22 m/1000 years (deep fore reef at —60 m) was also determined (Land, 1974).

Theoretical calculations by Chave *et al.* (1972) suggest that the highest gross productivity takes place on the upper- and lower-reef slopes, a con-clusion which conflicts with the observation that a very high production rate is characteristic of reef flats. Most dated reef flat sediments are less than 2000 years old whereas unconsolidated sediments of the deep sea-ward slopes may contain significant proportions of relict sediments which increase in importance with depth (Milliman, 1974, p. 171).

D. Mineralogical and Geochemical Differentiation of Facies

In view of the differences which exist between groups of organisms in terms of the mineralogy and chemistry of their skeletal material, and the zonation of sedimentary facies in a reef complex which reflects eco-logical factors, it is to be expected that mineralogical and geochemical investigations of the sediments will reveal distribution patterns related to these factors.

Foraminiferans and calcareous red algae consist chiefly of calcite whereas *Halimeda* and hermatypic corals consist chiefly of aragonite. Further, aragonite is considered to be a carrier of strontium, and there is a correlation between aragonite and relatively low-magnesium oxide. Chave (1954) showed that magnesium is abundant in calcite skeletons. Consequently a basic mineralogical and geochemical differentiation of facies may be established. Thus analyses of sediments from the beaches and lagoon floor of Bikini Atoll (Emery *et al.*, 1954, p. 67) show a de-crease in MgO and an increase in CaO and SrO with a transition from calcite to aragonite; such a transition accompanies a change from forami-niferan-rich to *Halimeda*-rich sediment.

Friedman (1968) recognized low-magnesium aragonite, high-mag-nesium aragonite, low-strontium aragonite, and high-strontium aragonite in shells and sediments of the Gulf of Aqaba, and showed that the mag-nesium content of the sediments reflects the relative proportion of mol-luscs, coralline algae, and corals present. Hoskin (1968) found that the magnesium and strontium content of muds from Alacran Reef, Mexico, varied according to environment and sediment source.

At Lady Musgrave Reef (a closed-ring-type platform reef of the Great Barrier Reef Province), a general trend towards the separation of fine, predominantly aragonitic deposits of the lagoon from predominantly coarse, calcitic debris of the reef flat accompanies a marked change in biogenic constituents, and the geochemical data suggest magnesium and strontium distribution patterns similar to the concentric horizontal arrangement of the physiographic zonation of the reef (Orme et al., 1974), see Fig. 3, and Plate I.

A knowledge of the relationship between the elemental composition of biogenically synthesized carbonate and the environment has an obvious application to palaeoenvironment analysis and may help in understanding certain diagenetic trends. Studies of that relationship have received considerable attention in recent years (see, e.g., Wolf et al., 1967; Bathurst, 1971). Chave (1954), for example, considered that the magnesium content of skeletal material is not only related to mineralogic form, to type, and to phylogenetic level, but also to water temperature. The relationship of temperature and other factors to the aragonite/calcite ratios in carbonate secreting marine organisms was considered by Lowenstam (1954). St. John (1973) carried out an investigation of the trace elements Fe, Cu, Cd, Co, Zn, Pb, and Ni in scleratinian corals of the Coral Sea Region, and concluded that their abundances resulted from ingestion and incorporation of particulate matter by the corals. IIe demonstrated that the trace element distribution patterns "reflect in an inefficient way, those of the aqueous chemistry of their environment rather than those of the carbonate solid state."

A relationship between the geochemical composition of carbonate sediments and the aqueous chemistry of their depositional environment is suggested by the studies of Friedman (1969) who investigated the barium, iron, and manganese content of molluscan shells and sediments found in fresh water, marine, and lagoonal (brackish) environments.

V. Sedimentation: Processes and Products

Petrographic criteria for the interpretation of the physical parameters of sedimentary environments are largely based on studies of terrigenous (detrital) sediments. Such grain characteristics as form, sphericity, roundness, surface texture, and mean grain size, together with grain size distribution data, convey valuable information regarding sediment provenance, transportation, and deposition. However, physical processes overshadow all others in the formation and dispersal of terrigenous particles, and the contrasting intrabasinal provenance and largely biogenic nature of most carbonate deposits casts doubt on the validity of the application of such granulometric criteria to sedimentation processes in the carbonate environment. Such speculation and dissent is fostered by an appreciation of the importance of organisms not only as providers of sediment, but also in the breakdown and dispersal of sedimentary particles; it is also fostered by an understanding of the powerful ecological controls on sediment facies through bottom community influences on the deposition and stabilization of carbonate sediments. One of the most important differences between the physical characteristics of carbonate and terrigenous particles is durability, indeed ". . . some skeletal carbonate might be described as having no durability, falling apart to extremely fine particles upon removal of organic integuments without the influence of abrasion or agitation" (Chave, 1960, p. 18). Morphological and bulk density differences are also important.

Sedimentation in the coral reef environment is primarily concerned with the interaction between biological and physical processes. We are concerned with the source factor, the availability of sedimentary particles, i.e., the availability by transportation and by *in situ* postmortem contribution, factors in which ecology is of great significance; with the durability of sedimentary particles which is a factor of microarchitecture and fabric; with bulk density and the morphology of grains which affect hydraulic behavior during transportation and deposition; and with chemical instability which becomes particularly important in postdepositional processes. The processes responsible for the development of the sedimentary characteristics of a reef complex may be considered broadly in terms of those involved in sediment production and in the modification and breakdown of sedimentary grains, those which disperse the sediment, i.e., those involved in sediment transportation and deposition, and those which influence stabilization and postdepositional modification of sedimentary deposits (Table II).[1]

Physical (mechanical), biological, and chemical processes and agents are all involved in the development of the reef complex; organisms play particularly important roles, both constructive and destructive. However, in the absence of adequate quantitative data, the grouping of agents involved according to their relative importance in these processes is somewhat arbitrary, for some agents have multiple roles in sedimentation, as carbonate producers, sediment disintegrators, and sediment transporters, and consequently the interrelationship between processes, agents, and products is manifold (see Schroeder and Zankl, 1974 p. 419).

---

[1] See foldout for Table II.

A. Formation and Breakdown of Sedimentary Particles

In a reef complex discrete sedimentary particles, the components of the facies outlined in Section IV, result from the *in situ* postmortem decay and disintegration of organisms, from erosion, from the activities of sediment-ingesting organisms, and locally from chemical and biochemical carbonate precipitation. Erosion of the reef is accomplished by the hydraulic action of waves and surf, by corrasion, by the activities of boring and rasping organisms, and by corrosion. Size reduction of sedimentary particles is also brought about by biological, mechanical, and chemical means. It is therefore convenient to consider sedimentary particle formation and breakdown in terms of biological, physical (mechanical), and chemical agents.

1. Biological Agents

Organisms which bore and rasp reef rock, beach rock, and skeletons are effective not only in erosion, but also in sediment production. The general groups of rock-destroying organisms, together with an indication of the methods they employ to erode substrates, is presented by Neumann (1968); calcibiocavities are tabulated by Carricker and Smith (1969) who also note their distribution; see also Scoffin and Garrett (1974, p. 435). The borers include endolithic algae, pelecypods, polychaetes, siphunculids, echinoids, barnacles, and sponges; the effectiveness of such organisms in eroding Bermuda and Bahama reefs, where, "by riddling the rock intensively, some borers such as pelecypods and sponges are able to virtually remove it layer by layer," is discussed by Schroeder and Zankl (1974, p. 415). Scoffin and Garrett (1974, p. 433) showed that most living borers on the tops of Bermuda patch reefs preferred a substrate of dead skeleton and consequently the under surfaces of corals are attacked during reef growth; thus the detachment of massive corals from the reef is facilitated.

The study of the effects of the barnacle *Lithotrya* on Puerto Rican beach rock (Ahr and Stanton, 1973) gives an indication of the sedimentological importance of rock borers. The sediment produced is bimodal consisting of calcilutite (removed from the tube) and cobble- to boulder-sized clasts formed by the crumbling of weakened, intensively bored beach rock ledges.

Rasping organisms such as turtles, crustaceans, and chitons produce furrows on hard surfaces and are likely to remove and partially triturate considerable quantities of matter (Bardach, 1961). The chiton *Acanthozostera gemmata* produces sediment covering a wide range of grade sizes (McLean, 1974, p. 407). Faecal pellets ("average length 4.5 mm and maximum diameter 1.0 mm") were composed mainly of carbonate material (60%–80%) and on disaggregation yielded carbonate particles of all sizes between clay grade and coarse sand grade. Trituration of sediment grains may result from the grazing activities of fish and sediment ingestion by bottom feeders such as holothurians. Large quantities of sediment are ingested by the latter; a total of 230 tons/km²/year was estimated by Crozier (1918) for *Stichopus moebii* in Harrington Sound, Bermuda, but there is a difference of opinion regarding the size reduction of sediment grains during passage through the gut of holothurians—see Gardiner (1931), and Stoddart (1969a, p. 479). However, Emery *et al.* (1954) and Emery (1962) showed that solution of fine grains occurred during sediment ingestion by holothurians on Guam. The decrease in sediment size from coarse sand to mud during its passage through the gut of the dominant polychaete of the reefs of north Florida was described by Ebbs (1966).

Microorganisms have an important role in sedimentation. They are abundant in the waters and sands of the reef environment and have a role in carbonate precipitation as well as in the breakdown of reef skeletons. The dysphotic regions of the reef, are favorable for the formation of microorganism communities and the highest counts of aerobic heterotrophic bacteria on reefs come from the internal reef sediments (Di Salvo, 1973). Bacteriological activity within such reef regenerative spaces may result in the precipitation of calcium carbonate. Bacteria may cause breakdown of coral skeletons (Di Salvo, 1969), and bacterial action on organic tissue which invests or is incorporated in carbonate grains will result in grain disintegration.

The disintegration of nearly every form of calcareous matter is assisted by boring filamentous algae (Duerden, 1902; Otter, 1931; Bertram, 1936; Bathurst, 1966) and their effect is well documented from modern and ancient deposits. Boring blue-green algae are believed to attack skeletal debris differentially (Ginsburg, 1957), the denser *Lithothamion* being less susceptible than the coral skeleton. Although their penetration of carbonate material appears to be almost ubiquitous, as has been pointed out by Cribb (1973, p. 67), "there is still little precise information on their quantitative importance or on the mechanism by which they effect erosion."

2. *Physical Agents*

Biological erosion makes reef rock and skeletal material more susceptible to mechanical erosion. Mechanical erosion due to the hydraulic action of waves and surf dislodges reef rock, detaches reef organisms from their substrates, and through agitation of debris causes attrition by rubbing and impact of particles. Powerful tidal currents may assist in this process.

The sediments formed on reefs depend primarily on the characteristics of contributing organisms, on the structure and microarchitecture of their skeletons, and on the influence of these characteristics on the breakdown and transportation of biogenic debris. The form of the skeleton is important in determining its resistance to mechanical erosion and also the sizes

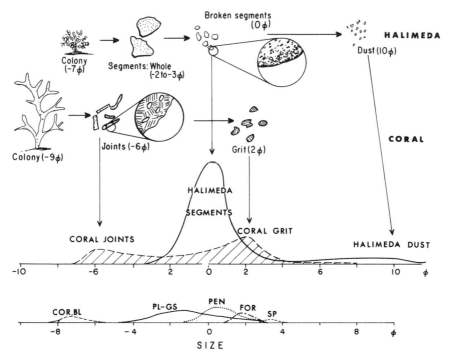

Fig. 4.A. Illustrates the relationship between particle size and type produced by breakdown characteristics of the two most abundant organic constituents of the beach sediments of Isla Perez, Alacran Reef (after Folk and Robles, 1964), staghorn coral, and *Halimeda*, each of which breaks down into two favored grain sizes. Consequently the gravels are composed of coral. *Halimeda* mostly constitutes the 0φ sand, the 2φ sand is mostly coral, and the 10φ dust is derived largely from the aragonite microcrystals of the calcareous skeleton of *Halimeda*. The lower scale shows the minor constituents: COR.BL: brain coral blocks, elkhorn coral blades, large conch shells; PEN: peneroplid Foraminifera; FOR: smaller Foraminifera; PL-GS: pelecypods, gastropods, and fragments; SP: gorgonian spicules, coralline algae stems. The relative volume of each constituent is indicated by the area of its frequency curve.

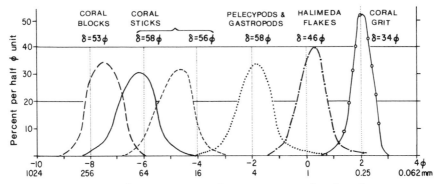

Fig. 4.B. Grain-size frequency curves for representative beach sediment types on Isla Perez, after Folk and Robles (1964). The curves illustrate the normality of the grain-size distributions and the uniform sorting over a wide grain-size range including a variety of grain shapes.

and shapes of the grains initially produced. The forms present are: (1) massive, e.g., varieties of coral, encrusting algae, and large mollusc shells; (2) segmented, e.g., *Halimeda*, echinoderms, crinoids, asteroids, and ophiuroids, in which carbonate segments form plates held together by organic tissue which, after the death and decay of the organisms, releases the plates; (3) spicular, e.g., sponges, tunicates, holothurians, alcyonareans, gorgonaceans, and some algae which have disseminated spicular particles; and (4) small, whole, hollow tests, e.g., foraminiferans, gastropods, and pelecypods.

Reef-rock and massive skeletons are relatively resistant to mechanical breakdown. Consequently, during storms, boulder deposits consisting of blocks of reef rock and unbroken coral colonies may be formed, whereas initially, branching organisms break at nodes into sticks, and coarse gravel deposits may result; such remain as a lag deposit on reef flats. The next stage of mechanical breakdown may be largely controlled by the micro-architecture of the shell. Thus, after the branching coral *Acropora cervicornis* has broken down into sticks of a size approximating the original segments, forming coral stick gravel (Plate III,C), it breaks down further into 2ϕ grit, "the approximate size of component crystal packets," without passing through intermediate stages of grain size reduction. This gives a distinct sinusoidal curve when the relationship between sorting and mean grain size is considered (Folk, 1962; Folk and Robles, 1964); see Fig. 4. This control has been named the Sorby Principle (Folk and Robles, 1964, p. 287) after Henry Clifton Sorby who, in what has become a classic study (Sorby, 1879), was the first to indicate the relationship between the breakdown of carbonate skeletons and their microstructure.

On Yucatan beaches Folk and Robles (1964) found the size distribution of *Halimeda* debris to be concentrated at two size grades, namely 0ϕ (1 mm) and about 10.0ϕ (1–3 μm), the latter representing the final disintegration size of aragonite crystals; they therefore assumed the operation of the Sorby Principle. Initially, sedimentary particles the size of the discrete segments are released by the decay and disintegration of the organic binding; these plates may then break into two or three segments and continued breakdown is controlled by the system of tubules which characterizes the internal structure of the algal plates. However, the modal sizes determined for the *Halimeda* sediments of the Florida Keys by Jindrich (1969) and by Basan (1973) depart from those observed by Folk and Robles (1964). Jindrich (1969, p. 542) expresses the view that "the breakdown of *Halimeda* and its concentration in certain size fractions is at least partly a function of the sorting potential of a given environment rather than wholly the result of structural characteristics."

Chave (1960, 1964) in tumbling-barrel experiments investigated the physical breakdown of carbonate skeletons. He concluded that skeletal durability varies widely, that skeletal mineralogy has no influence on physical breakdown, and that, although shell size may affect the rate of breakdown, "the smaller shells being less effective as abrasive agents," the "micro-architecture of the shell and the disposition of the organic matrix among the crystals of carbonate" are the major factors controlling resistance of skeletal material to physical destruction (Chave, 1964, p. 381). Consequently the most durable skeletons are dense and fine grained —*Nerita, Mytilus* etc., the hard porous coral structures; the organic-rich, coarsely crystalline shells of oysters are of intermediate durability, and the least durable forms are those like echinoderms, bryozoa, and algae with much open work and organic matrix. Chave's experiments also showed that an abundance of fine grains relative to coarse grains resulted from the mechanical breakdown of skeletal material (especially in the case of *Acropora*) and in this way a large crop of aragonite needles may result from the abrasion of aragonitic forms such as corals, algae, or molluscs.

In a more recent experimental study Force (1969) showed that, by the mechanical wearing of the organic binder, mollusc shells tend to break down into three particle sizes representing three basic microarchitectural units, namely layers 500–250 μm (1.0ϕ–2.0ϕ), sublayers 32–4 μm (5.0ϕ–8.0ϕ) and unit crystals 0.5–0.125 μm (11.0ϕ–13.0ϕ).

The durability of pelecypod valves was investigated experimentally in three natural marine environments by Driscoll (1967). He found that the variables which influence shell abrasion are grain size, surface area in relation to weight, energy of the environment, shell orientation, and the nature of the associated sediment. Thus, lighter delicate shells are easily picked up and transported by strong surf and the cushioning effect of water protects them from breakage. The larger valves tend to maintain a convex upward orientation in surf zones which results in selective abrasion of their margins by sediment transported as traction load. Furthermore, initial weight loss due to removal of surface ornamentation by the "sand blast" action of sandy sediment subjected to surf action is more strikingly shown in larger and heavier shells. The surface texture resulting from abrasion varies according to the nature of the sediment constituting the traction load, gravelly sand leading to a rougher shell surface than sand.

Although Driscoll's study was carried out in the nonreef environment off the Massachusetts coast and involved the reaction between pelecypod shells and terrigenous sediment, the processes involved have a wider application and in a qualitative sense are applicable to sedimentation in the coral reef environment.

3. Chemical Agents

Chemical solution may be significant in grain size reduction and removal (Chave, 1962, 1964; Chave et al., 1962), especially in the intertidal zone and below the sediment/water interface. Diurnal cycles in pH, dissolved CO_2, and carbonate saturation in marine waters of the coral reef environment have been observed. They are believed to result from changing rates of respiration and photosynthesis (Schmalz and Swanson, 1969) and are considered to induce the nighttime solution and daytime precipitation of solid carbonate phases. However, Davies (1974), who demonstrated the importance of solution in the intertidal zone of reefs of the Capricorn Group (Great Barrier Reef Province), presents evidence for the daytime solution of calcium and magnesium, and suggests that some solubility variations may be related to halmyrolitic reactions resulting from organic attack on carbonate grains. Below the sediment/water interface solution of shells may be important (Driscoll, 1970). Solution of some sediment particles may occur during their passage through the gut of deposit-feeding organisms (Emery et al., 1954; Emery, 1962).

B. Dispersal of Sedimentary Particles

The reef tract being the site of most intensive carbonate production and organic activity, and the locus of powerful wave and surf action, is the provenance of a wide range of grain types covering a broad spectrum of grade sizes. Here, at the source, an initial tendency of grain separation reflecting ecology, mineral composition, and differential resistance to destructive agencies may be effected which is further developed under suitable physical conditions during transportation (Maxwell *et al.*, 1964; Orme *et al.*, 1974).

Some grade sizes may be favored by specific skeletal types and their breakdown products, e.g., *Acropora cervicornis* (see p. 153) whereas debris of skeletons which show a more continuous breakdown may be present in a range of grade sizes. However, the Sorby Principle, though significant, is only one of several factors controlling grain size distribution, and its relative importance may be overshadowed by others involved in the formation of texturally more mature sediment (Jindrich, 1969). In some cases ecological factors may dominate the character of sediment which may reflect death assemblages, e.g., the distribution of Foraminifera on Heron Reef is maintained, with little dispersal, by the protected nature of their favored habitats.

Bioturbation is common in coral reef sediments and some organisms are effective sediment transporters, but the most significant transporting agents are waves and currents which carry sediment in suspension and disperse it also by surface creep or rolling, and by saltation. Deposition brought about by the reduction of energy of the transporting medium may be due to the influence of bottom communities (e.g., sea grasses) on the hydraulic regime. A successful interpretation of texture must be based on an appreciation of the mechanisms of transportation and deposition in relation to grain size, an aspect of carbonate studies which has been neglected.

1. Agents and Mechanisms of Transportation and Deposition

Organisms may be important as transporting agents in terms of the overall sediment budget of the reef, e.g., fish may rival waves and currents as transporting and triturating agents (Emery, 1956, p. 1513). Estimates made by Cloud (1959) at Saipon, and Bardach (1961) in Bermuda, of the fine sand contributed to the reef surface by fish indicate how effective this mode of transport is, and suggest a deposition rate of 1–3 mm/year (Stoddart, 1969a, p. 479).

Physical agents, viz., translatory waves and currents, are effective in transporting and sorting sediment. Separation of components through winnowing is particularly effective in the surf zone and gravitational influence on sediment movement is especially effective on seaward reef slopes. In the high-energy regions of the reef, finer grades of sediment are removed and carried in suspension to quieter regions, leaving behind lag gravels and sands. Folk (1962, p. 243) found, in beach deposits associated with the Alacran reef complex, that the mean grain size is affected by wave energy, but for a given size the sorting values are nearly constant. "Thus the process (surf action) seems to have an inherent sorting capacity that is constant although the intensity of surf action varies enormously."

Some workers, e.g., Stoddart (1969a, p. 447), believe that size, shape, and sorting in sediments on reefs are more dependent on the skeletal characteristics of contributing organisms than on waves and currents. The considerable range of initial sizes, shapes, and bulk densities of the biogenic particles, and their relatively low durability impose controls which clearly differentiate carbonate sedimentation in a reef environment from noncarbonate sedimentation. These factors, together with the potentially high incidence of *in situ* postmortem contributions, may produce complex sediment dispersal patterns, and the resulting sediment textures may be difficult to interpret in terms of sedimentary environment and sedimentary processes, especially when ancient carbonate deposits are being considered.

Some Fundamental Hydrodynamic Aspects of Sediment Dispersal. Many variables influence the transportation and deposition of sedimentary particles by a fluid medium. These include the nature of the current flow (direction, stability, turbulence, speed, and duration), the character of the surface over which the current is moving, the nature of the sediment being transported (grain size, shape, bulk, density, and the quantity of sedimentary particles), and the volume of sediment supply. Some factors are difficult to determine, and, especially in the coral reef environment with its attendant powerful biological influences, a bewildering diversity of variables is present. An appreciation of the responses of sedimentary particles to different physical conditions according to basic hydrodynamic principles is a prerequisite to understanding the implications of grain size distribution in bioclastic carbonates.

The classic works of Rubey (1933), Hjulström (1939), and Inman (1949) have lead to a clarification of some fundamental aspects of sediment transport and deposition, particularly with respect to spherical quartz grains. It is appropriate to consider these aspects here and to indicate the complexities introduced when the sediment consists of an assortment of biogenic allochems.

The relationship between three parameters, *roughness velocity, threshold velocity,* and *settling velocity* is illustrated in Fig. 5. The upper levels of water flows are inherently turbulent, but if the surface over which they move is hydrodynamically smooth a thin layer of laminar flow will occur at the base. On the other hand, if the bottom surface is hydrodynamically

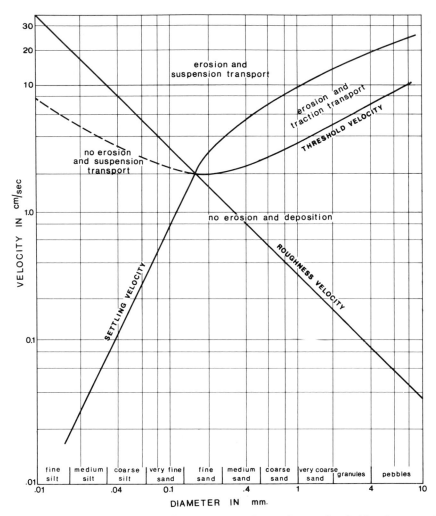

Fig. 5. Relationship of grain diameter to settling velocity, threshold velocity and roughness velocity. After Inman (1949).

rough, i.e., when roughness elements such as sedimentary particles protrude above the laminar layer, turbulence prevails to the bottom of the water flow. The *roughness velocity* is the critical friction velocity at which the character of the bottom changes from smooth to rough, and a criterion between rough and smooth bottoms may be "some constant value of the ratio of the bottom grain diameter to the thickness of the laminar layer" (Inman, 1949, p. 55). Experimental studies have shown that sediment particles will not move until a critical drag force, which is a function of grain size, is exceeded. Such a force may be expressed in terms of critical drag velocity, which is the point at which movement of particles begins and is called the *threshold velocity* (Bagnold, 1942). For spherical quartz grains it has been shown that this parameter decreases with grain size until a diameter of 0.18 mm is reached when the bottom becomes effectively smooth. As grain diameter decreases below 0.18 mm the velocity necessary to start surface movement, i.e., at the sediment/water interface, increases (Hjulström 1939).

The *settling velocity* is the velocity at which the frictional drag of the fluid approaches the value of the impelling force (the difference between the force due to gravity and the buoyancy force). Two modes of settling are recognized, *viscous* and *impact settling* (Rubey, 1933). Rubey (1933) demonstrated that for quartz particles smaller than .18 mm Stokes Law applies and viscous settling occurs, but larger particles create turbulence, settling velocities depart from Stokes Law, and impact settling occurs.

The settling behavior of sedimentary particles varies considerably according to shape, size, and surface texture. Thus plates and irregular angular fragments offer much greater assistance than spheres and rods (Rubey, 1933, p. 329). These basic physical properties of sedimentary particles cover a very broad spectrum in sediments of substantially biological origin such as characterise the coral reef environment. The specific gravities (bulk densities) of bioclastic materials aragonite and calcite (Fig. 6) may depart considerably from the specific gravity of solid, equidimensional aragonite and calcite grains (see Maiklem, 1968) which, owing to their greater specific gravities (calcite, 2.72; aragonite, 2.95), have settling velocities slightly greater than the 2 cm/sec calculated by Rubey for spherical quartz grains (specific gravity, 2.65) of .18 mm diameter. Furthermore, there is a divergence from that velocity according to shape, size, and bulk density of the particles of a size greater than approximately 0.125 mm (3ϕ) diameter. Fig. 7 suggests that departures in settling behavior between different grain types become very pronounced in the coarser sand grades. Maiklem (1968) examined experimentally the behavior of bioclastic sedimentary particles from the Great Barrier Reef Province, where corals and coralline algae provide blocks (mollusc shells may be a source of smaller blocks); rods are provided mainly by coral sticks, spines, turreted gastropods, and the foraminiferan *Alveolinella;* plates are supplied by *Halimeda* segments, pelecypod valves, and by *Marginopora;* and spheres are formed either of particles rounded by abrasion or from foraminiferans such as *Calcarina* and *Buculogypsina.*

Settling motion is related to particle size and shape, the most direct

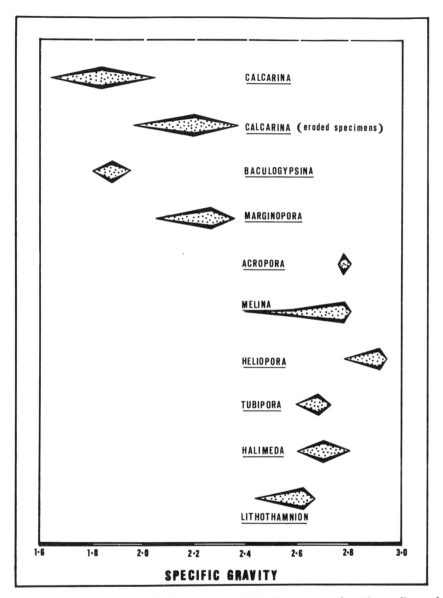

Fig. 6. Specific gravity (bulk density) of bioclastic materials. After Jell *et al.* (1965).

path being taken by equidimensional particles (spheres and small blocks); larger blocks and plates oscillate, the degree of oscillation varying with angularity and surface texture. The slowest grains to settle are plates, and

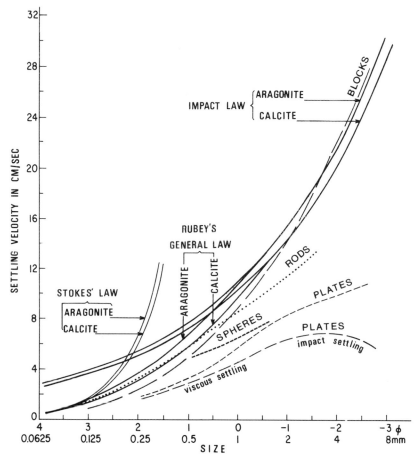

Fig. 7. Settling velocity curves for bioclastic grains (grouped according to shape). Curves calculated for aragonite and calcite are included. After Maiklem (1968).

within this category there is a difference imposed by different bulk densities; e.g., the very porous *Marginopora* plates settle more slowly than *Halimeda* plates or pelecypods (see Fig. 9). There are clearly two parts to each of the two curves for plates, and this is interpreted by Maiklem (1968, p. 106) as a reflection of the change from general to impact settling above −1φ (2 mm). The divergence between these two plate curves above −1.6φ (3 mm) is considered to be due to differences in bulk density between *Halimeda* plates and pelecypod valves on the one hand and *Marginopora* tests on the other, and it suggests that with regard to settling velocity at larger sizes, bulk density is of greater importance than shape.

The decrease in velocity indicated above -1.6ϕ by the lower curve "appears to be due to a change in settling motion, the larger grains oscillating as they settle and thus lengthening the distance travelled" (Maiklem, 1968, p. 107). The importance of shape and bulk density on settling behavior as grain size increases is also illustrated by Braithwaite's (1973) experiments with bioclastic sediments from the Seychelles. The difference in behavior of small particles where viscous settling occurs, and coarser grains where inertia forces and form drag dominate, and higher Reynolds numbers apply, is endorsed. Further, he shows that as grain size increases different grain types pass through a series of fall regimes (straight fall, spinning, spiral, and erratic tumbling) "at rates dependent upon their shape and effective density" (Braithwaite, op. cit. p. 260).

The texture of a sedimentary deposit due to sorting accomplished by differential transportation and deposition of grains may reflect the relative importance of traction and suspension loads; consequently their differentiation through the detection of distinct log-normal grain size subpopulations within individual log-normal grain size distributions may be possible. The traction load may move by surface creep or saltation. Surface creep, the sliding or rolling of particles along the bottom at a velocity less than that of the fluid flow, begins when the threshold velocity is reached, and saltation occurs when sedimentary particles are lifted off the bed into the fluid stream to be carried along with it at the same velocity until gravity brings them back to the surface (sediment/water interface). Saltation is not significant for fine-grained sediments (Inman, 1949, p. 58). If suspension is to be significant, threshold velocity must be equal to or greater than settling velocity, which, if the particles in suspension are spherical and of equal specific gravity, is a function of grain diameter.

Thus whether surface creep or suspension is important will depend on the interrelationship of these factors (roughness velocity, threshold velocity, settling velocity, and grain size); see Fig. 5. The threshold velocity is much greater than the settling velocity for small grains and if once put into suspension they will tend to be transported in suspension rather than by surface creep. Thus, the fine aragonite debris resulting from coral breakdown in the high-energy zones of a reef may be transported in suspension for a considerable distance before being finally deposited. For larger grain sizes the threshold velocity is greater than the roughness velocity and less than the settling velocity, and movement will be by surface creep rather than suspension. As current strength increases, larger particles will be moved by surface creep and saltation, and larger grains will be carried in suspension.

Inman (1949, p. 61) showed that for spherical grains of specific gravity

2.65 those of approximately 0.18 mm diameter are the most easily transported. These are hydrodynamically unique, they are moved by weaker currents than larger or smaller grains, and once moved they do not have as great a tendency to go into suspension as do smaller grains, but they are more readily carried in suspension than larger grains. Therefore fine sand marks the transition zone between transportation in suspension and transportation by surface creep, where friction velocity does not greatly exceed the threshold velocity. However, bioclastic grains with their wide range of shapes and porosities (bulk densities) create a more complex relationship between grain size and ease of transportation. For example, *Halimeda* plates, because of their shape and high porosity, tend to outrun more equidimensional calcite aragonite grains (Jindrich, 1969, p. 548) in transportation by unidirectional currents. This is similar to the response of mica flakes in terrigenous sediment which are commonly separated in transportation from terrigenous sand grains of equal volume, eventually to be deposited with quartz of silt grade.

Nevertheless, as the size of bioclastic particles is reduced, the hydrodynamic behavior of different grain types becomes more uniform, and it is not unreasonable to assume that finer grades of bioclastic sediment may show grain size distributions which reflect different modes of transportation and deposition characteristic of particular environments.

2. The Development of Clastic Carbonate Textures

Strong tidal currents and powerful translatory waves affecting a reef may transport relatively coarse-grained sediment in suspension. In the intertidal zone strong surf action effectively winnows away the more readily suspended particles leaving behind coarse-grained lag deposits on the reef flat. The sediment in suspension may be transported considerable distances, depending on local circumstances, into regions of quieter water where it settles and may accumulate. Therefore, owing to the particular morphological and hydrological nature of the reef environment with its characteristic energy gradient from reef front to back-reef and lagoon, "currents of removal" will be most effective on the reef while "currents of delivery" may be most influential in determining the texture of deposits in back reef, interreef, and lagoonal situations. However, this is a generalization which may be considerably modified through the influence of ecological and bathymetric controls (Orme *et al.*, 1977). It might be anticipated that, especially near the source, an abundance of certain grade sizes and a scarcity of others will be controlled not only by the type of organism supplying the sediment, but also by the dominance of the Sorby

Principle. Bramlette (1962), Stoddart (1962), and Folk and Robles (1964) found an abundance and scarcity of certain size grades for Samoa, British Honduras, and for carbonate beaches on the Alacran Reef, Yucatan, respectively. Under these circumstances the compositionally mature (biologically uniform) sediments, such as pure *Halimeda* sand, or pure coral sand, show the best sorting (Folk and Robles, 1964).

Sorting may be purely local through selective transportation of those grains having hydraulic characteristics which favor easy removal. Sorting is improved progressively in transportation and grain size populations which are more readily transported in suspension than by surface creep or saltation may become differentiated. Under conditions of continuous supply comprising all sediment grades, where the average velocity of a unidirectional current gradually decreases in magnitude in a down-current direction, and where the initial stress on the bottom is sufficient to cause movement of all particles, there will be a progressive decrease in the median diameter of the bottom samples (Inman, 1949, p. 63), but the wide range of property variations characteristic of biogenic particles (see page 159) imposes constraints on the application of this general concept in carbonate sedimentation.

In many of the contemporary back reef situations *Halimeda* plates are quantitatively the most important sediment, followed by molluscs and foraminiferans, and they provide an illustration of the influence of shape and bulk density on hydraulic behavior. They behave hydraulically like much smaller grains and tend to outrun larger, denser, more spherical grains in transportation. "The coarse grained *Halimeda* sediment presents a rough surface to the current and, having practically negligible cohesive forces between individual grains, can be easily resuspended after it has settled out" (Jindrich, 1969, p. 548).

Thus, Jindrich (1969) and Basan (1973), in their studies of the Florida Shelf, found that *Halimeda* plates and cellular foraminiferans tend to be transported mostly in suspension or saltation, and undergo less fragmentation than particles in the traction load (Jindrich, 1969, p. 542; Basan, 1973, pp. 50–51). The mode of transport of mollusc shells depends on the ratio of their surface area to weight (Driscoll, 1967). However, because of their greater density (as compared to algal plates and cellular foraminiferans), Basan (1973) found that they tended to be transported by bottom traction (surface creep) and consequently became more abraded. Heavier molluscan fragments which, through prolonged abrasion, reached 1.ϕ ($\frac{1}{2}$ mm) diameter became more spherical and behaved hydraulically like rock fragments (Pleistocene limestone fragments); due to their homogeneity and relatively high sphericity such particles showed the least departure between "sieving size" and "hydraulic size." The effective distance and method of transportation of sedimentary particles therefore influences the compositional and textural maturity of deposits which are carried some distance from their source, and may indeed become of greater importance than the influence of the Sorby Principle.

Jindrich (1969, pp. 542–543) demonstrated the textural changes that occur during "channel" transportation in the lower Florida Keys, where *Halimeda*-rich sediment became progressively enriched in molluscan and rock fragments during transport. In this case he showed that, despite diverse particle type and grain morphology, with decreasing size and increasing sorting, "a hydraulically well balanced mixture of all components represented in equal proportions" formed the texturally mature sediment (Fig. 8). Further, the effect of sorting in transportation is reflected in the frequency curves, and their skewness is indicative of environmental contrasts. Thus, some materials are trimodal "with a distinct mode commonly at -1.5ϕ (2.8 mm), signalling a high content of coarse *Halimeda* grains. Moderately sorted channel and delta sands tend to approach normal frequency distribution but are strongly negatively skewed," the skewness being caused by the introduction of "coarse-grained debris entering the channel from places along its course" (see Jindrich, 1969, Fig. 7, p. 541). On the other hand, Jindrich (1969) found that "plateau" sands have an almost normal symmetrical distribution with almost no negative skewness which he considered to be due to the slow uniform rate of sediment supply and the greater duration of current sorting of the carbonate sands.

Strong negative skewness of the sediments of the Lady Musgrave Reef lagoon is believed to be due to the contribution of coarse debris from patch reefs and the occasional transport of coarse debris during periods of increase in the strength of waves moving over the reef flat (Orme *et al.*, 1974). Kurtosis is closely related to source factors and leptokurtic curves may result from the mixing of sediments from multiple sources.

Jindrich (1969) found that a sinusoidal trend "resulted from mixing of four major constituents which tend to attain different size proportions during the sorting process. These mixtures behave at certain mean grain sizes almost as single populations and represent the best sorted sediment recorded." The best sorted sediments coincide with the lower deflections on the sinusoidal trend (see Jindrich, 1969, Fig. 10, p. 545).

The relationship of mean, sorting, dispersion (Friedman, 1962; Spencer, 1963), skewness, and kurtosis to environmental factors may be complex but the influence of reef morphology, source, and water movement on them may be reflected in the distribution of these parameters (Orme *et al.*, 1974).

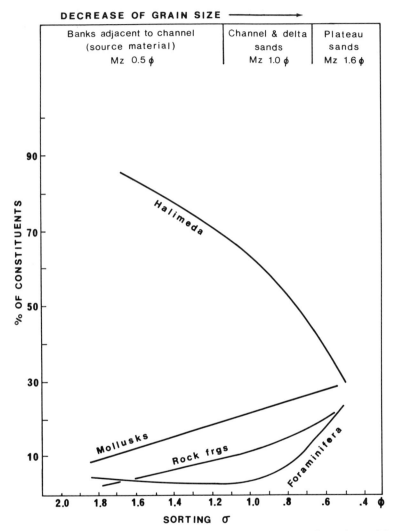

DECREASE OF GRAIN SIZE ⟶

| Banks adjacent to channel (source material) Mz 0.5 φ | Channel & delta sands Mz 1.0 φ | Plateau sands Mz 1.6 φ |

Fig. 8. Compositional changes of sediments during sorting and mechanical break-down, lower Florida Keys. After Jindrich (1969).

The polymodality of reef complex sediments is often explained by a mechanism of mixing distinct sediment populations—see Maiklem (1970), Frankel (1974), and see Fig. 9, and while mixing of sediment from different sources may be the dominating textural influence in some en-vironments, e.g. interreef areas of the Great Barrier Reef Province, differ-

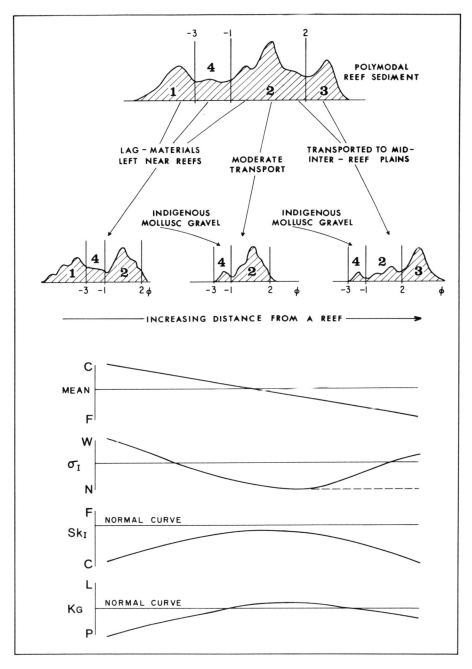

Fig. 9. Diagram showing four-sediment populations and their sorting with distance from a reef. The variation in textural parameters in this direction are also shown. After Maiklem (1970).

ent modes of sediment transportation might be reflected by some of the polymodal characteristics of some reef flat and backreef sediments (Flood and Scoffin, 1977, Flood *et al.*, 1977, Orme *et al.*, 1977).

Collective movement of the traction load as opposed to individual motion of particles, under appropriate conditions, results in bed forms such as ripple marks which characterize sandy areas on reef flats.

C. Interpretation of Grain Size Distribution Data

The characteristics of sedimentary environments reflect grain-size distribution, and consequently granulometric analyses are often used as environmental indicators. However, though parameters such as mean grain size, standard deviation, and skewness of detrital (terrigenous) sands are independent of mineralogy (Friedman, 1961), for carbonate sands they have been considered to be somewhat unreliable for estimating depositional conditions, because of the exceptionally wide range of characteristics of the sedimentary particles in the coral reef environment, the considerable range of bottom conditions, and the fact that the biogenic particles may form lag deposits or undergo only short transportation. The diverse morphology coupled with differences in bulk density of skeletal grains is basically responsible for their unequal hydraulic behavior, and may produce considerable discrepancies between "sieving size" and "hydraulic size" (see e.g., Braithwaite, 1973); the more homogenous and more nearly spherical the particles, the closer the agreement between these two parameters. Thus, the texture of an immature deposit may be dominated by source factors and the Sorby Principle rather than by transportation factors.

Visher (1969) recognizes distinct grain size subpopulations in detrital sediments which reflect different modes of transportation, the subpopulations occupying different parts of cumulative curves (Fig. 10). Thus suspension, saltation, and surface creep populations can be identified, and "the number, amount, size-range, mixing, and sorting of these populations vary systematically in relation to provenance, sedimentary processes and sedimentary dynamics" (Visher, 1969). There is clearly a relationship of sedimentary processes to textural responses, but the variability in the coarse fraction of biogenic carbonates and the lag nature of many reef sediments suggests that such textural responses might be absent or difficult to recognize, especially in coarse-grained, relatively immature deposits. Nevertheless, work in progress at Queensland University suggests that in some carbonate sediments of the Great Barrier Reef Province statistical separation of grain size populations which may reflect different

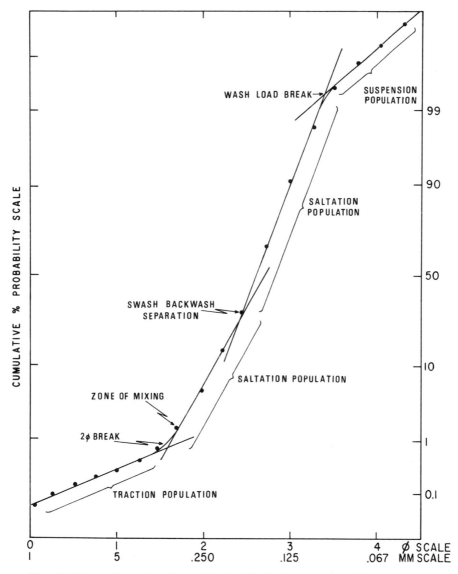

Fig. 10. Truncation points in grain size distribution curves reflecting grain size populations related to sediment transport modes. After Visher (1969).

modes of transportation is possible. The recognition of such textural responses and their correlation with specific environmental situations would provide a valuable aid to the interpretation of the genesis of ancient clastic carbonates.

Multivariate statistical techniques offer a method of dealing with the complex of variables involved in carbonate sedimentation. Klovan (1966) demonstrated the advantage of such an approach in dealing with recent sediments of Barataria Bay, Louisiana, where grain size distribution is determined by the relative amounts of three energy types active at the site of deposition, viz., wind and wave energy, current energy, and gravitational energy.

However, the possible influences of bottom communities on current strength and sediment transportation and deposition may be considerable, and should be borne in mind when considering the implications of sediment textures. The need to examine "interfacial biological communities when examining depositional and erosional processes at the sediment-water interface, or when making interpretations from the products of these processes in ancient rock" is emphasized by Neumann et al. (1970, p. 274). The kind of anomaly which might be encountered in interpreting clastic carbonate deposits is illustrated by recent studies of carbonate bank sedimentation. Thus Basan (1973, p. 51) points out that it might be difficult to determine bank environments from textural parameters, and Davies (1970, p. 85) maintains that "where the organic cover, like sea-grasses is not preserved in the sediment body, its former existence may be inferred by the presence of detached encrusting organisms and a large proportion of silt- and clay-size skeletal fragments."

The common assertion that grain size distribution data in biogenic carbonate sediments are of no value in determining environmental parameters probably reflects the fact that relatively few studies have been concerned in a quantitative way with the processes of sedimentation in the coral reef environment. It is also a reflection of the complexity of such studies which becomes apparent if some basic differences between carbonate sediments and terrigenous sediments are considered: (1) particles in carbonate sediments, which are formed and deposited in the same depositional basin, exhibit great ranges in bulk densities and morphologies, especially at larger sizes, and the net distance of transportation may be short in contrast to terrigenous sediments, which may travel great distances from their provenance to their depositional site; (2) appreciable additions to the carbonate sediment may be made by postmortem contributions of skeletal material *in situ*, or while the sediment is *en route* to its site of accumulation; such contributions may detract from the maturity of the deposit and cause negative skewness; (3) although grain size populations corresponding with different modes of transport may be detected in granulometric analyses of carbonate sediments, their presence may be masked by *in situ* contributions especially where only short transportation has taken place. Consequently the argument that currents of removal are more important in biogenic carbonate deposits than currents of delivery receives some support. This concept is one of the main arguments advanced against total acceptance of the genetic implications of many textural classification of limestones. However it is too simple a concept to account for the complexities inherent in the various aspects of sedimentation in the coral reef environment.

VI. Some Aspects of the Interrelationship of Organisms and Sedimentation

A. Influence of Bottom Communities on the Depositional Environment

Whether sediment after deposition will actually accumulate, thereby preserving an environmental record, is a matter of current strength, sediment supply, and stability of the relationship between sea level and sea bed (the sediment/water interface). Conditions may change within an environment so that erosion of the sea bed occurs, thus removing the record of a previous episode of sedimentation. However, organisms have the power to modify the physical environment just as changes in physical environment may bring about changes in biota (Ginsburg and Lowenstam, 1958). Thus, the development and morphological expression of a coral reef controls bathymetry and influences water circulation over a wide area. At the other end of the scale organic films, algal mats, and marine grasses trap and bind loose sediments. Basan (1973, p. 51) noted the association of brown, gelatinous algae with floating sheets of sand-grade carbonate sediment. Red, green, and blue-green algae may constitute subtidal biological mats with a rigid, open network in which "mucilaginous sections of blue-green algae and diatoms in association with the fine filaments of blue-green algae bind the grains to each other and to the mat network" (Neumann et al., 1970). Considerable sediment stabilization is effected by this means for Neumann et al. (1970) found that sediment bound in this manner could withstand current velocities much greater than that required to move unbound sediment, and furthermore, the intact mat surface could withstand direct current velocities greatly in excess of those of normal tidal currents in the mat environment. Neumann et al. (1970) also observed that "grain size, sorting, packing, structure, and sediment surface morphology, are influenced by mat formation."

Marine grasses such as *Thalassia,* which are conspicuous in many back reef and lagoonal environments, form a "bafflelike" carpet which reduces current flow, and traps and binds silt and clay grade particles (Ginsburg and Lowenstam, 1958), so that fine grades of sediment not usually associated with intertidal regions may occur as a consequence of stabilization by plants. Scoffin (1970) studied the influence of plants on sedimentation in the Bimini lagoon and ascertained that *Rhizophora* roots and *Thalassia* blades formed very effective baffles so that current velocity of 30 cm/sec (sufficient to "transport loose sand grains in clear areas" was reduced to zero at the sediment/water interface). Sea grasses also have a major role in the development and stabilization of carbonate banks (Davies, 1970; Basan, 1973), and also have the effect of modifying the chemical environment above and below the sediment/water interface.

B. Influence of Sedimentation on Bottom Communities

Gebelein (1969, p. 49) showed that current velocity and rates of sediment movement over the bottom controlled the distribution, morphology, and abundance of stromatolitic structures in Bermuda.

Firm substrates, e.g., rock or large shells, are necessary for the attachment of sessile benthonic organisms. Mobile sandy bottoms are unsuitable for either sessile benthonic organisms or the establishment of an infauna, but beyond the broad physical substrate controls there is a correlation between sediment texture and distribution patterns of benthos which is a reflection of feeding habits (see Purdy, 1964; Driscoll, 1969).

Sediment texture is related to current strength, all other things being equal, and the ease of transportation of sedimentary particles is controlled by the factors outlined in Section V.

This interrelationship determines the stability of sediment substrates. Sessile benthonic organisms which rise above the influence of the "traction carpet" are adversely affected only when the volume of sediment in transport becomes so great as to engulf them or when the current detaches them from their foundation. Vagile forms can also tolerate high-current velocities which bring increased food supply, but infaunal suspension feeders are commonly associated with weaker currents and fine sediments (Driscoll, 1969, p. 340). Low-current velocities associated with fine sediment grades permit the deposition of particulate organic matter from suspension, which may be extracted by sediment-ingesting organisms. Furthermore, some coral forms are better adapted to higher sedimentation rates than others (see Hubbard and Pocock, 1972). Therefore, it is to be anticipated that the general ecological zonation in a reef complex will reflect in a broad sense those physical parameters of sedimentation which may change progressively from reef front to deep lagoon.

VII. Summary and Conclusions

In the coral reef environment the role of organisms is manifold. They act as frame builders, sediment contributors, sediment movers (Kornicker and Boyd, 1962, p. 415), sediment transporters, and sediment stabilizers and binders, they are active in the breakdown of sedimentary particles, and they act as cementing agents and have the ability to inhibit reef growth (Fig. 11). An appreciation of their importance as agents of sedimentation, and of the diversity of their skeletal components, fosters doubts concerning the application of principles for the interpretation of sedimentary texture which stem from studies of terrigenous sedimentation.

In coral reef complexes there is a broad correlation between morphological and ecological zones and sedimentary facies, particularly in the vicinity of the reef itself. Such a correlation would be regarded as evidence of little sediment transportation. Indeed, although sedimentary particles will tend to separate into transportation populations according to intrinsic hydraulic properties, the net distance travelled under the influence of flood- and ebb-tidal currents may be short. However, some sediment may travel considerable distances from its provenance, especially when carried in suspension, and then the transportation history will dominate the texture of the deposit; to this may be added the influence of interfacial communities at the site of eventual deposition. The genetic implications of the texture of bioclastic carbonate accumulations are often difficult to decipher, and many anomalies may arise which can be appreciated only through quantitative studies of sedimentation processes.

The problem of the "true size" of sedimentary particles has been raised by a number of authors. Folk and Robles (1964) found great differences in settling velocities between skeletal particles and quartz grains of the same size, and Maiklem (1968) pointed out the differences in the settling behavior of skeletal grains and near-spherical grains of calcite and aragonite. The importance of the microarchitectural control on skeletal breakdown (the Sorby Principle) is emphasized by Folk and Robles (1964), while Jindrich (1969) stresses the importance of other factors, such as sorting, in determining grain-size distributions (Basan, 1973, p. 50). It may be concluded that the grain-size characteristics of bioclastic carbonates are the result of inherited and acquired features (Maiklem, 1968, p. 102). Thus, the former is controlled chiefly by the type and abun-

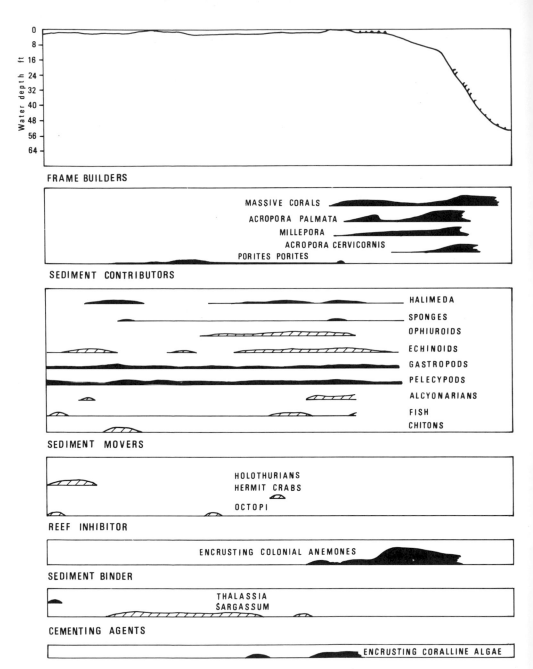

Fig. 11. Windward reef transect profile at "the boilers," Alacran Reef Complex, Mexico (after Kornicker and Boyd, 1962), illustrating the diverse roles of reef organisms.

dance of contributing organisms, and by the microarchitecture, fabric, and the method of breakdown of skeletal material. The acquired features are induced by hydraulic modification of the inherited characteristics. These may occur during transportation when the mode (surface creep, saltation, suspension) of physical transport will strongly influence grain size distribution by effecting a separation of grain size populations especially within the finer grade sizes; selective transportation of a particular grain size population will increase sorting (dispersion), producing differential abrasion of skeletal fragments, the traction load suffering more corrasion than the suspension load. The effective extent of time, during which transportation occurs, or the effective distance of transportation are naturally of supreme importance in the development of acquired features. Other important attributes may be due to the mixing of grain size populations during transportation and/or deposition, the acquisition of significant postmortem contributions *en route* or at the site of deposition, or to the introduction of sediment from a different source, the rapidity of deposition and stabilization of the sediment, the influence of interfacial communities (Section VI), or the physical reworking of the deposit and bioturbation.

Therefore a generalization may be stated to the effect that very coarse, immature reef sediments have inherited characteristics indicative of a source in which the influence of the Sorby Principle is marked, whereas a relatively mature sediment has a modified grain-size distribution which may be due to features acquired chiefly during transportation, and consequently may convey more information regarding the depositional environment than about the source.

From Inman's (1949, p. 68) consideration of the interrelationship of four parameters, viz., threshold velocity, roughness velocity, settling velocity, and grain diameter, certain fundamental observations are made: (1) for spherical grains the fine sand is the easiest transported; (2) in settling, the fine grains behave differently from the coarse, and the shape controls the settling path; (3) the relationship between suspension and traction transport depends on the interrelationship of settling velocity to threshold velocity and grain size; (4) sorting (dispersion) and skewness depend on variations in all of these factors and there may be sorting under way in addition to sorting *in situ.* Separation of grain size populations occurs according to the mode of transport; a fast flow will effect all three modes of transport, i.e., surface creep, saltation, and suspension. Fine-grade sizes may be rapidly removed from a reef flat and carried some distance in suspension so that the suspension load may not be apparent in the granulometric characteristics of reef flat sediment. The recognition of environments and sedimentary processes is based on sediment composition and texture. Thus, it is possible to distinguish fore reef, reef, and back reef deposits, and, for example, Folk and Robles (1964) were able to differentiate carbonate beach sands from adjacent "submerged bare sands," and from sands covered with marine grasses at Alacran reef. It is conceivable that a wide range of environments within a reef complex can be defined on the basis of grain-size distribution as well as by grain types which reflect ecological factors, but further study is needed to define the complex relationship between sediment textures, sedimentary processes, and environments in coral reef complexes.

References

Ahr, W. M., and Stanton, R. J., Jr. (1973). The sedimentological and paleoecologic significance of *Lithotrya*, a rock-boring barnacle. *J. Sediment. Petrol.* **43**, 20–33.

Bagnold, R. A. (1942). "Physics of Blown Sands and Desert Dunes." Methuen, London.

Baines, G. B. K., Beveridge, P. J., and Maragos, J. E. (1974). Storms and Island Building at Funafuti Atoll, Ellice Islands. *Proc. Int. Symp. Coral Reefs, 2nd, 1973* Vol. 2, pp. 485–486.

Bardach, J. E. (1961). Transport of calcareous fragments by reef fishes. *Science* **133**, 98–99.

Barnes, J., Bellamy, D. J., Jones, D. J., Whitton, B. A., Drew, E. A., Kenyon, L., Lythgoe, J. H., and Rosen, B. R. (1971). Morphology and ecology of the reef front of Aldabra. *Symp. Zool. Soc. London* **28**, 87–114.

Barrell, J. (1917). Rhythms and the measurement of geologic time. *Geol. Soc. Am. Bull.* **28**, 745–904.

Basan, P. B. (1973). Aspects of sedimentation and development of a carbonate banks in the Barracuda Keys, south Florida. *J. Sediment. Petrol.* **43**, 42–53.

Bathurst, R. G. C. (1966). Boring algae, micrite envelopes and lithification of molluscan biosparites. *Geol. J.* **54**, 15–32.

Bathurst, R. G. C. (1971). "Carbonate Sediments and Their Diagenesis." Elsevier, Amsterdam.

Bertram, G. C. L. (1936). Some aspects of the breakdown of coral at Ghardaqa, Red Sea. *Proc. Zool. Soc. London* pp. 1011–1026.

Blumenstock, D. I. (1958). Typhoon effects at Jaluit Atoll in the Marshall Islands. *Nature (London)* **182**, 1267–1269.

Blumenstock, D. I. (1961). A report on typhoon effects upon Jaluit Atoll. *Atoll Res. Bull.* **75**, 1–105.

Blumenstock, D. I., Fosberg, F. R., and Johnson, C. G. (1961). The re-survey of typhoon effects on Jaluit Atoll in the Marshall Islands. *Nature (London)* **18**, 618–620.

Braithwaite, C. J. R. (1973). Settling behaviour related to sieve analysis of skeletal sands. *Sedimentology* **20**, 251–262.

Bramlette, M. N. (1926). Some marine bottom samples from Pago-Pago Harbor, Samoa. *Carnegie Inst. Washington, Pap. Dep. Mar. Biol.* **23**, 1–35.

Carricker, M. R., and Smith, E. H. (1969). Comparative calcibiocavitology: Summary and conclusions. *Am. Zool.* **9**, 1011–1020.

Chave, K. E. (1954). Aspects of the biogeochemistry of Magnesium 1. Calcareous marine organisms. *J. Geol.* **62**, 266–283.

Chave, K. E. (1960). Carbonate skeletons to limestones: Problems, *Trans. N.Y. Acad. Sci.* [2] **23**, 14–24.

Chave, K. E. (1962). Factors influencing the mineralogy of carbonate sediments. *Limnol. Oceanogr.* **7**, 218–233.

Chave, K. E. (1964). Skeletal durability and preservation. *In* "Approaches to Palaeoecology" (J. Imbrie and N. D. Newell, eds.), pp. 377–387. Wiley, New York.

Chave, K. E., Deffeyes, K. S., Garrels, R. M., Thompson, M. E., and Weyl, P. K. (1962). Observation on the solubility of skeletal carbonates in aqueous solutions. *Science* **137**, 33–34.

Chave, K. E., Smith, S. V., and Roy, K. J. (1972). Carbonate production by coral reefs. *Mar. Geol.* **12**, 123–140.

Cloud, P. E., Jr. (1952). Preliminary report on the geology and marine environments of Onotoa Atoll, Gilbert Islands. *Atoll Research Bull.* **12**, 1–73.

Cloud, P. E., Jr. (1959). Geology of Saipon, Mariana Islands. 4. Submarine topography and shoal water ecology. *U.S., Geol. Surv., Prof. Pap.* **280-K**, 361–445.

Cribb, A. B. (1973). The algae of the Great Barrier Reef. *In* "Biology and Geology of Coral Reefs" (O. A. Jones and R. Endean, eds.), Vol. 2, pp. 47–73. Academic Press, New York.

Crozier, W. J. (1918). The amount of bottom material ingested by holothurians. *J. Exp. Zool.* **26**, 379–389.

Cuffey, R. J. (1972). The roles of bryozoans in modern coral reefs. *Geol. Rundsch.* **61**, 542–550.

Davies, G. R. (1970). Carbonate bank sedimentation, eastern Shark Bay, Western Australia. *Mem., Am. Assoc. Pet. Geol.* (1970) **13**, 85–168.

Davies, P. J. (1974). Cation electrode measurements in the Capricorn Area, Southern Great Barrier Reef Province. *Proc. Int. Symp. Coral Reefs, 2nd, 1973* Vol. 2, pp. 449–455.

Di Salvo, L. H. (1969). Isolation of bacteria from the Corallum of *Porites lobata* (Vaughn) and its possible significance. *Am. Zool.* **9**, 735–740.

Di Salvo, L. H. (1973). Microbial ecology. *In* "Biology and Geology of Coral Reefs" (O. A. Jones and R. Endean, eds.), Vol. 2, pp. 1–15. Academic Press, New York.

Driscoll, E. G. (1967). Experimental field study of shell abrasion. *J. Sediment Petrol.* **37**, 1117–1123.

Driscoll, E. G. (1969). Animal-sediment relationships of the Coldwater and Marshall formations of Michigan. *In* "Stratigraphy and Palaeontology" (K. W. S. Campbell, ed.), pp. 337–352. Australian National University Press, Canberra.

Driscoll, E. G. (1970). Selective bivalve shell destruction in marine environments, a field study. *J. Sediment. Petrol.* **40**, 898–905.

Duerden, J. E. (1902). Boring algae as agents in the disintegration of corals. *Bull. Am. Mus. Nat. Hist.* **16**, 323–332.

Ebbs, N. K., Jr. (1966). The coral-inhabiting polychaetes of the northern Florida reef tract. Part I. Aphroditidae, Polynoidae, Amphinomidae, Eunicidae and Lysaretidae. *Bull. Mar. Sci.* **16**, 485–554.

Eden, R. A., Orme, G. R., Mitchell, M., and Shirley, J. (1964). A study of part of the margin of the Carboniferous Limestone "Massif" in the Pin Dale area of Derbyshire. *Bull. Geol. Surv. G. B* **21**, 73–118.

Emery, K. O. (1956). The geology of Johnston Island and its surrounding shallows, central Pacific ocean. *Geol. Soc. Am. Bull.* **67**, 1505–20.

Emery, K. O. (1962). Marine geology of Guam. *U.S. Geol. Surv., Prof. Pap.* **403-B**, 1–76.

Emery, K. O., Tracey, J. I., Jr., and Ladd, H. S. (1954). Geology of Bikini and nearby atolls. 1. Geology. *U.S., Geol. Surv., Prof. Pap.* **260-A**, 1–265.

Endean, R. (1973). Population explosions of *Acanthaster planci* and associated destruction of hermatypic corals in the Indo-West Pacific region. *In* "Biology and Geology of Coral Reefs" (O. A. Jones and R. Endean, eds.), Vol. 2, pp. 381–438. Academic Press, New York.

Fairbridge, R. W. (1950). Recent and Pleistocene coral reefs of Australia. *J. Geol.* **58**, 330–401.

Flood, P. G. (1974). Sand movements on Heron Island—a vegetated sand cay. *Proc. Int. Symp. Coral Reefs, 2nd, 1973* Vol. 2, pp. 387–394.

Flood, P. G., Orme, G. R., and Scoffin, T. P. (1977). An analysis of the textural variability displayed by inter-reef sediments of the "Impure Carbonate Facies" in the vicinity of the Howick Group, Great Barrier Reef, Australia, *Philos. Trans. R. Soc.* London (in press).

Flood, P. G. and Scoffin, T. P. (1977). Reefal sediments from the northern Great Barrier Reef. *Philos. Trans. R. Soc.* London (in press).

Folk, R. L. (1962). Sorting in some carbonate beaches of Mexico. *Trans. N.Y. Acad. Sci.* [2] **25**, 222–244.

Folk, R. L., and Robles, R. (1964). Carbonate sands of Isla Perez, Alacran Reef complex, Yucatan. *J. Geol.* **72**, 255–292.

Force, L. M. (1969). Calcium carbonate size distribution on the west Florida shelf and experimental studies on the microarchitectural control of skeletal breakdown. *J. Sediment. Petrol.* **39**, 902–934.

Frankel, E. (1974). Recent sedimentation in the Princess Charlotte Bay area, Great Barrier Reef Province. *Proc. Int. Symp. Coral Reefs, 2nd, 1973* Vol. 2, pp. 355–369.

Frankel, E. (1975). *Acanthaster* in the past: Evidence from the Great Barrier Reef. *In* "Crown-of-Thorns Starfish Seminar Proceedings," pp. 159–165. Aust. Gov. Publ. Ser., Canberra.

Friedman, G. M. (1961). Distinction between dune, beach, and river sands from the textural characteristics. *J. Sediment. Petrol.* **31**, 514–529.

Friedman, G. M. (1962). Comparison of moment measures for sieving and thin-section data in sedimentary petrological studies. *J. Sediment. Petrol.* **32**, 15–25.

Friedman, G. M. (1968). Geology and geochemistry of reefs, carbonate sediments, and waters, Gulf of Aqaba (Elat), Red Sea. *J. Sediment. Petrol.* **28**, 895–919.

Friedman, G. M. (1969). Trace elements as possible environmental indicators of carbonate sediments. *Soc. Econ. Paleontol. Mineral., Spec. Publ.* **14**, 193–198.

Gardiner, J. S. (1931). "Coral Reefs and Atolls." Macmillan, New York.

Garrett, P., Smith, D. C., Wilson, A. O., and Patriquin, D. (1971). Physiography, ecology and sediments of two Bermuda patch reefs. *J. Geol.* **79**, 647–668.

Gary, M., McAfee, R., Jr., and Wolf, C. L., eds. (1972). "Glossary of Geology." Am. Geol. Inst., Washington, D.C.

Gebelein, C. D. (1969). Distribution, morphology, and accretion rate of Recent subtidal algal stromatolites, Bermuda. *J. Sediment. Petrol.* **39**, 49–69.

Ginsburg, R. N. (1957). Early diagenesis and lithification of shallow-water carbonate sediments in south Florida. *Soc. Econ. Paleontol. Mineral., Spec. Publ.* 5, 80–99.

Ginsburg, R. N., and Lowenstam, H. A. (1958). The influence of marine bottom communities on the depositional environment of sediments *J. Geol.* 66, 310–318.

Ginsburg, R. N., and Schroeder, J. H. (1973). Growth and submarine fossilization of algal cup reefs, Bermuda. *Sedimentology* 20, 575–614.

Glynn, P. W. (1973). Aspects of the ecology of coral reefs, in the western Atlantic region. *In* "Biology and Geology of Coral Reefs" (O. A. Jones and R. Endean, eds.), Vol. 2, pp. 271–324. Academic Press, New York.

Goreau, T. F., and Goreau, N. I. (1973). The ecology of Jamaican Coral Reefs. II. Geomorphology, zonation, and sedimentary phases. *Bull. Mar. Sci.* 23, 399–404.

Henson, F. R. S. (1950). Cretaceous and Tertiary reef formations and associated

Hjulström, F. (1939). Transportation of detritus by moving water. *In* "Recent Marine Sediments" (P. D. Trask, ed.), pp. 5–31. Am. Assoc. Petrol. Geol., Tulsa, Oklahoma.

Hoffmeister, J. E., and Ladd, H. S. (1944). The antecedent-platform theory. *J. Geol.* 52, 388–402.

Hoskin, C. M. (1968). Magnesium and strontium in mud fraction of Recent carbonate sediment, Alacran Reef, Mexico. *Am. Assoc. Pet. Geol. Bull.* 52, 2170–2177.

Hubbard, A. E. B., and Pocock, Y. P. (1972). Sediment rejection by recent scleractinian corals; a key to palaeo-environmental reconstruction. *Geol. Rundschou.* 61, 598–626.

Humm, H. F. (1964). Epiphytes of the sea grass *Thalassia testudinum* in Florida. *Bull. Mar. Sci. Gulf Caribb.* 14, 306–341.

Inman, D. L. (1949). Sorting of sediments in the light of fluid mechanics. *J. Sediment. Petrol.* 19, 51–70.

Jell, J. S., Maxwell, W. H. G., and McKellar, R. G. (1965). The significance of larger Foraminifera in the Heron Island reef sediments. *J. Paleontol.* 39, 273–279.

Jindrich, V. (1969). Recent carbonate sedimentation by tidal channels in the lower Florida Keys. *J. Sediment. Petrol.* 39, 531–553.

Klovan, J. E. (1966). The use of factor analysis in determining depositional environments from grain-size distributions. *J. Sediment. Petrol.* 36, 115–125.

Kornicker, L. S., and Boyd, D. W. (1962). Shallow water geology and environments of Alacran reef complex. Campeche Bank, Mexico. *Am. Assoc. Petrol. Geol. Bull.* 46, 640–73.

Kukal, Z. (1971). "Geology of Recent Sediments," p. 366. Academic Press, New York.

Ladd, H. S. (1961). Reef building. *Science* 134, 703–715.

Ladd, H. S., and Schlanger, S. O. (1960). Drilling operations on Eniwetok Atoll. *U.S., Geol. Surv., Prof. Pap.* 260-Y, 863–903.

Land, L. S. (1974). Growth rate of a West Indian (Jamaican) reef. *Proc. Int. Symp. Coral Reefs, 2nd, 1973* Vol. 2, pp. 408–412.

Lang, J. C. (1974). Biological zonation at the base of a reef. *Am. Sci.* 62, 273–281.

Logan, B. W. (1969). Coral reefs and banks: Yucatan shelf, Mexico. *Mem., Am. Assoc. Petrol. Geol.* 11, 129–198.

Lowenstam, H. A. (1950). Niagaran reefs of the Great Lakes area. *J. Geol.* 58, 430–487.

Lowenstam, H. A. (1954). Factors affecting the aragonite-calcite ratios in carbonate-secreting marine organisms. *J. Geol.* **62**, 284–322.

Loya, Y. (1975). Possible effects of water pollution on the community structure of Red Sea corals. *Mar. Biol.* **29**, 177–185.

McKee, E. D. (1956). Geology of Kapingamarangi Atoll, Caroline Islands. *Atoll Res. Bull.* **50**, 1–38.

McKee, E. D., Chronic, J., and Leopold, E. B. (1959). Sedimentary belts in the lagoon of Kapingamarangi Atoll. *Am. Assoc. Petrol. Geol. Bull.* **43**, 501–562.

McLean, R. F. (1974). Geologic significance of bioerosion of beachrock. *Proc. Int. Symp. Coral Reefs, 2nd, 1973* Vol. 2, pp. 401–412.

Macnae, W. (1968). A general account of the fauna and flora of mangrove swamps and forests in the Indo-West-Pacific region. *Adv. Mar. Biol.* **6**, 73–270.

Maiklem, W. R. (1968). Some hydraulic properties of bioclastic carbonate grains. *Sedimentology* **10**, 101–109.

Maiklem, W. R. (1970). Carbonate sediments in the Capricorn reef complex, Great Barrier Reef, Australia. *J. Sediment. Petrol.* **40**, 55–80.

Maxwell, W. G. H. (1968a). "Atlas of the Great Barrier Reef." Elsevier, Amsterdam.

Maxwell, W. G. H. (1968b). Relict sediments, Queensland continental shelf. *Aust. J. Sci.* **31**, 85–86.

Maxwell, W. G. H., Jell, J. S., and McKeller, R. G. (1964). Differentiation of carbonate sediments on the Heron Island Reef. *J. Sediment. Petrol.* **34**, 294–308.

Milliman, J. D. (1973). Caribbean Coral Reefs. *In* "Biology and Geology of Coral Reefs" (O. A. Jones and R. Endean, eds.), Vol. 1, pp. 1–50. Academic Press, New York.

Milliman, J. D. (1974). "Marine Carbonates." Springer-Verlag, Berlin and New York.

Nelson, H. F., Brown, C. W., and Brineman, J. H. (1962). Skeletal limestone classification. *Mem., Am. Assoc. Pet. Geol.*, **1**, 224–252.

Neumann, A. C. (1968). Biological erosion of limestone coasts. *In* "Encyclopaedia of Geomorphology" (R. W. Fairbridge, ed.), pp. 75–81. Van Nostrand-Reinholt, Princeton, New Jersey.

Neumann, A. C., Gebelein, C. D., and Scoffin, T. P. (1970). The composition, structure and erodability of subtidal mats, Abaco, Bahamas. *J. Sediment. Petrol.* **40**, 274–297.

Newell, N. D. (1971). An outline history of tropical organic reefs. *Am. Mus. Novit.* **2465**, 1–37.

Newell, N. D., Rigby, J. K., Fischer, A. G., Whiteman, A. J., Hickox, J. E., and Bradley, J. S. (1953). "The Permian Reef Complex of the Guadalupe Mountains Region, Texas and New Mexico." Freeman, San Francisco, California.

Odum, H. T., and Odum, E. T. (1955). Trophic structure and productivity of a windward coral reef community on Eniwetok Atoll. *Ecol. Monogr.* **25**, 291–320.

Orme, G. R. (1971). The D_2-P_1 "reefs" and associated limestones of the Pin Dale-Bradwell Moor area of Derbyshire. *C. R. Congr. Int. Stratigr. Geol. Carbonif. Sheffield* **111**, 1249–1262.

Orme, G. R., Flood, P. G., and Ewart, A. (1974). An investigation of the sediments and physiography of Lady Musgrave Reef—a preliminary account. *Proc. Int. Symp. Coral Reefs, 2nd, 1973* Vol. 2, pp. 371–386.

Orme, G. R., Flood, P. G., and Sargent, G. E. G. (1977). Sedimentation trends in the lee of outer (ribbon) reefs, northern region of the Great Barrier Reef Province. *Philos. Trans. R. Soc. London* (in press).

Otter, G. W. (1931). Rock-destroying organisms in relation to coral reefs. *Sci. Rep. Great Barrier Reef. Exped.* 1, 323–352.

Plumley, W. G., Risley, G. A., Graves, R. W., Jr., and Kaley, M. E. (1962). Energy Index for Limestone Interpretation and Classification. *In* "Classification of Carbonate Rocks" (W. E. Ham, ed.), pp. 85–107. Am. Assoc. Pet. Geol., Tulsa, Oklahoma.

Purdy, E. G. (1963). Recent calcium carbonate facies of the Great Bahama Bank. *J. Geol.* 71, 472–97.

Purdy, E. G. (1964). Sediments as substrates. *In* "Approaches to Paleoecology" (J. Imbrie and N. B. Newell, eds.), pp. 238–269. Wiley, New York.

Purdy, E. G. (1974a). Reef configurations: Cause and effect: *Soc. Econ. Paleontol. Mineral., Spec. Publ.* 18, 9–76.

Purdy, E. G. (1974b). Karst-determined facies patterns in British Honduras: Holocene carbonate sedimentation model. *Am. Assoc. Pet. Geol. Bull.* 58, 825–855.

Rubey, W. W. (1933). Settling velocities of gravel, sand, and silt particles. *Am. J. Sci.* 25, 325–338.

Schmalz, R. F., and Swanson, F. J. (1969). Diurnal variations in the carbonate saturation of seawater. *J. Sediment. Petrol.* 39, 255–267.

Schroeder, J. H., and Zankl, H. (1974). Dynamic reef-formation: A sedimentological concept based on studies of recent Bermuda and Bahama Reefs. *Proc. Int. Symp. Coral Reefs, 2nd, 1973* Vol. 2, pp. 413–428.

Scoffin, T. P. (1970). The trapping and binding of subtidal carbonate sediments by marine vegetation in Bimini Lagoon, Bahamas. *J. Sediment. Petrol.* 40, 248–273.

Scoffin, T. P., and Garrett, P. (1974). Processes in the formation and preservation of internal structure in Bermuda patch reefs. *Proc. Int. Symp. Coral Reefs, 2nd, 1973* Vol. 2, pp. 429–448.

Sorby, H. C. (1879). Anniversary Address of the President (Structure and Origin of Limestones). *Geol. Soc. London Proc.* 35, 56–95.

Sorokin, Y. T. (1973). Microbiological aspects of the productivity of coral reefs. *In* "Biology and Geology of Coral Reefs" (O. A. Jones and R. Endean, eds.), Vol. 2, pp. 17–45. Academic Press, New York.

Spencer, D. W. (1963). The interpretation of grain size distribution curves of clastic sediments. *J. Sediment. Petrol.* 33, 180–190.

St. John, B. E. (1973). Trace elements in corals of the Coral Sea: Their relationship to oceanographic factors. *In* "Oceanography of the South Pacific (1972)" (R. Fraser, compiler), pp. 148–158. New Zealand National Commission for UNESCO, Wellington.

Stoddart, D. R. (1962). Three Caribbean atolls: Turneffe Islands, Lighthouse Reef, and Glover's Reef, British Honduras. *Atoll Res. Bull.* 87, 151.

Stoddart, D. R. (1963). Effects of Hurricane Hattie on the British Honduras reefs and cays October 30–31, 1961. *Atoll Res. Bull.* 95, 1–142.

Stoddart, D. R. (1965). Re-survey of Hurricane effects on the British Honduras reefs and cays. *Nature (London)* 207, 588–592.

Stoddart, D. R. (1968). Castastrophic human interference with coral atoll ecosystems. *Geography* 53, 25–40.

Stoddart, D. R. (1969a). Ecology and morphology of Recent coral reefs. *Biol. Rev. Cambridge Philos. Soc.* 44, 433–498.

Stoddart, D. R. (1969b). Post-hurricane changes on the British Honduras reefs and cays: Re-survey of 1965. *Atoll Res. Bull.* 131, 1–25.

Stoddart, D. R. (1972). Catastrophic damage to coral reef communities by earthquake. *Nature (London)* **239**, 51–52.

Stoddart, D. R. (1974). Post-hurricane changes on the British Honduras Reefs: Resurvey of 1972. *Proc. Int. Symp. Coral Reefs, 2nd, 1973* Vol. 2, pp. 473–483.

Taylor, J. D. (1968). Coral reef and associated invertebrate communities (mainly molluscan) around Mahe, Seychelles. *Philos. Trans. R. Soc. London, Ser. B* **254**, 129–206.

Tracey, J. I., Jr., Cloud, P. E., Jr., and Emery, K. O. (1955). Conspicuous features of organic reefs. *Atoll Res. Bull.* **46**, 1–3.

Visher, G. S. (1969). Grain size distributions and depositional processes. *J. Sediment. Petrol.* **39**, 1074–1106.

Wells, J. W. (1954). Recent corals of the Marshall Islands, Bikini and nearby atolls. *U.S. Geol. Surv., Prof. Pap.* **260-I**, 385–486.

Wells, J. W. (1957). Coral reefs. *Mem., Geol. Soc. Am.* **67**, 609–631.

Wolf, K. H., Chilinger, G. V., and Beales, F. W. (1967). Elemental composition of carbonate skeletons, minerals, and sediments. *Dev. Sedimentol.* **9B**, 24–148.

Wolfenden, E. B. (1958). Paleoecology of the Carboniferous reef complex and shelf limestones in northwest Derbyshire, England. *Geol. Soc. Am. Bull.* **69**, 871–898.

6

RADIOMETRIC GEOCHRONOLOGY OF
CORAL REEFS

H. H. Veeh and D. C. Green

I. Absolute Age Determinations

Numerous studies have shown that coral reefs, due to their tendency to develop into broad platforms close to sea level, are among the best "geologic tide gauges" available, and as such offer much insight into the geologic processes of a dynamic earth (Kuenen, 1955). Locally, relative sea level shifts, as revealed by emerged coral reef terraces hundreds of meters above the present sea level, can be interpreted in terms of crust deformation, especially if these terraces are tilted as well. Similarly, drowned reef faunas on guyots and thick sequences of shallow-water reef deposits underlying coral atolls attest to the net subsidence of individual islands or of larger sections of the ocean floor. On a global scale, widespread emerged coral reefs of similar elevation are attributed by many to eustatic sea level shifts, which are at least partially controlled by the amount of water tied up as glacial ice; hence they have considerable potential as paleoclimatic indicators. More likely, however, eustatic sea level fluctuations and crustal unrest have, at a given location, combined to produce a complex pattern of fossil strand lines, impossible to unravel without the aid of adequate age determination.

For these reasons the development of an absolute chronology of coral reefs in different parts of the world is highly desirable. The only currently valid methods for the absolute dating of geologic events are those based on the radioactive decay of naturally occurring radioisotopes. Several such methods are suitable for coral reefs and have found increasing application in recent years. This chapter presents a brief review of these methods, together with significant results obtained so far.

A. THE RADIOCARBON METHOD

Since Libby (1955) first demonstrated that ^{14}C, produced by cosmic rays and having a half-life of 5730 years, could be used for dating organic and carbonate material, this method has found widespread applications with phenomenal success. In principle it is assumed that (1) the initial $^{14}C/^{12}C$ ratio of carbon in a given material has always been the same, and (2) there has been no chemical or isotopic exchange of the carbon in this material with its surroundings from the time it was deposited until it was dated in the laboratory.

It is not possible to review here the substantial literature devoted to the verification of these assumptions and to the various corrections that have to be applied to individual radiocarbon measurements. In a recent symposium edited by Olsson (1970), evidence and possible causes for small but significant variations in the $^{14}C/^{12}C$ ratio during the last 7000 years were discussed. Very little is known at present about the extent and magnitude of such variations beyond 7000 years B.P.

A source of uncertainty most relevant to the present topic, however, is the possibility of contamination of old carbonate material with modern carbon. The error caused by such contamination increases rapidly with greater sample age (Broecker and Kulp, 1956). Thus the incorporation of 5% of contemporary carbonate by weight into a 10,000-year-old sample will yield a ^{14}C age only slightly less than the true age, but the same amount of contamination in a 100,000-year-old sample would result in an apparent age of 24,500 years.[1] It is this fact, coupled with the difficulty in recognizing small amounts of contamination by modern carbonate, that makes all radiocarbon ages of carbonate material older than about 20,000 years suspect. Such ages should be regarded as

---

[1] These figures are conservative, as they are based on natural radiocarbon levels, and do not take into consideration the additional amounts of radiocarbon produced by bomb tests, which would make old carbonates even more sensitive to contamination with modern carbon.

minimum ages only, unless absence of contamination with modern carbonate can be demonstrated by other means. The various effects of diagenesis on ^{14}C dating of corals and molluscs have been discussed by Chappell and Polach (1972).

With proper regard of these limitations, the application of radiocarbon dating to the chronology of coral reefs has been very fruitful. Significant results pertaining to the detailed history of the sea level during the Holocene transgression and its effect on the morphological development of coral reefs are reviewed by W. G. H. Maxwell in Chapter 9 of Volume I.

B. URANIUM-SERIES METHODS

The radioactive decay series headed by ^{238}U, ^{235}U, and ^{232}Th (Figs. 1–3) provide unique methods of determining the age of formation of natural materials. If a particular intermediate nuclide is initially absent, its concentration will build up as its parent decays. The daughter product will itself decay at a rate proportional to its concentration until its rate of decay equals its rate of formation, at which point radioactive equilibrium is established for that particular step. Similarly, the decay of an unsupported daughter nuclide in any of the decay series will proceed until radioactive equilibrium with its parent is reattained. The attainment of secular equilibrium through the entire ^{238}U series, for example, requires about 10^6 years. For this reason methods of geochronology based on the disturbance of secular equilibrium are limited to the last million years or less.

But, in addition, the helium produced by the alpha emitters in the uranium decay series provides a useful accumulation time clock with theoretically unlimited range. An excellent review of the principles and assumptions underlying the various uranium-series methods has been given by Ku (1976). Those methods with specific applications to marine carbonates and hence to coral reefs are outlined below.

1. The ^{234}U/^{238}U Method

Thurber (1962) discovered an excess of ^{234}U activity relative to the activity of its parent nuclide ^{238}U in recently formed carbonates and suggested that this phenomenon might be used for age determinations. This method is based, in principle, on the premise that the excess ^{234}U activity (15% in marine carbonates) decays with a half-life of 2.48×10^5 years in returning to secular equilibrium with its parent ^{238}U, and on the assumptions that (1) the initial excess of ^{234}U in the water from

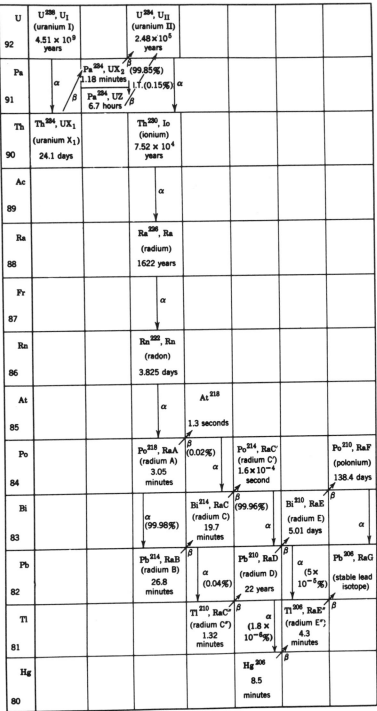

Fig. 1. The decay series of ^{238}U. From Friedlander *et al.* (1964), with permission of John Wiley and Sons, Inc.

186

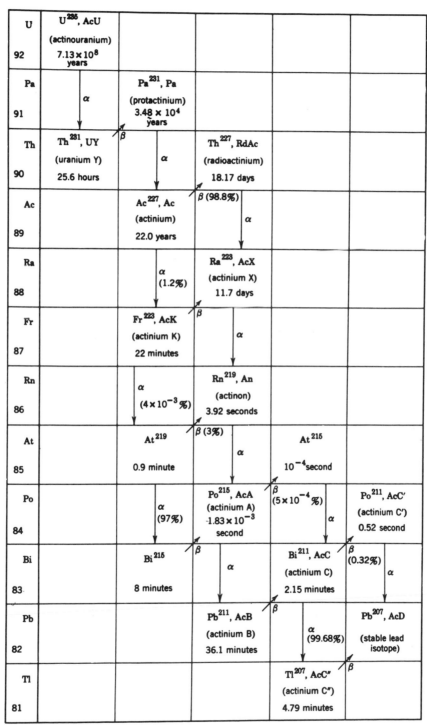

Fig. 2. The decay series of ^{235}U. From Friedlander *et al.* (1964), with permission of John Wiley and Sons, Inc.

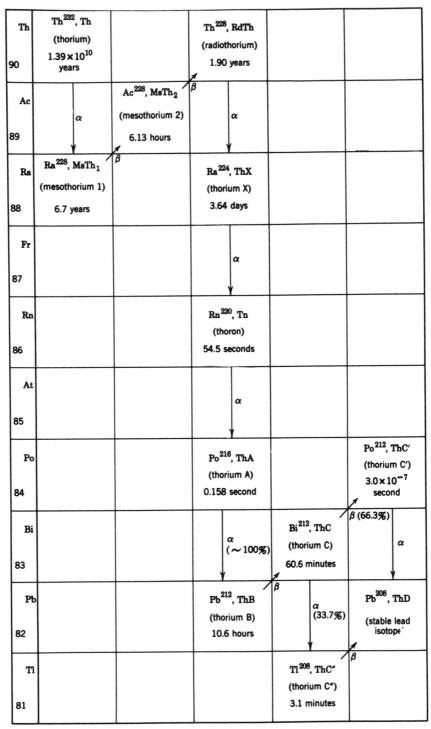

Fig. 3. The decay series of ^{232}Th. From Friedlander *et al.* (1964), with permission of John Wiley and Sons, Inc.

which the carbonate was derived is maintained at a constant value, so that it can be readily assessed; and (2) after its formation the carbonate has remained a closed system with respect to the uranium isotopes.

Experience has shown (Kaufman and Broecker, 1965; Blanchard *et al.*, 1967; Kaufman *et al.*, 1971) that, with the notable exception of corals, the second of these assumptions is frequently violated. Increasing uranium content with age, and failure of the $^{234}U/^{238}U$ ratio in the carbonate to decrease with time, indicate that many mollusc shells have received sizeable contributions of uranium isotopes from their surroundings subsequent to their formation.

On the other hand, corals, and to a lesser extent aragonitic oolites grown in an open marine environment, have yielded $^{234}U/^{238}U$ ratios which are consistent with independent age determinations (Thurber, 1962; Thurber *et al.*, 1965; Osmond *et al.*, 1965; Veeh, 1966; Sakanuoe *et al.*, 1967; Broecker and Thurber, 1965; Broecker *et al.*, 1968; Veeh and Chappell, 1970; Bloom *et al.*, 1974; and Ku *et al.*, 1974). Although this method has a potential range of about one million years, practical considerations arising from relatively large analytical errors limit it to the last 800,000 years.

The main application of this method so far has been in conjunction with the $^{230}Th/^{234}U$ method discussed below, and for age estimates of corals beyond the range of $^{230}Th/^{234}U$ dating.

2. $^{230}Th/^{234}U$ (*Ionium Deficiency*) *Method*

The applicability of this method to marine carbonates was first demonstrated by Barnes *et al.* (1956) who discovered that reef-building corals contain several parts per million uranium, but are essentially free of ^{230}Th at the time of formation. The growth of ^{230}Th towards a state of radioactive equilibrium with uranium can be used as a measure of age, subject to the conditions that (1) only uranium, but no thorium, enters the carbonate at the time of its formation; (2) the carbonate remains a closed system with respect to uranium and its decay product; and (3) the initial $^{234}U/^{238}U$ ratio has a fixed value which can be readily assessed (see Section 1, above).

The various criteria for evaluation of analytical results on samples of unknown age have been discussed by Tatsumoto and Goldberg (1959), Sackett (1958), Thurber *et al.* (1965), Stearns and Thurber (1965), Kaufman and Broecker (1965), Blanchard *et al.* (1967), and Bloom *et al.* (1974). The most important of these criteria are summarized below:

1. There should be no evidence of recrystallization and/or deposition of void-filling cement, since such processes facilitate chemical and isotopic exchange of uranium and thorium between the sample and its surroundings. It has been found that fossil corals containing significant amounts of calcite have lost part of their original uranium to their surroundings. Inasmuch as corals deposit their skeletons as aragonite, the onset of recrystallization to calcite can be readily recognized by x-ray diffraction analysis. Void filling with secondary aragonite needles is detectable by microscopic examination of thin sections.

2. The uranium concentration in the sample should not differ markedly from that in a contemporary equivalent of the carbonate in question. In modern corals, the uranium contents vary from about 2 to 4 ppm, depending on species and location, while in other biogenic carbonates such as mollusc shells, the uranium contents generally do not exceed 0.5 ppm. Uranium contents in fossil carbonate material falling outside these respective ranges may indicate loss or gain of uranium.

3. The $^{234}U/^{238}U$ ratio should be consistent with the $^{230}Th/^{234}U$ age of the sample, i.e., it exhibits a systematic decrease with sample age.

4. The sample should be free of ^{232}Th, since the presence of ^{232}Th would indicate addition of unknown proportions of external ^{230}Th to the sample. Under favorable circumstances, a correction for this "nonradiogenic" or common ^{230}Th is possible. In order to make such a correction, the proportion of $^{230}Th/^{232}Th$ in the environment of the sample and the time of addition of these isotopes to the sample have to be known.

5. The $^{230}Th/^{234}U$ ages should be consistent with stratigraphic position and, if available, with independent age determinations.

Many measurements of uranium series isotopes in carbonates of various types and ages have been made since the original discovery by Barnes et al. (1956), permitting a reassessment of the general validity of the $^{230}Th/^{234}U$ method as applied to marine carbonates and, in particular, to corals (Broecker, 1963; Sackett, 1958; Osmond et al., 1965; Broecker and Thurber, 1965; Stearns and Thurber, 1965; Thurber et al., 1965; Veeh, 1966; Blanchard et al., 1967; Sakanuoe et al., 1967; Veeh and Valentine, 1967; Broecker et al., 1968; Valentine and Veeh, 1969; Veeh and Giegengack, 1970; Veeh and Veevers, 1970; Veeh and Chappell,

1970; Kaufman *et al.*, 1971; Bloom *et al.*, 1974; Ku *et al.*, 1974; Chappell and Veeh, 1977).

From all accounts so far it appears that reliable ages can be obtained for corals, provided the guidelines listed above are strictly observed. Thus the uranium concentrations in well-preserved (i.e., unrecrystallized) fossil corals of Pleistocene age do not differ greatly from those of contemporary species equivalents, indicating that, to a first approximation, the assumption that the corals have remained a closed system is justified. This is further supported by the uranium isotope data. The $^{234}U/^{238}U$ ratios in unrecrystallized corals of all ages are generally consistent with the time-dependent decrease of this ratio, demonstrating that no significant secondary addition of uranium isotopes from the surroundings of the coral sample has taken place subsequent to its formation. Finally, mutual cross checks between ^{230}Th ages and independently determined ages on identical coral samples have been satisfactory. Thus a comparison between ^{230}Th ages and semiindependent ^{231}Pa ages of corals has given concordant results (see below). Similarly, cross checks between ^{230}Th and ^{14}C ages are in reasonably good agreement, at least for the last 10,000 years. Cross checks beyond 10,000 years would be highly desirable, but are difficult to perform due to the scarcity of suitable samples within the interval 10,000–20,000 years,[2] and because ^{14}C ages beyond about 20,000 years become increasingly sensitive to contamination with modern carbon, as previously discussed. In a few isolated instances, a comparison between ^{230}Th ages of coral reefs and K/Ar ages of stratigraphically related lava flows has been possible. Again, the results have so far been in satisfactory agreement (see below).

In contrast to that of corals, the present status of uranium-series dating of molluscs is not encouraging. On the basis of some 400 available analyses of uranium-series isotopes in mollusc shells from diverse locations and ages, Kaufman *et al.* (1971) have concluded that the assumption of a closed system for this material is violated so frequently that uranium-series ages of mollusc samples cannot be accepted as valid unless there is independent proof that postdepositional migration of uranium and/or its daughters has not taken place. The evidence presented for migration of uranium and/or its daughters includes: (1) increase of ^{238}U with time, (2) increase of $^{234}U/^{238}U$ with time, (3) disagreement between $^{230}Th/^{234}U$ and $^{231}Pa/^{235}U$ ages, and (4) disagree-

[2] During this time interval, the sea level had reached its maximum depression in response to the peak of the Wisconsin glaciation (Curray, 1965; Milliman and Emery, 1968; and Veeh and Veevers, 1970) so that corals living at that time are not easily accessible for sampling.

ment between ^{230}Th/^{234}U and ^{14}C ages within the range in which ^{14}C dates are considered reliable (i.e., up to about 20,000 years).

Several attempts have been made to correct for such postdepositional migration of uranium-series isotopes in mollusc shells by devising various "open system models" in order to calculate absolute ages from existing analytical data (Rosholt, 1967; Szabo and Rosholt, 1969; Osmond *et al.*, 1970; and Kaufman *et al.*, 1971). Although such an approach to the problem holds some promise, none of the models devised so far has yielded reliable results and hence all uranium-series ages of molluscs should be viewed with reservation. Pending further investigations of this nature, only uranium-series ages of unrecrystallized corals meeting all the conditions outlined above should be considered as absolute ages.

In summary, until the complicated migration processes of uranium-series isotopes in mollusc shells are better understood, the extra time spent in the field searching for well-preserved and well-placed corals goes a long way towards obtaining reliable uranium-series ages of coral reefs.

3. The ^{231}Pa/^{235}U (^{231}Pa deficiency) Method

In this technique, first suggested by Sackett (1958) and first attempted by Rosholt and Antal (1962), the age of the sample is calculated from the relative amount of ^{231}Pa which has accumulated during growth towards radioactive equilibrium with the decay series headed by ^{235}U (see Fig. 2). Provided that the carbonate has remained a closed system with respect to uranium and protactinium, absolute ages to about 120,000 years can be obtained. The method is very similar in principle and assumptions to the ^{230}Th/^{234}U method discussed above and therefore is subject to the same limitations.

Good agreement with the ^{230}Th/^{234}U method has been reported for unaltered coral samples (Sakanuoe *et al.*, 1967; Ku, 1968; Ku *et al.*, 1974), but not for molluscs (Kaufman *et al.*, 1971). The method is useful as an independent check on ^{230}Th/^{234}U and ^{14}C ages. Drawbacks of the method are the shorter range as compared to the ^{230}Th/^{234}U method, the relatively large sample requirements due to the very low concentration of ^{235}U and hence of ^{231}Pa, and the fact that there is no independent isotope of protactinium to serve as an indicator of possible contamination with external ^{231}Pa.

4. The Helium-Uranium Method

Although the helium method is one of the earliest methods in the history of radiometric dating of rocks, the possibility of applying it to carbonates has only recently been seriously explored (Fanale and Kulp, 1962; Fanale and Schaeffer, 1965; Bender, 1973). In principle, this method makes use of the generation of helium nuclei (alpha particles) within a carbonate material containing alpha emitters. The age of a given carbonate sample can be calculated by adding together the integrated helium production rates of the principal alpha sources such as: (1) ^{238}U, which is essentially constant over the time period considered; (2) the excess ^{234}U decaying from its initial concentration to its ^{238}U-supported concentration; and (3) ^{230}Th and its daughters as they grow into transient equilibrium with ^{234}U (see Fig. 1). The contribution of ^{235}U together with its alpha-emitting daughters adds a minor increment to the total helium (see Fig. 2).

As this method involves the uranium series isotopes, it is, *a priori,* subject to many of the conditions and limitations discussed previously. Additional limitations arise from the loss of alpha particles shot out of crystal lattices, possible diffusive loss of radon, and from the unknown initial helium content observed in a number of samples, probably the result of incorporation into the carbonate of old detritus containing alpha emitters, and hence helium. Nevertheless, agreement between the ages of samples dated by the helium-uranium and those obtained by other methods has generally been good for unrecrystallized corals, but predictably less so for mollusc samples (Fanale and Schaeffer, 1965; Bender, 1973). The main value of this method lies in its theoretically unlimited range, subject only to the ability of aragonite corals to remain a closed system with respect to uranium and its various decay products.

C. The Potassium-Argon Method

This technique, widely used for dating minerals and rocks of Precambrian to Pleistocene age, has so far not been applied directly to carbonates. However, under favorable circumstances, indirect dating of coral reefs would be possible by analyzing volcanic material with known stratigraphic relationship to the reef deposits, provided the time interval between formation of the reef and the volcanic eruption has been short.

An example can be cited from the island of Oahu, Hawaii. Here a massive reef limestone, evidently formed during the interglacial Kaena (Yarmouth ?) stand of the sea (Stearns, 1966) is overlain by a basalt flow with a potassium-argon age of 297,000 years (Gramlich *et al.,* 1971), thus providing a lower age limit for this reef. This is consistent with ^{230}Th/^{234}U and ^{234}U/^{238}U age estimates of greater than 200,000 years and 600,000 years, respectively, for reef deposits formed during the Kaena stand on Oahu (Veeh, unpublished data).

Similar associations between lava flows and coral reefs can be expected on many volcanic islands within the coral reef belt, and should provide opportunities for badly needed mutual cross checks between uranium-series and potassium-argon ages.

D. RESULTS

Using the methods outlined above, the following significant results concerning the absolute chronology of coral reefs have been obtained so far.

First, the maximum lowering of sea level in response to the last glacial advance, and corresponding to the peak of the Wisconsin glaciation, was about 17,000 years ago (Veeh and Veevers, 1970). This compares favorably with a ^{14}C date of 18,000 years for the maximum ice advance in the northern hemisphere (Flint, 1955).

Second, a worldwide occurrence of emerged coral reefs between 2 and 10 m above the present sea level in tectonically "stable" areas[3] with a mean age of about 120,000 years (Broecker and Thurber, 1965; Osmond et al., 1965; Veeh, 1966; Land et al., 1967; Ku et al., 1974). This sea stand is tentatively correlated with the end of the Sangamon interglacial and probably represents the last time at which, in such areas, the sea stood higher than it does today.

Third, elevated coral reef terraces in active orogenic belts and island arcs along lithosphere plate boundaries display a systematic increase in age with height above sea level, with one or more of the following age groups persistently recurring: 5000–9000 years, 35,000–42,000 years, 48,000–66,000 years, 82,000–86,000 years, 102,000–107,000 years, 118,000–125,000 years, 130,000–140,000 years, and 180,000–190,000 years (Broecker et al., 1968; Mesolella et al., 1969; Konishi et al., 1970, 1974; Veeh and Chappell, 1970; Bloom et al., 1974; and Chappell, 1974; Chappell and Veeh, 1977). These results strongly support the hypothesis that the combined effects of continuous tectonic uplift of a given coast and rapid glacioeustatic sea level fluctuations inevitably produce a sequence of transgressive-regressive cycles which appear as a series of coral reef terraces that increase in age with height above sea level and that are

---

[3] Tectonically "stable" areas in this context are defined as areas far removed from Lithosphere plate boundaries and include locations such as Florida, the Bahamas, Bermuda, Hawaii, the Cook Islands, the Tuamotu Islands, Western Australia, and the Seychelles. It should be noted, however, that in the dynamic earth model of modern Plate Tectonics the concept of crustal stability has only relative meaning, and that the possibility of minor vertical movements of the crust at such locations, either positive or negative, cannot be excluded.

separated by steep scarps (Mitchell, 1968; Mesolella *et al.*, 1969; Chappell, 1974).

This chronology of implied sea level fluctuations shows features remarkably similar to those of the ^{18}O record in the deep sea cores (Broecker and van Donk, 1970; Shackleton and Opdyke, 1973). Although originally interpreted in terms of the paleotemperature variations of surface ocean water (Emiliani, 1955, 1966), it has recently been suggested that at least 50% of the observed oxygen isotope variations in the shells of planktonic foraminifera are directly related to the glacially controlled ocean volume (Shackleton, 1967; Dansgaard and Tauber, 1969). If this is correct, both the ^{18}O record of deep sea cores and the record of sea level fluctuations in the form of raised coral reefs should have one principal component in common, namely the extent of continental glaciations. In this regard the geochronology of coral reefs becomes a powerful tool in the investigation of ice ages and their causes (Broecker and van Donk, 1970; Veeh and Chappell, 1970; Chappell, 1973).

Fourth, the thick sequence of shallow water reef carbonates underlying many coral atolls, revealed by drilling at Eniwetok and Mururoa, shows a systematic increase in age with depth below sea level, with a persistently recurring gap in the age sequence between 7,000 and about 100,000 years (Thurber *et al.*, 1965; Lalou *et al.*, 1966). Another age gap, though less well defined, occurs between about 130,000 years and more than 200,000 years at Eniwetok (Thurber *et al.*, 1965) and between about 160,000 years and more than 200,000 at Mururoa (Lalou *et al.*, 1966). These results can be interpreted as reef growth on a slowly subsiding foundation, interrupted only during times when glacially lowered sea levels had transformed these atolls into temporary high islands.

Fifth, several attempts have been made to use the absolute ages of coral reefs discussed above to quantitatively assess neotectonic movements of the crust along zones of lithosphere convergence. Such attempts are at present seriously limited by the lack of an "absolute datum" of reference, against which vertical tectonic movements can be measured. Using an assumed datum of +6 m for the sea level of 120,000 years ago[4], based on widespread evidence for the approximate position of the sea level at that time (see above), mean uplift rates have been calculated for the crust at several locations along active orogenic belts and island arcs (Table I). In these calculations the simplifying assumption

[4] An uncertainty of a few meters in the actual position of sea level at that time is not critical, considering the relative magnitude of uplift in tectonically active areas.

TABLE I

NEOTECTONIC UPLIFT RATES OF THE CRUST IN ACTIVE
OROGENIC BELTS AND ISLAND ARCS

Location	Uplift rate[a] (m/10³ years)	Reference
Barbados	0.24–0.38	Broecker *et al.* (1968) Mesolella *et al.* (1969)
Ryukyu Islands	0.4–1.5	Konishi *et al.* (1970) Konishi *et al.* (1974)
Huon Peninsula, New Guinea	0.91–2.62	Bloom *et al.* (1974) Chappell (1974)
Atauro, Inner Banda Arc	0.5	Chappell and Veeh (1977)
East Timor, Outer Banda Arc	0.03–0.5	Chappell and Veeh (1977)

[a] Calculated from radiometric ages of emerged coral reefs, and based on a datum of +6 m for the sea level of 120,000 years ago (see text).

was made that the tectonic uplift rate has been uniform with time along any measured sequence of emerged coral reefs, at least for the last 120,000 years. This assumption appears to be well based on Atauro (Chappell and Veeh, 1977), but could be argued on Huon Peninsula, New Guinea (Chappell 1974). Clearly, many more age determinations of coral reefs in carefully selected locations will have to be made, before the full potential of coral reefs as "geologic tide gauges" can be fully exploited.

II. Growth Rate Studies

An area of special interest under the general topic of radiometric age determinations of corals is the possibility of using radiometric methods to determine the growth rates of individual coral species. Such information is useful to the coral ecologist and to the carbonate sedimentologist seeking quantitative data relevant to the carbonate budget within the coral reef ecosystem. It is necessary that the radiometric techniques used should have sufficient resolution to distinguish between the ages of different sections within individual coral heads.

A. THE ²²⁸Ra METHOD

The decay of unsupported ^{228}Ra (^{232}Th series, see Fig. 3) has been used by Moore and Krishnaswami (1972, 1974), and Dodge and Thomson (1974) to determine growth rates of individual coral heads. The excess ^{228}Ra is introduced into sea water by diffusion from sediments, where it is generated by decay of ^{232}Th. Corals have been found to incorporate radium in approximately the same proportion to calcium as these two elements exist in sea water (Broecker, 1963), but they incorporate very little, if any, thorium. Excess ^{228}Ra (unsupported by ^{232}Th) decays with a half-life of 6.7 years and hence can be used for growth rate studies over a 30-year interval. Growth rates measured by this technique in several coral specimens from different locations were found to be consistent with growth rates based on alternating radial density bands revealed by x-ray radiography in the same coral specimens, and believed to represent annual growth increments (see below).

B. The ^{210}Pb Method

The isotope ^{210}Pb in the ^{238}U series is mainly generated in the atmosphere by decay of the noble gas ^{222}Rn and is introduced into the ocean by rain water, permitting geochronological studies in marine sediments over periods as long as a century (Rama *et al.*, 1961; Koide *et al.*, 1972).

In contrast to the ^{228}Ra method discussed above, attempts to determine coral growth rates on the basis of ^{210}Pb measurements have not always been successful. This may be due in part to the longer half-life of ^{210}Pb (22 years), which would require much larger coral specimens in order to define a logarithmic decay with depth in the coral accurately (Dodge and Thomson, 1974), but could also be caused by variable uptake of ^{210}Pb from sea water by the corals, especially if they are growing in turbid near shore waters (Moore and Krishnaswami, 1972, 1974).

C. The ^{90}Sr Method

The explosion of nuclear devices has provided an unexpected application to growth rate studies of corals. Aside from the sudden increase in ^{14}C which may be used as a recognizable "pulse" in the residual activity across the section of a given coral head, a number of other short-lived fission products such as ^{90}Sr (half-life 28.1 years) have been introduced into the ocean-atmosphere system. Like radium, strontium is incorporated into coral skeletons in proportions, which, compared with those of calcium, are roughly similar to the ratio of the two elements in sea water. Due to its dilution by natural strontium in the ocean, ^{90}Sr is generally not detected in sufficient concentrations for accurate measurements. In the vicinity of a former nuclear test site such as the Eniwetok lagoon, however, the sudden onset of ^{90}Sr activity is faithfully recorded by corals which were growing at the time of the explosion, and can be clearly detected by radiometric methods, as well as by autoradiography of coral sections (Knutson *et al.*, 1972).

The bomb-produced activity appears to be concentrated along definite contrasting-density bands, possibly growth bands, as revealed by x-ray radiography of the same coral sections. Since the times of nuclear testing at Eniwetok are known, the coral growth rates can be calculated. In many of the corals analyzed (Knutson *et al.*, 1972) the number of contrasting-density bands between successive radioactive bands coincided exactly with the number of years that had elapsed between successive nuclear tests at Eniwetok. Similarly, the total number of growth bands between the radioactive bands and the present coral surface matches the total number of years elapsed since the respective nuclear tests. These results provide strong support for the idea that the contrasting-density bands, as revealed by x-ray radiography, are in fact annual growth bands, a discovery which opens exciting possibilities for reconstructing microclimatic fluctuations in the coral reef environments of the geologic past (Knutson *et al.*, 1972; Buddemeier, 1974).

References

Barnes, J. W., Lang, E. J., and Potratz, H. A. (1956). *Science* **124**, 175.

Bender, M. (1973). *Geochim. Cosmochim. Acta* **37**, 1229.

Blanchard, R. L., Cheng, M. H., and Potratz, H. A. (1967). *J. Geophys. Res.* **70**, 4055.

Bloom, A. L., Broecker, W. S., Chappell, J., Matthews, R. K., and Mesolella, K. J. (1974). *Quat. Res.* (*N.Y.*) **4**, 185.

Broecker, W. S. (1963). *J. Geophys. Res.* **68**, 2817.

Broecker, W. S., and Kulp, J. L. (1956). *Am. Antiq.* **22**, 1.

Broecker, W. S., and Thurber, D. L. (1965). *Science* **149**, 58.

Broecker, W. S., and van Donk, J. (1970). *Rev. Geophys. Space Phys.* **8**, 169.

Broecker, W. S., Thurber, D. L., Goddard, J., Ku, T. L., Matthews, R. K., and Mesolella, K. J. (1968). *Science* **159**, 297.

Buddemeier, R. W. (1974). *Proc. Second Internat. Coral Reef Symposium* **2**, 259.

Chappell, J. (1973). *Quat. Res.* (*N.Y.*) **3**, 221.

Chappell, J. (1974). *Geol. Soc. Am. Bull.* **85**, 553.

Chappell, J., and Polach, H. A. (1972). *Quat. Res.* **2**, 244.

Chappell, J., and Veeh, H. H. (1977). *Geol. Soc. Am. Bull.* (in press).

Curray, J. R. (1965). *In* "The Quaternary of the United States" (W. H. E. Wright, Jr., and D. G. Frey, eds.), pp. 723–735. Princeton Univ. Press, Princeton, New Jersey.

Dansgaard, W., and Tauber, H. (1969). *Science* **166**, 499.

Dodge, R. E., and Thomson, J. (1974). *Earth Planet. Sci. Lett.* **23**, 313.

Emiliani, C. (1955). *J. Geol.* **63**, 538.

Emiliani, C. (1966). *J. Geol.* **74**, 109.

Fanale, F. P., and Kulp, J. L. (1962). *Econ. Geol.* **57**, 735.

Fanale, F. P., and Schaeffer, O. A. (1965). *Science* **149**, 312.

Flint, R. F. (1955). *Am. J. Sci.* **253**, 249.

Friedlander, G. Kennedy, J. W., and Miller, J. M. (1964). "Nuclear and Radiochemistry" (2nd edition). Wiley, New York.

Gramlich, J. W., Lewis, V. A., and Naughton, J. J. (1971). *Geol. Soc. Am. Bull.* **82**, 1399.

Kaufman, A., and Broecker, W. S. (1965). *J. Geophys. Res.* **70**, 4039.

Kaufman, A., Broecker, W. S., Ku, T. L., and Thurber, D. L. (1971). *Geochim. Cosmochim. Acta* **35**, 1155.

Knutson, D. W., Buddemeier, R. W., and Smith, S. V. (1972). *Science* **177**, 270.

Koide, M., Soutar, A., and Goldberg, E. D. (1972). *Earth Planet. Sci. Lett.* **14**, No. 3, 442.

Konishi, K., Schlanger, S. O., and Omura, A. (1970). *Mar. Geol.* **9**, 225.

Konishi, K., Omura, A., and Nakamichi, O. (1974). *Proc. Second. Internat. Coral Reef Symp.* **2**, 595.

Ku, T. L. (1968). *J. Geophys. Res.* **73**, 2271.

Ku, T. L. (1976). *Ann. Rev. Earth Planet. Sci.* **4**, 347–379.

Ku, T. L., Kimmel, M. A., Easton, W. H., and O'Neil, T. J. (1974). *Science* **183**, 959.

Kuenen, P. H. (1955). *Geol., Soc. Am. Spec. Pap.* **62**, 193.

Lalou, C., Labeyrie, J., and Delibrias, G. (1966). *C. R. Habd. Seances Acad. Sci.* **263**, 1946.

Land, L. S., MacKenzie, F. T., and Gould, S. J. (1967). *Geol. Soc. Am. Bull.* **78**, 993.

Libby, W. F. (1955). "Radiocarbon Dating," 2nd ed. Univ. of Chicago Press, Chicago, Illinois.

Mesolella, K. J., Matthews, R. K., Broecker, W. S., and Thurber, D. L. (1969). *J. Geol.* **77**, 250.

Milliman, J. D., and Emery, K. O. (1968). *Science* **162**, 1121.

Mitchell, A. H. G. (1968). *J. Geol.* **76**, 56.

Moore, W. S., and Krishnaswami, S. (1972). *Earth Planet. Sci. Lett.* **15**, 187.

Moore, W. S., and Krishnaswami, S. (1974). *Proc. Second Internat. Coral Reef Symposium* **2**, 269.

Olsson, I. U. (1970). "Radiocarbon Variations and Absolute Chronology," *12th Nobel Symp.* Wiley, New York.

Osmond, J. K., Carpenter, J. R., and Windom, H. L. (1965). *J. Geophys. Res.*, **70**, 1843.

Osmond, J. K., May, J. P., and Tanner, W. F. (1970). *J. Geophys. Res.* **75**, 469.

Rama Koide, M., and Goldberg, E. D. (1961). *Science* **134**, 98.

Rosholt, J. N. (1967). *Radioact. Dating Methods Low-Level Counting, Proc. Symp., 1967 IAEA Publ.* SM-87/50, pp. 299–311.

Rosholt, J. N., and Antal, P. S. (1962). *U.S., Geol. Surv., Prof. Pap.* **450-E**, (209) E108.

Sackett, W. M. (1958). Ph.D. Dissertation, Washington University, St. Louis, Missouri.

Sakanuoe, M., Konishi, K., and Komura, K. (1967). *Radioact. Dating Methods Lew-Level Counting, Proc. Symp., 1967 IAEA, Publ.* SM-87/28, 313–329.
Shackleton, N. J. (1967). *Nature (London)* **215**, 15.
Shackleton, N. J., and Opdyke, N. D. (1973). *Quat. Res.* **3**, 39.
Stearns, C. E., and Thurber, D. L. (1965). *Quaternaria* **7**, 29.
Stearns, H. T. (1966). "Geology of the State of Hawaii." Pacific Books, Palo Alto, California.
Szabo, B. J., and Rosholt, J. N. (1969). *J. Geophys. Res.* **74**, 3253.
Tatsumoto, M., and Goldberg, E. D. (1959). *Geochim. Cosmochim Acta* **17**, 201.
Thurber, D. L. (1962). *J. Geophys. Res.* **67**, 4518.
Thurber, D. L., Broecker, W. S., Potratz, H. A., and Blanchard, R. L. (1965). *Science* **149**, 55.
Valentine, J. W., and Veeh, H. H. (1969). *Geol. Soc. Am. Bull.* **80**, 1415.
Veeh, H. H. (1966). *J. Geophys. Res.* **71**, 3379.
Veeh, H. H., and Chappell, J. (1970). *Science* **167**, 862.
Veeh, H. H., and Giegengack, R. (1970). *Nature (London)* **226**, 155.
Veeh, H. H., and Valentine, J. W. (1967). *Geol. Soc. Am. Bull.* **78**, 547.
Veeh, H. H., and Veevers, J. J. (1970). *Nature (London)* **226**, 536.

7

SOME NOTES ON THE CORAL REEFS OF THE SOLOMON ISLANDS TOGETHER WITH A REFERENCE TO NEW GUINEA

O. A. Jones

The late F. W. Whitehouse, in his article on the Coral Reefs of the New Guinea area (Volume 1, Chapter 6), omitted reference to three very important papers on the reefs of the Solomon Islands (Stoddart, 1969a,b,c) and one relating to New Guinea (Veeh and Chappell, 1970), an omission which the editors failed to rectify. All these are certainly pertinent to Whitehouse's topic, so brief reference is made to them here.

The Solomon Islands area is of importance in relation to New Guinea, first, because, since Guppy's series of papers between 1884 and 1890, it has been regarded as a type area for theories of reef development that are opposed to Darwin's theory, and second because, like New Guinea, it is a mobile area showing evidence of uplift in the region as a whole, although it has areas with local subsidence and others with tilting, as well as eustatic movements of sea level.

As Stoddart has pointed out (1969a, p. 357), although recent deep drilling on atolls such as Bikini, Mururoa, and others has provided proof of Darwin's theory for open-ocean atolls, it does not follow that the theory is applicable to reefs in tectonically mobile belts (nor, indeed, to reef growth on relatively stable continental shelves).

In both these kind of areas one finds fringing reefs, barrier reefs, patch reefs, elevated reefs, banks of sediment probably concealing reef formations, elevated coral reefs, an atoll or two, and submerged barriers; in the New Georgia group of the Solomons there are elevated barriers, and double and even triple barriers, and at Ugi there are thin veneers of coral overlying deep-sea sediments.

To quote from Stoddart (1969a, p. 279), "uplift has not been uniform and many of the first-order reef features are clearly related to differential local movements." Examples which he cites include Gizo Island, near New Georgia, which has been tilted to the north, so that on the south side there is a steep coast with a fringing reef, but on the north a drowned coast with a barrier reef; the Russel Islands, which are also tilted to the north, displaying abrupt, reefless southern coasts and lower, drowned northern coasts rimmed with a barrier reef; the east coast of Malaita, which is submergent, with inlets and mangroves while the west coast is emergent with a raised barrier; and the reef complex of Northwest Santa Isabel, which is formed around the faulted and foundered end of the island. But "the great areas of reefs on the New Georgia group appear to be formed round volcanoes of different ages in a classical 'Darwinian manner.'" However, even in that group some local features call for a tectonic factor in addition to vertical movements; such is the double barrier around the Marovo lagoon, where the outer barrier is higher than the inner—difficult to reconcile with Darwin's theory without an additional tectonic factor. Grover (1958) wrote on "Rennell—the great uplifted atoll on the edge of the Coral Sea."

There are as yet no absolute dates for either reefs or volcanic flows in the Solomons. Stoddart (1969a, pp. 357–358) emphasized the difficulty of geomorphological interpretation because of the lack of absolute dates, little knowledge of the geomorphic effects of Pleistocene climatic changes in low latitudes, and eustatic fluctuations of the sea level with an amplitude of not less than 150 m during a period of 1–2×10^6 years, which may have been superimposed on a general Quaternary regression of about 200 m.

In this respect and this respect only, New Guinea is slightly better known, for in 1970 Veeh and Chappell, incidental to writing to lend support to Milankovitch's astronomical theory of climatic change, briefly described a complex flight of terraces rising to nearly 700 m on the northeast coast of the Huon Peninsula (on the north coast of New Guinea). The terraces include over 20 uplifted reefs (both fringing and ancient barrier reefs and lagoons) of a wide range of widths and thicknesses. They list 29 ^{230}Th and/or ^{14}C dates on unrecrystallized corals or *Tridacna* shells from these reefs and use the absolute ages with measured heights above sea level of the samples to establish a curve of eustatic sea level fluctuations for the last 2×10^5 years.

Furthermore, in 1963 Christiansen described the morphology of some coral cliffs in the Bismarck Archipelago, suggesting that they are indicators of the eustatic 1.5 m sea level.

Other problems discussed by Stoddart (1969a) include that of the forms created by erosion and solution on both steep and flat shores on elevated reefs and of the widespread mortality of modern corals growing close to low tide levels.

In a second paper, Stoddart (1969b) included descriptions of the exceptionally well-developed double (in part triple) barrier and in a third paper, (Stoddart, 1969c), descriptions are given of the previously undescribed sand cays of Marovo Sound. These almost unique, very youthful single cays were developed on relatively old topographic reef features and, because of the restricted range of features developed, they do not fit easily into earlier proposed schemes of cay morphology.

Following the general study by members of the Royal Society Expedition, 1965, Morton (1974) made a detailed study of the ecology of eight groups of coral and associated reef building organisms. This is an important, essentially biological, paper.

References

Christiansen, S. (1963). *Geogr. Tidsskr.* 62, 1–23.
Grover, J. C. (1958). *Geol. Surv. Br. Solomon Isl.* 2, 134–139.
Morton, J. (1974). *Proc. Int. Symp. Coral Reefs, 2nd, 1973* Vol. 2, pp. 31–53.
Stoddart, D. R. (1969a). *Philos. Trans. R. Soc. London, Ser. B* 255, 355–382.
Stoddart, D. R. (1969b). *Philos. Trans. R. Soc. London, Ser. B* 255, 383–402.
Stoddart, D. R. (1969c). *Philos. Trans. R. Soc. London, Ser. B* 255, 403–432.
Veeh, H. H., and Chappell, J. (1970). *Science* 167, 862–865.

8

THE GREAT BARRIER REEFS PROVINCE, AUSTRALIA

O. A. Jones

PART I. A SUMMARY OF THE GEOLOGY AND HYDROLOGY OF THE QUEENSLAND COAST AND THE CONTINENTAL SHELF

I. The Regional Setting

To the north, the Great Barrier Reef Province[1] is bounded by the Gulf of Papua and to the east and west of that Gulf by Papua itself. To the east of the Province lies, from south to north, Hervey Bay and Fraser Island with its northerly extension Breaksea Spit, the Capricorn Channel (or Southern Shelf Embayment), the Coral Sea Platform (or Plateau), the Queensland Trench[2] and the Papuan Trough, the last four being western elements of the Coral Sea. To the west of the Province lies the land mass of Queensland; Queensland east of the Main Divide should, for structural and historical reasons, be included as part of the Province.

Defined in this way, the Province has a length, from Bramble Cay (9°13′S) to Lady Musgrave Island (23°50′S), of about 1600 km; but the main aggregation of reefs itself (the Great Barrier Reef or Reefs), not a straight line, is approximately 2000 km in length. The shelf, from which the reefs arise, varies in width from 36 km at Cape Melville (14°10′S) to about 412 km near Broad Sound (22°S).

However, the outer barrier, which is the most easterly fringe of reefs and fronts the Pacific Ocean, is set back from the edge of the shelf by up to 140 km in the north and up to 170 km south of Mackay. Only between about 10°40′S and 17°S does it follow closely the edge of the shelf. The least distance between the outer edge of the reef system and the coast is 28 km at Cape Melville and the greatest about 260 km at Cape Clinton (approximately 22°26′S). The area of the Province, i.e., the area between the outer reefs and the coast, is about 215,000 km².

[1] The term "Province" is used in a geographical sense. Zoologists divide the area into Faunal Provinces.

[2] Chapter 11 "The Basement beneath the Queensland Continental Shelf" by A. R. Lloyd should be read in conjunction with this; it includes deductions from geophysical work not available when this was written.

Fig. 1. Map of the Great Barrier Reefs Province, showing the main reefs only, the sedimentary basins, and the three main regions.

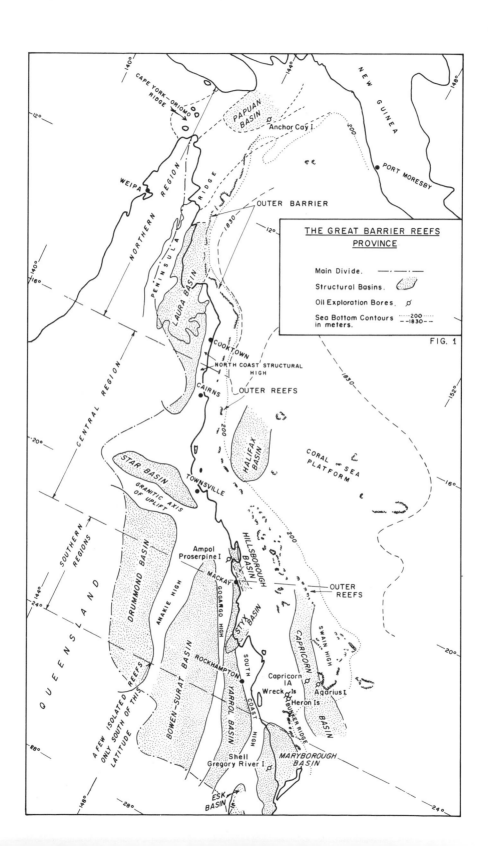

FIG. 1

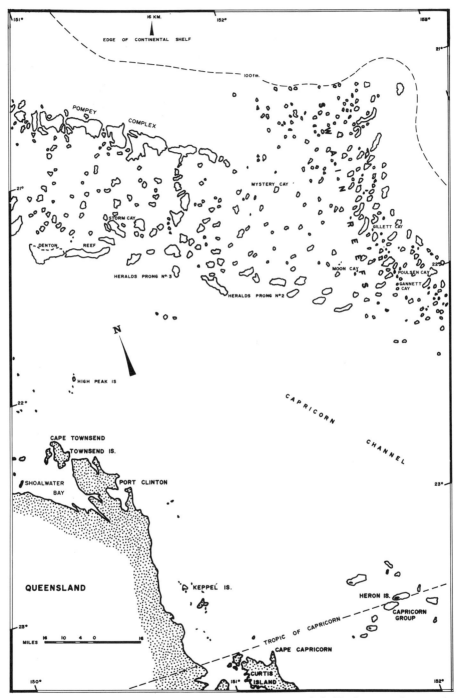

Fig. 2. The southern end, set far in from the continental shelf edge. Broad channels and closely spaced reefs.

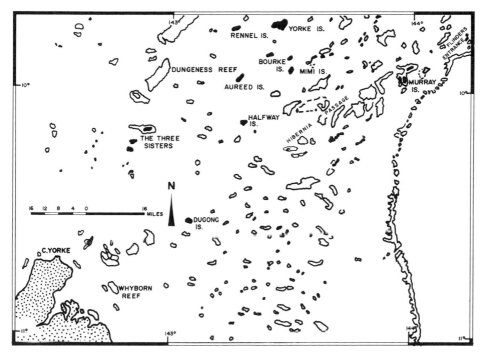

Fig. 3. The barrierlike character of the northern end.

The Main Divide (the so-called "Great Dividing Range," see Fig. 1) is a significant feature in the present context, for all rivers flowing east from it carry sediment onto the Continental Shelf. Furthermore, it would appear probable that the geological features, formations, and structures, etc., of the coastal strip (the area between the Divide and the coast) would continue onto the Continental Shelf. The Divide, almost coincident with the coast line north of Cairns, reaches its maximum distance inland, 480 km, not far south of the tropic. Thus, although only about a third of Queensland lies east of the Divide, there is still ample area to provide heavy loads of sediment for the rivers during the tens of millions of years that it has been subject to erosion.

Structurally the area east of the Divide falls into three divisions—structural (not necessarily topographic) highs consisting mainly of Palaeozoic metamorphic rocks; basins (not necessarily topographic lows) containing Middle and Upper Palaeozoic greywacke, shale, chert, and limestone, Triassic and Jurassic sandstones and shales, and Cretaceous shales; and, confined to the hinterland in the north, a complex of Palaeozoic igneous and possibly Precambrian metamorphic rocks. Igne-

sediments and coral rock) would be, for the most part, seaward extensions of the rocks of coast and nearby hinterland. This expectation is supported too, by the nature of the many continental islands bordering the coast.

These islands are, as the name implies, built of continental rocks—granitic varieties, tuffaceous volcanics, metamorphics and, rarely, sandstones and related sedimentaries. Their composition suggests a continuation of the land mass beneath the shelf, but they are, with few exceptions, such as Lizard Island, situated relatively close to the continent, so again their evidence must not be overweighted.

The small total of this visible evidence emphasizes the importance of bottom sampling, boring, and especially of deep boring and geophysical surveys.

Only four bores have been sunk on the shelf to basement[3] rock—one in Anchor Cay in the far north (120 miles northeast of Cape York) and three in the south, one on Wreck Island in the Capricorn Group and, nearby, one on either side of the Capricorn Channel (Capricorn No. 1A and Aquarius No. 1). Two of lesser depth were sunk by the Great Barrier Reef Committee, one on Michaelmas Cay in the north to 183 m and one on Heron Island in the south to 223.26 m. The data from these bores will be discussed below.

2. Geophysical Evidence

Incomplete though the geophysical survey (aeromagnetic, gravimetric, and seismic) of the shelf is, it has helped to define "highs" of older rocks, "basins" of younger rocks, and areas of relatively thick sedimentary sequences overlying basement rocks. These, together with the geophysical surveys, indicate the pattern of Highs and Basins as shown on Fig. 1.

The shelf, then, is essentially an easterly, submarine extension of the Eastern Highlands region of Queensland. But, when examined in greater detail, some important differences emerge.

Aeromagnetic and seismic surveys have given some indication of the thickness of sedimentary rocks overlying the "basement" in the Reef Province (see Ericson's submission to the Royal Commission, Ericson, 1972). As is apparent in Ericson's figure (Royal Commission exhibit No. 123), south of lattitude 17°S the thickness is less than 3000 ft (915 m) over the greater part of the area, and sediments in excess of 5000 ft

---

[3] The term "basement" is used in the sense of economic basement in respect of possible oil occurrence. It includes, in addition to igneous and metamorphic rocks, highly indurated, but little-metamorphosed strata.

(1524 m) are limited to two small areas northeast of Townsville and to a much larger area east of Rockhampton and Gladstone. North of latitude 17°S sedimentary strata may be in excess of 5000 ft (1500 m) in the area of Princess Charlotte Bay, and northeast of Cape York the thickness is more than 7000 ft (2134 m).

3. *Highs and Basins; Deep Bores*

The limited areas over which the basement is deep, or moderately deep, are with one exception directly related to onshore sedimentary basins of Mesozoic age and hence may be taken to define seaward extensions of these basins.

Geophysical work has shown that the Maryborough Basin is bounded to the east by the Bunker Ridge (or High), a northwesterly extension of Fraser Island continuing to Lady Elliot Island and the Bunker and Capricorn Groups. The Wreck Island (Capricorn Group) well penetrated this ridge at 1795 feet. Magnetic and seismic surveys suggest a second high farther still to the northeast under the Swain Reefs—the Swains High. Between the two ridges is the Capricorn Basin (or "Southern Shelf Embayment"). This Basin has been penetrated by two wells, Capricorn No. 1A in the west and Aquarius No. 1 towards its eastern margin. These bores proved about 30 m of Recent and Pleistocene beds and 1400 m of Tertiary (of which all but the lowermost 160 m are nonmarine) unconformably overlying almost 950 m of conglomerates and red beds, probably Cretaceous in age, and in turn resting unconformably on the basement of possibly Palaeozoic metamorphics.

The Styx Basin is small with about 400 m of freshwater Cretaceous sediments onshore. A marine seismic survey indicated that offshore it is restricted to a part of Broad Sound and contains not more than 460 m of sediments.

Farther north again is the Hillsborough Basin, also small, but with 1280 m of freshwater sediments and volcanics onshore. Seismic evidence has shown that offshore, in Repulse Bay, there is a thickness of 3000–4500 m of probably Teritary sediments.

One aeromagnetic survey has suggested the existence of the deep Halifax Basin. Apparently of considerable extent, most of it is seaward of the continental shelf on which however, it impinges to a width of 30 to 40 km between the Palm and Grafton Passages.

West of Cooktown the Laura Basin onshore contains about 380 m of marine lower Cretaceous rocks overlying freshwater Jurassic sandstones. Marine seismic surveys show that the basin extends northeastward under Princess Charlotte Bay and beyond, with an area of some 10,000 km².

In the far north the Papuan Basin appears to extend into the northern Great Barrier Reef area (Peninsula Trough of Woods). Marine seismic surveys show that throughout most of the area the thickness of Cainozoic sediments is less than 180 m, but that there is a sharp increase in thickness to the northeast. In this northeast area Anchor Cay No. 1 bore, of 3625 m depth, revealed some 1372 m of Upper Jurassic and lower Cretaceous marine strata overlaid unconformably by nearly 1830 m of marine strata ranging in age from Eocene to Pliocene. This is by far the thickest and most complete marine section yet found beneath the shelf. The Anchor Cay well established the presence of a Pliocene coral reef in this northern part of the Province. The only other indication of "pre-Great Barrier Reef" coral reef development is in the Miocene of the Wreck Island bore.

Lloyd (see Volume I, Chapter 11, p. 365) summed up as follows: "Conditions similar to those existing today along the inner parts of the Great Barrier Reefs first came into existence along the northeastern Australian shelf during the early middle Miocene. There was a break in these conditions along most of the shelf during the late middle and upper Miocene . . . In the lower Pliocene, conditions similar to the area of the outer Great Barrier Reefs came into existence and continued to the present time, possibly, however, with interruptions and even erosion at times, in the Pleistocene."

4. Eustatic Changes of Sea Level

The gentle eastward deepening of the Queensland continental shelf to the continental slope is interrupted not only by the innumerable reefs, but also by features which have been interpreted as due to changes of sea level and in some cases as still stands. Maxwell's (1968, 1969a,d) careful study of the bathymetric charts, supplemented by some 40 traverses with echo sounding apparatus, indicated features at depths of 18.3, 29.3, 36.6, and 58.5 m (10, 16, 20, and 32 fm) and, less certainly, others at 65.9, 87.8, and 102.5 m (36, 48, and 56 fm). In addition, there is the bench recognized by many workers on numerous islands and reefs, as well as at many points along the coast at 1.8–3 m (6–10 ft) above present sea level.

There is difference of opinion regarding Holocene fluctuations of sea level, particularly a high level still stand, and the interpretation of radiocarbon dates. Stoddart (1971, pp. 25 and 31) summarized the pertinent radiocarbon evidence and concluded that it is against a still stand, in which case the benches would have to be explained in terms of local tectonic events, or else as survivals of a feature or features formed in pre-Holocene time.

Maxwell (Volume I, pp. 251–264) reexamined the evidence and came to conclusions supporting fluctuations and still stands.

The —29-m surface is deeply dissected (suggesting subaerial erosion) and the —36-m surface is widespread and persistent. These features were interpreted by Maxwell as summarized below and correlated with fluctuations of the sea level during the last 18,000 years.

He determined, based on evidence of the age of the —119–128-m terrace at the edge of the shelf off New South Wales and in the Gulf of Carpentaria, that the Queensland shelf may have been cut some 24,000 years B.P., during the Würm emergence. From that depth the sea level rose successively to —102 m (not 92 m as shown on Maxwell's fig. 6, p. 262) at 18,000 years B.P; to —88 m at 15,100, —66 m at 13,400; —58 m at 13,000; —37 m at 11,400; —29 m at 10,900; fell to —36 m at 10,300; rose again to —18 m at 9,250; and, after a minor rise, fell again to —18 m at 7250 B.P., and then rose to above the present sea level, with the most recent movement being a fall of about 3 m.

However, Hopley (1974) identified three distinct Holocene shorelines. The oldest, with radiogenic ages of between 6000 and 5000 B.P., consists of terrigenous material (boulder beaches), reworked by a transgressive sea which appears to have reached its maximum between 4500 and 4000 B.P., leaving shorelines of raised reefs, peat, shell beds, and beach rock. More recent shore lines, with radiogenic ages of between 3800 and 3200 B.P., are also biogenic. All these shore lines display strong warping, but earth movements seem to have ceased about 2500 B.P.

Hopley concludes that "the hypothesis of present sea level being reached between 4500 and 4000 years B.P. and not fluctuating to any marked degree since appears to fit the evidence best."

It is difficult to separate the eustatic from the tectonic factors but the latter appear to have been significant.

By no means has the last word been said (or written) on this subject and, in particular, publication of dates obtained by the 1973 Stoddart Expedition to the northern section of the reef is awaited with interest.

Maxwell envisaged that reef growth continued during the fluctuations, with some reefs being killed during periods of regression of the sea and with recolonization taking place with each advance of the sea. Evidence in support of this thesis is to be seen in the thick veneer of dead reef material and sediments observed on many of the Near Shore and Inner Shelf reefs.

A corollary of this interpretation is that no post-Tertiary cross warping of the shelf took place, but, it does not preclude Recent faulting in central and northern coastal regions to give the typical drowned coast-line appearance and the corridors of those areas (for an analysis of the type of tectonics involved, see Heidecker, 1973, Volume I, Chapter 9). Maxwell (1968) has suggested that the warping which produced the hypothetical Longreach, Forsayth and Arnhem Swells did not finally cease until early Tertiary times. Such movements, if they did occur, are of importance, for when the shelf was elevated above the sea, the swells helped to channel the drainage from the mouths of present-day rivers through the now submarine valleys in the shelf to some of the deeper passes in the outer barrier. Saville-Kent (1893, p. 132) noted that the mouths of many of the large present-day rivers are opposite passes in the Reef. Others (Richards and Hedley, 1923; Jardine, 1925a,c; Bryan, 1928a, Steers, 1929; and Fairbridge, 1950b) looked for confirmation of Saville-Kent's suggestion, but found the available data insufficient. However, Maxwell's (1968, pp. 51–58) bathymetric data, additional to that on the Admiralty charts, does lend a large measure of support to the idea, although the picture is complicated and to some extent obscured by the later growth of reefs and deposition of sediment, and it would be easy to carry deductions too far.

An additional important feature of the swells is that they modified the longitudinal highs which, in Maxwell's view, are the foundations for the reef zones. They also modified the basins, the source of sediments, accumulations of which also provide foundations for reefs in some areas.

Maxwell's scheme of development of the Great Barrier Reefs does not require slow, long-continuing subsidence; tectonic adjustments and eustatic changes of sea-level are, in his view, sufficient to allow growth of the thicknesses observed and deduced. He maintains also that the main reef developments rest upon crustal belts which have *risen rather than subsided*.

Heidecker (Volume I, Chapter 9) expresses a different view. "Subsidence outstripped reef development on certain of the segments, e.g., the Coral Sea Plateau. Along the inner shelf, steady foundering at an optimum rate provided an ideal tectonic setting for growth of the Great Barrier Reef."

However, Whitehouse (Volume I, Chapter 6, p. 184) has written "not only, as Darwin postulated, are the great coral reefs of the world formed upon subsiding foundations, but also that there is a favourable rate of subsidence, and that such reefs are rarely formed if the rate of vertical movement is above or below that favourable value."

Thus one of the major questions regarding the Great Barrier Reefs—was slow, long-continued subsidence involved in its formation—is still not unequivocally answered.

B. The Present Topography of the Shelf

1. *Latitudinal Subdivisions*

The surface of the shelf is far from being a plane, showing, in fact, a surprising degree of relief.

Maxwell (1968) has studied its bathymetry from existing charts and although some of these, especially in areas towards the outer reefs, are based on old and sometimes very old observations and contain many inaccuracies, the picture he has presented must be regarded as accurate in general though perhaps erroneous in some details.

The shelf deepens progressively from north to south and it may be usefully divided into a north, a central, and a southern region (see Figs. 1, 2, 3, 4). In the northern region, north of latitude 16°S, depths are, with few exceptions, 30 m or less. The shelf, as narrow as 33 km in the south of the region, widens to the north to as much as 160 km off Cape York and further north still to about 350 km. From lat. 16°S to 10°30′S reef development is highly distinctive, rising on the edge of the shelf as a narrow, almost continuous belt of linear reefs with recurved ends, (variously termed ribbon, linear, or cuspate reefs. West (leeward) of these are large, but dispersed, reefs and abundant near-shore reefs (including fringing reefs). North of lat. 10°30′S the reef is set back a distance of up to 60 km but the linear character continues northward to about 9°50′S.

The central region, from latitude 16°S to just south of latitude 21°S, ranges in depth from 9–16 m over a shelf width of about 40–290 km. From 16°–19°S the reefs are increasingly sparsely distributed but from there on are rather more numerous as far south as 20°S. From that latitude south to 21°10′S the outer reefs are densely packed and again take the form of a barrier. They are, however, set well back from the 100-fm (180-m) line and the predominant reef forms are not the ribbon reefs of the north, but patch and platform reefs, many of the latter with lagoons.

In the southern region, with depths ranging from 64–128m, the shelf varies in width from about 200–390 km and exhibits large, rather dispersed areas of reefs. Thus, looking at the reef system as a whole, one can see remarkable changes in its nature from north to south. North of latitude 16°S the shelf edge reef system is frequently termed the "outer barrier," and it does form a barrier of reefs, though possibly not a barrier reef in the Darwinian sense; it protects a reef studded "lagoon." South of 16°S ribbon reefs are few, the outer reef zone retreats from the shelf edge to a distance of more than 100 km, and except between latitudes 20° and 21°S, the barrierlike character has disappeared. At about latitude 22.5°S the reef system is interrupted by the Capricorn Channel (or "Southern Shelf Embayment"), but is found again to the west as the Capricorn and Bunker Groups of coral cays and reefs, a mere 80 km from the coast. This is the southern end of the Great Barrier Reefs system.

2. *Longitudinal Zones: Types of Sediments*

The division into longitudinal zones, subparallel to the coast line, is based on the reefs themselves, their types and abundance, and on the several kinds of sediment being deposited.

In the near-shore zone (down to a depth of 9 m) reefs are relatively few and are mainly fringing reefs around rocky headlands, near river mouths, and bordering long, straight stretches of coast adjacent to relatively deep water. The sediments of this zone are terrigenous—beaches of fine quartz sand, coarse quartz feldspar sand or gravel, and, on tidal flats, mud and sand.

The inner self zone lines between the 9- and 36-m depth contours, forming a belt about 37 km wide with little relief, but with a rather uniform slope. Reefs are relatively few and the sediments being deposited are terrigenous sands and muds with localized patches of carbonate. This zone with its comparatively few reefs corresponds approximately with the "inner" or "steamer channel" of some earlier writers.

The outer (or marginal) shelf is the zone between the inner shelf and the shelf edge, between the 36- and 90-m depth contours. This is the zone of greatest reef growth and of maximum concentration of reefs with fine, muddy, carbonate sands between the reefs. On the reef surfaces carbonate gravels and coarse sands are dominant on the reef flats, finer sands and silts are typical of the lagoons, and boulder accumulations and shingle banks may be found on reef crests and rims.

Maxwell (1968, p. 106) drew attention to the heavy detrital blankets of coral debris, in some cases cemented to form raised platforms, which occur on many of the Inner Shelf and Near Shore reefs in the northern and north central regions. He suggested that such features are the result of erosion of reefs which had grown some 2–3 m higher during the time of highest sea level. He further suggested that in these regions the outer reefs are sufficiently close to the coast to provide protection from the heavy seas of the open ocean, so that much of the debris is preserved on the reef flats.

These suggestions raise two questions—one concerning the rate at which erosion of coral reefs proceeds and the other concerning the length of time reefs may have been exposed above sea level during the Holocene.

Looking first at the possible length of time of exposure during the

Holocene, Fairbridge has suggested multiple fluctuations above and below the present sea level but, even if one accepts this, the sum total of exposure of reefs since the first Holocene high stand almost 6000 years ago would be little more than 3000 years.

Measurements by Stoddart (1969) of subaerial and intertidal weathering and corrosion processes gave a figure of 1 or 2 mm per year for intertidal erosion, and he suggested that rates of surface solution by rainwater are probably two or three orders of magnitude lower.

Stoddart also cited elevated reefs with a radiometric age of 40,000 years or more which still stand 2 m or more above present sea level; they survived the last major low stand of the sea with only minor topographic changes.

Unless rates of erosion in the Great Barrier Reef Province were very much greater, it would seem that there was insufficient time of exposure between the upward fluctuations of the sea for the described features to be formed.

Maxwell (1968, pp. 106 and 107) also described "corroded reefs"— "embayed corroded relicts or clusters rising from large submerged platforms," the platforms being the planed surfaces of earlier reefs. Again there would seem to have been insufficient time between the hypothetical high sea levels of the last 6000 years for the production of the submerged platforms.

The alternatives would appear to be either that there were local tectonic adjustments producing submergence and emergence in addition to and preceding eustatic fluctuations within the Holocene, or else that the structure of the corroded reefs is a karst structure inherited from the Pleistocene when erosion periods were significantly longer.

C. The Waters of the Shelf

The two major surface currents which affect the waters of the Queensland Continental Shelf (the Great Barrier Reef Province) are the swift-flowing South Equatorial Current in the north, and, farther south, the broad, westward-flowing Trade Wind Drift which divides and diverges in the Coral Sea, a northern stream flowing into the Solomon Sea and washing the north Queensland shelf, and a southern stream forming the strong, but narrow East Australian Current. Maxwell (1968, Figs. 4 and 5, pp. 10 and 13) reproduced data from Wyrtki (1960) showing a southerly flow of surface water over the whole of the Queensland shelf in January, and in July a divergence to the north and to the south at about latitude 16°S.

Rochford (1959, quoted by Maxwell, 1969d, p. 357 and Fig. 63; see also Maxwell, 1968, Fig. 6, p. 13) showed that four water masses make contributions to the waters of the Great Barrier Reef Province. One of these, the Coral Sea mass, spreads under a second and a third mass in the West Central South Pacific, and has been identified at the surface only in the northwest of the Province. The fourth mass is that of the Gulf of Carpentaria and the Arafura Sea.

Brandon (Volume I, Chapter 7) set out the following generalizations: (1) isothermal and isohaline conditions prevail over the whole Shelf, except during the rainy season; (2) the southern shelf shows the smallest seasonal variation in salinity and the largest in temperature; (3) the northern Shelf shows clearly defined temperature and salinity gradients across the shelf during spring and summer; (4) the seasonal variation in temperature is smallest in the north; (5) the central region is most complex owing to high rainfall and great runoff, but the northern and the southern parts of that region have anomalously low rainfall—the high rainfall reduces the surface salinity during the wet season, but the waters of the whole area are isohaline and isothermal while the southeast winds are blowing; (6) around the north of Cape York and in the Torres Strait conditions are influenced by large quantities of water from the Fly River (Papua) and by West Irian water. There is a sharp boundary between the lower-salinity water to the north and the higher-salinity water to the south. This boundary migrates southward during the northwest monsoon.

These conclusions are based mainly on studies carried out during one year. There must of course be many temporary variations during abnormal seasons and there is a need for the collection of such data on a continuing basis.

The most important application of hydrological data is perhaps in the zoological field since variation in hydrological conditions affects the nature and rate of coral growth, and these in turn affect the shape of reefs, the size, nature, and position of coral islands, and ultimately the geological history of the reefs.

III. Types of Reefs

Recent workers in this field have drawn a sharp distinction between oceanic coral reefs (the Darwinian fringing and barrier reefs and atolls) and coral reefs of the continental shelves, the latter category, of course, including the Great Barrier Reefs.

The forms of the reefs of the Great Barrier have been most closely studied by Maxwell (1968), who, in addition to fringing reefs which border the coast in a few places, mostly around rocky headlands, and surround many of the Continental Islands, distinguished some 13 types which he suggests (1968, pp. 98–107, Fig. 72; 1969d, p. 364, Fig. 66) developed from embryonic colonies according to the scheme as shown in the figures quoted.

Maxwell and Swinchatt (1970, pp. 700–706, Fig. 6) set down arguments in support of evolutionary relationships of the 13 types. These arguments are summed up in the caption to their Fig. 6—"Reef shape is controlled by differential growth in response to bathymetric and hydrologic influences. Where these are uniform around the reef, symmetrical growth results. Where favorable conditions are restricted to one side of the reef, elongation, cuspation, and closing result. Unfavorable conditions cause 'resorbtion'." More detail is given in the text, but comments by Stoddart (1969a, p. 453) and Ladd (see Chapter 1) do not give unqualified approval, suggesting, as indeed I think is the case, that much more detailed study is required before such a scheme can be regarded as established.

IV. Reef Structure

A number of workers have described the structure of various reefs, but it has remained for Maxwell to summarize and generalize the structure into five types (1968, pp. 98–107, Fig. 72; 1969a, pp. 11–13, Fig. on p. 12). From the windward side the *off-reef floor*, with a depth of 29–58 m, becomes more irregular near the reef and marginal ridges rising 13–18 m may occur. The *reef slope*, with flourishing coral, rises steeply (average gradient about 30°) to the *reef surface;* one or more terraces (19, 13, and 3.6 m) may be seen interrupting the slope. The upper part of the reef slope is the *reef front* with a *spur-and-groove system.* The rising reef front culminates in the *reef rim* or *algal rim*, the highest part of the reef surface, a height not infrequently enhanced by large dark-colored masses of dead coral and/or coral shingle. From the rim one descends slightly to the main *coral zone*, a dense mass of living coral with many "coral pools," one of the most beautiful features of coral reefs. The living coral grades first into predominantly dead coral and then into the *sand flat.*

The *sand flat* may pass to leeward into a repetition, in reverse order, of the zones already listed; it may grade into a shallow lagoon or fall sharply into a deep lagoon (ring and mesh reefs); or, on open ring and composite apron reefs, it is succeeded by a back-reef apron and back-reef shoals.

The algal rim or algal ridge is a feature of particular interest. According to Maxwell it is present on the seaward margins, both windward and leeward, of all closed reefs except in the Low Wooded Islands. Reefs which are open to leeward (open ring and composite apron reefs) develop an algal rim on the windward side only. But some closed ring reefs and mesh reefs have a second algal rim forming the crest of the lagoon slope (see Maxwell, 1968, p. 106, Fig. 72), a feature which appears to be unique to this area.

Maxwell studied a wealth of aerial photographs of reefs and visited and studied a large number. Nevertheless, it is possible that he may have overgeneralized the position and that this feature is one which requires much more study and documentation, especially in the far north of the Province, the area affected by the northwest monsoon.

Stoddart (1969a, p. 455) drew a fundamental distinction between reefs with an algal ridge and those without. He pointed out that it is best developed on reefs which lie in the tradewind belts of the Pacific Ocean. It is absent as a topographic feature from reefs in monsoonal and doldrum areas in Indonesia, New Caledonia, New Hebrides, the Solomon Islands, and atolls in the Gilbert Islands.

In the southern two thirds of the Great Barrier Reef Province the southeast trades are dominant for the whole of the year; in the northern third the trades are dominant from about mid-March to mid-December and the northwest monsoon for the other three months.

In general, then, one would expect the development of an algal ridge because of the dominance of the southeast over the whole area for most of the year. But the northernmost third, which experiences relief from the trades and is affected by the northwest monsoon for three months of the year, exhibits some interesting features. It is in precisely this part that the outer, the shelf-edge reefs, assume the character of an almost continuous barrier—a line of linear, ribbon, wall, and cuspate reefs running for hundreds of kilometers with only narrow breaks between individual reefs. Because of the extreme difficulty of access, and because weather which permits landing upon them is rare, these reefs are not well known, but such descriptions as do exist (as, for example, of Yonge Reef) indicate the presence of a very well-developed algal rim; its growth is not inhibited by the brief yearly visit of the northwest monsoon.

Another distinctive feature of the Northern Region is the development, in the "steamer channel" between latitudes 14° and 16°S, of the variously designated Low Wooded Island Islands (Steers, 1929, 1937), Island Reefs (Spender, 1930), Moat Islands (Stoddart, 1965) or High Reefs of the Inner Shelf (Maxwell, 1968). This distinctive type of reef was first recognized and described by Steers from this area. It has since been found to occur, though not commonly in some other parts of the world.

Low Isles, the headquarters of the Yonge Expedition, is an island (reef) of this type, though not completely typical. Naturally it was described in detail by members of that Expedition and has been surveyed in detail twice since then (Fairbridge and Teichert, 1947, 1948). In it the place of the algal rim or ridge is taken by a system of coral shingle ramparts, one, or in some cases two, with a "moat" between. Other features include mangroves, perhaps consisting of only a few stunted trees, but ranging to a well-developed mangrove swamp, and a vegetated sand cay; there may also be a coral "boulder tract" on the leeward side (see Isobel Bennett, 1971, p. 42, for an excellent diagram and illustrations).

In an important paper Stoddart (1965) concluded that these islands form under rather special conditions (and hence conditions which are likely to be local)—protection by barriers (reef or islands) to windward, but with sufficient fetch and water depth for moderate wave energy to be generated. An additional requirement is a moderate growth of mainly fragile coral species to yield an abundant supply of coral shingle. The energy needs to be sufficient to lift the shingle onto the reef flat (which is at a higher level than the Shelf Edge Reefs), but not to transport it across the reef flat. The shingle accumulates to form ridges on the reef crest and sand produced by abrasion accumulates by refraction, the locus of accumulation depending on the size and shape of the flat and the depth of water upon it.

V. Types of Islands

The Continental or High Islands fall in a number of fairly well-defined groups, with only a few scattered examples between the groups. The main groups comprise the Keppel Islands, 23°15′S; the Northumberland Isles, 22°S; the Cumberland Isles (including Whitsunday Island), 20°–21°S; the Palm Isles, 18°40′S; the Family Group, 18°S; the Barnard Islands, 17°42′S; the Frankland Islands, 17°12′S; the Howick Group, 14°30′S; the Flinders Group, 14°10′S; and the islands of the Torres Strait. To these may be added Magnetic Island and Hinchinbrook Island, which are the largest and highest (Mt. Bowen, 1113 m). Peaks exceeding 270 m are not uncommon. All these are much closer to the mainland than to the outer barrier. The furthest from the coast are islands in the Great North East Channel, running northeast from Cape York. Of these, Bramble Cay is about 190 km distant, Darnley Island, 180 km, and the Murray Islands, in latitude 10°S, about 185 km distant; all are just within the outer barrier. Isolated Lizard Island, 14°40′S, is nearer the outer barrier, 19 km, than to the coast, 27 km.

Most of the Continental Islands are of granitic or metamorphic rocks, though there is an impressive series of ignimbrites cropping out on many of the islands in Torres Strait and a few, such as Albany off the tip of Cape York, are capped with Mesozoic lacustrine sediments. The Murray Islands, Bramble Cay, and Darnley Island are also exceptional in being composed of andesitic lavas and ash beds. At the Murray Islands there is evidence of the interplay of Recent uplift and Recent eruption, masses of coral reef rock from elevated reefs being included in the lavas.

In contrast to the "Continental Islands" are the "true coral islands" or "low islands" formed almost solely from fragments of the skeletons of coral and associated organisms which make up the reefs—"coral sand." Several classifications have been set up for such islands in the Great Barrier Reefs Province, notably by Steers (1929, 1937), Spender (1930), and Fairbridge (1950b).

Stoddart and Steers (Chapter 3, this volume), writing on the nature and origin of such islands in general, recognize seven types, five of which are known within the area of The Great Barrier Reefs.

A. SAND CAY

The first of these, the sand cay, may be either unvegetated or vegetated. The former are commonly unstable and migrate seasonally and are found in all areas and on all types of reefs except fringing reefs. The vegetated sand cays are moderately stabilized, generally include beach rock, and are widely distributed, although seldom found on the outer, more exposed reefs.

B. SAND CAY WITH SHINGLE RIDGES (MOTUS)

The sand cay with shingle ridges are generally vegetated. The windward shingle ridge is usually of coarser material than that of the leeward ridge. Between the two ridges the surfaces may fall gently or there may be an interior enclosed depression with standing water. One Tree Island in the Capricorn Group, described by Steers (1937) and described and figured by Steers (1938) appears to fit into this category.

C. SHINGLE CAY: UNVEGETATED OR VEGETATED

This type is moderately stabilized, widely distributed, and generally found on smaller, more exposed reefs. It is not common and its stability probably depends on vegetation obtaining a hold. Shingle cays usually occur on exposed reefs or on the exposed side of reefs; they may be the first stage in the formation of a motu.

D. Low Wooded Islands

Low wooded islands, (Steers, 1929), have also been called *island reefs* by Spender (1930), *low wooded island reefs* by Fairbridge and Teichert (1947), and *moat islands* by Stoddart (1965). An essential feature of the low wooded island is that the platform reef bears on its surface: (1) a shingle ridge or rampart close to the windward edge; (2) a sand cay on the leeward side, in some cases joined to the shingle rampart; and (3) an area between the cay and the rampart of open water which may or may not be colonized by mangroves.

This type of island was first described by Steers (1929) in the Great Barrier Reefs. Low Isles (16°23'S), the headquarters of the Yonge expedition of 1928–1929, is the most southerly; it and Three Isles (15°17'S) were mapped in detail by Spender (1930). Later, Steers (1937, 1938) mapped a further 15 and proposed Bewick Island (14°26'S) as the type example. Maxwell (1968, pp. 101 and 106, Figs. 66 and 72) referred to them as "High reefs of the inner shelf (Northern Region)." These are reefs (with one or more islands) rather than islands and their structure and the conditions under which they form have been described briefly above, pp. 222–223.

E. Emerged Reef-Limestone Islands

A fringe of recent sand, or shingle beach ridges, or ramparts may be present in this type of island. It is not clear to what extent this type occurs in the Great Barrier Reefs province. Fairbridge (1950b) cited only Raine Island and "a reef some 12 miles southwest of Raine Island." Stoddart and Steers (this volume, Chapter 3) say only that it is represented by islands of the Great Barrier Reefs.

In discussing the origin of these various types, Stoddart and Steers emphasize that they are "equilibrium forms in adjustment with controlling processes in the reef environment at any given location in space, and time, and that they do not form stages in any evolutionary sequence, nor are they inherited from environmental conditions of the past."

Stoddart and Steers's discussion of the origin of the various types suggests that a vital factor in the formation of sand cays is a change in the smooth prolongation of the geometry of the reef such that refraction of waves takes place in such a way that two wave trains meet and deposit their loads of reef-derived sediment, usually near the leeward end of the reef, but in some cases centrally. Other factors are reef geometry, strength and direction of wave action, and tidal range. A small tidal range encourages their formation. Sand cays tend to form at gaps in linear reefs, on small arcuate or oval reef segments, or at pronounced elbows or bends in reefs.

In the case of sand shingle cays or motus, additional factors include: (1) exposure to severe wave action, predominantly from one direction so that coral shingle and/or boulders are lodged on the reef flat near the windward edge—the ridge or rampart so formed traps some of the finer reef fragments to yield coral sand; and (2) waves or currents from the leeward side of the reef, or if present from the lagoon, which neutralize the partially spent storm waves and lead to deposition of coral sand in the lee of the shingle ridge. (However, it is likely that in at least some cases the process is more complicated than is suggested above.)

Lacking waves from leeward to oppose the trains of storm waves, shingle cays might still form near the windward edge of the reef, but they would tend to be unstable unless vegetation was able to gain a foothold.

The Low Wooded Islands are the most specialized type in the Great Barrier Reefs Province and are most fully developed and most numerous within the outer barrier in the northern section of the Province. The special factors essential for their formation have been enumerated above, Section IV.

VI. Sediments

The major study of the sediments has been carried out by Maxwell (Maxwell et al., 1961; Maxwell, 1962, 1968, 1969b,c, 1970, 1973b; Maxwell et al., 1963, 1964; Maxwell and Maiklem, 1964; Maiklem, 1967, 1968, 1970; Maxwell and Swinchatt, 1970; Jell et al., 1965; Maxwell, Volume I, Chapter 10), but the work of the Stoddart Expedition of 1973, the results of which are as yet unpublished, will supplement that of Maxwell in the least-known region, north of Cooktown.

The major components of the sediments are, of course, mud and sand, but Maxwell emphasizes the importance of the carbonate-terrigenous (noncarbonate) ratio of each of these components. The terrigenous material has come almost solely from the continental mass to the west, whereas carbonate muds and sands are products of the breakdown of the reefs of the reef zone on the eastern flank of the Province.

The main components are quartz sand, terrigenous mud, and carbonate sand (with a little carbonate mud), but there are numerous minor components—contributions by forams, molluscs, *Halimeda,* bryozoans, corals, echinoderms, and *Lithothamnion.* The distribution of the coral, *Halimeda,* and *lithothamnion* components is essentially reef-controlled and consequently all three have a similar distribution.

There is considerable confusion in the use of the term "facies" in relation to the sediments of the Reefs Province. Maxwell in his Atlas (1968) rightly emphasized the contrast between carbonate-rich and silica-rich sediments and, using these factors and relative amounts of mud and sand, described 18 facies. Unfortunately, he also used the term facies in relation to components (quartz, bryozoan and mollusc fragments, etc.) content, thus adding no less than 35 to the above 18 "facies". In one paragraph, he even applied it to the proportion of carbonate present as aragonite—"weak aragonite facies." Thus he mentioned 54 "facies" in all.

Maxwell and Swinchatt (1970, p. 709) attempted to combine mineral nature (quartz, sand, mud, carbonate) with skeletal fragment (coral, *Halimeda,* forms, etc.) content, in this way obtaining six "facies":

1. Terrigenous sand >60% quartz, no mud
2. Terrigenous muddy sand and mud
3. Interreef sand >30%, foram >15%, mollusc >5%, coral, *Halimeda,* and bryozoa
4. Nonreef carbonate sand >15% bryozoa, >30% mollusc, and >5% coral and forams
5. Carbonate muddy sand
6. Carbonate mud

Maxwell (1973b, p. 327, Fig. 11) described five regional sedimentary facies:

1. Terrigenous sand <40% carbonate, 10% mud
2. Carbonate sand >80% carbonate
3. Mixed sand <80% carbonate, 60% terrigenous
4. Mud >40% (carbonate and terrigenous)
5. Muddy sand 10–40% mud

Frankel (1974), in his work on Princess Charlotte Bay, clarified the classification schemes considerably by separating the mainly mineralogical and the mainly textural factors thus:

a. Mineralogical (mainly carbonate and silica content)
   High carbonate facies (80% carbonate)
   Impure carbonate facies (60–80% carbonate)
   Transitional facies (40–60% carbonate)

Terrigenous facies (20–40% carbonate)
High terrigenous subfacies (<20% carbonate)
b. Textural
Terrigenous mud
Muddy carbonate sand
Recent terrigenous sand
Relict terrigenous sand
and a Reef facies which he did not describe.

Maxwell and Swinchatt (1970) and Frankel (1974) were dealing with restricted areas whereas Maxwell (1968, Volume I, pp. 299–345) was developing a classification for the whole region. The former two schemes are apparently not applicable to the whole province and it is not possible to correlate any one with any other. Clearly there is more work needed in this area.

With products of weathering and erosion (mainly noncalcareous) being shed by the continent via the rivers to the west and carbonate sands with some mud from the reef zone to the east, it is not unexpected to find, in general, muddy sediments in the axial region of the shelf, terrigenous sands to the west bordering the continent, and calcareous sand restricted to the reef zone.

This simple distribution, however, has been modified by local factors. Quartzose sand is restricted to that part of the shelf near major stream outlets in the lower rainfall region south of 19°S and to a small area around Princess Charlotte Bay. Much of this sand must come from the large and extensive dune systems of the present day, but earlier, now submerged, dune systems must have added their quota and erosion of the extensive outcrops of granitic rocks on the mainland and on Continental islands continues to contribute. To the north in the higher rainfall belts mud and, in the central region, muddy sand replace the sand facies in the near shore zone. A mixed-sand facies with up to 80% carbonate is characteristic of the coastline north of Princess Charlotte Bay.

The main reef zone is characterized throughout its whole length by the carbonate sand facies and this same facies appears between the axial mud and the near shore sand between 21° and 24°S. The reef here is at its maximum distance from the coast and many coral reefs flourish between the coast and the Capricorn Channel.

On the reef surfaces the few studies carried out suggest that the sediments consist solely of fragments of the skeletons of reef animals and algae, the major components being fragments of calcareous algae (*Halimeda* and *Lithothamnion*), Foraminifera, Mollusca, and corals.

VII. The Origin of the Reefs

Maxwell (1968, 1969d), Stoddart (1969a), and others have drawn a clear distinction between deep-sea oceanic reefs on the one hand and reefs of continental platforms and those associated with Continental islands on the other. The subsidence theory of Darwin is now regarded as established for the former type of reef.

The Great Barrier Reefs belong, of course, to the latter type and the work of the last 20 years has drawn attention to the difficulties of earlier attempts to apply the subsidence theory to them, but has not led to any unanimity of opinion on the extent to which subsidence has been involved.

Maxwell sees the reefs developing in post-Pleistocene times almost solely on tectonic highs on the continental shelf, that is, on areas which have risen rather than subsided. (The tectonic highs consist of ancient rocks and any uplift which has affected them occurred long before the reefs originated). In Maxwell's view, Holocene eustatic fluctuations in sea level did, however, have a marked effect on reef growth, growth being favored by rising sea level while low stands of sea level led to reef erosion and perhaps destruction.

Maxwell also sees the upwelling of deep, nutrient-rich water along the edge of the continental shelf as an important factor leading to three regions of differential growth and developments of the Reefs. But Brandon (Volume I, Chapter 7) believes that the amount of upwelling of nutrient-rich waters along the eastern margin of the reef zone would be quite insufficient to account for the marked differences in density and growth in the three regions of the Reefs.

There are, however, sharp differences of opinion on the number, if any, and the times of sea-level fluctuations during the last 20,000 years including the high stands during the last 6000 years. The eight levels which Maxwell believes are reflected in geomorphological and/or reef features on the shelf are all within this period, the oldest dating back to 18,000 B.P. and the youngest to 7250 B.P.

Both Heidecker (Volume I, Chapter 9) and Lloyd (Volume I, Chapter 11) argue for subsidence, though neither suggest an amount or a rate. However, Whitehouse (Volume I, Chapter 6) suggests that an optimum rate of subsidence is necessary and that coral will not flourish to the extent of forming reefs if subsidence proceeds either more slowly or more rapidly than that optimum rate.

It is obvious then, that in spite of the upsurge of work done on the Reefs during the last 20 years, much more work—more geophysical work, more bottom sampling, more boring through superficial sediment into the rocks beneath, more deep bores—is necessary before we can reach anything approaching unanimity of opinion on when and how the Reefs originated and grew to their present-day form.

In writing the above I have drawn heavily upon publications by Maxwell (1968, 1969d), Hill (1970), Steers (1929, 1937), and Stoddart (1969a). These works may be found listed in the Bibliography in Part II.

The results of the Royal Society of London, University of Queensland expedition to the northern section of the Reefs (north of Cooktown) are being published (1977). According to Farrow (1976), the use of high resolution seismic reflection equipment by Dr. Orme yielded profiles that demonstrate clearly the remnant nature of the present day reefs.

The information gained will change many of the earlier ideas on the age and origin of the whole reef system.

The following references have been used in the foregoing, but are not included in the Bibliography in Part II.

Stoddart, D. R. (1969c). *Phil. Trans. Roy. Soc. London*, Ser. B **255**, 355–382.
Stoddart, D. R. (1971). *Symp. Zool. Soc. London* **28**, 3–38.
Whitehouse, F. W. (1973). *In* "Biology and Geology of Coral Reefs" (O. A. Jones and R. Endean, eds.), Vol. I, 169–186. Academic Press, New York.

PART II. A BIBLIOGRAPHY OF THE GEOLOGY AND HYDROLOGY OF THE GREAT BARRIER REEFS AND THE ADJACENT QUEENSLAND COAST (INCLUDING A NUMBER OF TITLES BROUGHT TOGETHER BY THE LATE PROFESSOR EMERITUS W. H. BRYAN)

VIII. Introduction

In a review of Maxwell's *"Atlas of the Great Barrier Reef"* D. R. Stoddart wrote "It is a pity that the short list of references was not turned into a proper bibliography; no guide to the literature of the Great Barrier Reef exists . . ." Others have expressed similar opinions. The Great Barrier Reef Committee has at least twice in its 54 years of existence set out to produce a bibliography, but in each case the project was short lived, because the field to be covered embraces so many branches of science and specialists in those fields inevitably feel that their time is more usefully employed producing original work than in searching out and listing the works of others.

It is perhaps appropriate that I, retired and writing in a field outside my specialities, though one in which I have for very long been deeply interested, should attempt to remedy the deficiency, though at present only in part. It is hoped that at some future time, with the assistance of other members of the Great Barrier Reef Committee, this nucleus may be expanded to cover the whole range of fields involved.

When the present work was at a fairly advanced stage, the writer located the beginnings of a bibliography prepared by the late Professor Emeritus W. H. Bryan. This was the outcome of the first of the Great Barrier Reef Committee's bibliographic ventures. Though I had already listed many of the works included in Bryan's card index, the latter did add a considerable number of titles to my list. Furthermore, it encouraged me to include papers dealing with the geology of the neighboring coastline, which is a vital factor in reef studies. For these reasons reference has been made to Professor Bryan's name in the authorship of this chapter.

IX. Scope

The general limitations of the bibliography are indicated in the title— only works in the fields of geology, geomorphology, geophysics, hydrology, and oceanography are included. The listing of papers dealing with biological aspects of the Reefs is left to those with a more extensive and deeper knowledge of that field than the writer. Further limitations include the exclusion of: (1) nonscientific books and articles; (2) reviews of books and papers, with a very few exceptions; and (3) most works on coral reefs in general, despite the fact that these may contain incidental reference to the Great Barrier Reefs.

A great deal of difficulty has been experienced with two quite diverse sections of the field—the geology of the coastal areas fronting the reefs and the hydrology of the large ocean area involved. The difficulty in the first case arises from the very large number of papers involved. At first my intention was to select and list only the more important papers but, as the work continued, it became increasingly difficult to decide what geological details were important to workers in the past, which are important in our present state of knowledge, and which may be relevant to the work of future investigators. Indeed it became physically impossible to do the vast amount of reading necessary to attempt to make such decisions in the available time. In spite of quite intensive work, it is probable that a considerable number of papers in this field have been missed. In the other field, hydrology, the difficulty arose because my knowledge of this area is neither extensive nor deep; I can only hope that little of importance has been omitted.

In spite of these rather formidable restrictions and limitations it is hoped that the lists will prove valuable to workers in the field of coral reefs and will form a sound basis on which to build a more complete and more comprehensive bibliography.

X. Discussion

The geographical title used in this chapter is "The Great Barrier Reefs." Reasons for the use of the plural "Reefs" rather than the more commonly used singular "Reef" are several. Discovered by Captain James Cook on June 10, 1770, (Cook, 1955) this enormous system of coral reefs was not named by Cook but was designated "The Great Barrier Reef" by Flinders in 1814.

Beete Jukes, the naturalist aboard the HMS Fly 1842–1846 and the first scientist to discuss the origin of the Reefs, used the plural which is also used on the British Admiralty charts of the region, surely a weighty reason for using "Reefs" (Jukes, 1847).

The geographer J. A. Steers has argued cogently for the use of "Reefs" (1937, p. 4) and used it in the title of his two major papers on the reefs (1929, 1937). The plural has been used by a number of investigators since that time.

Finally and perhaps most importantly (as stressed by Steers) the coral structures do not form a barrier reef in the usual scientific sense of that term. Even if one confines one's attention to the "outer barrier," it is a Barrier Reef for less than a half of its length and in the whole of the complex of reefs many different types are present.

In addition to nearly 500 miles of reefs similar to a Barrier Reef in form, if not in origin, (the northern section) Fairbridge (1950b) recognized 5 types of reef and Maxwell (1968) 13. Two of Maxwell's types make up most of the Barrier, but also occur elsewhere in the system, as do other types.

To repeat Steers (1937, p. 4) own words, "A comprehensive name to cover all these varieties would be difficult to find. Following the lead of the Admiralty Charts, I would suggest that every one should use the phrase "Great Barrier Reefs."

XI. Bibliography

A. PUBLISHED BOOKS AND PAPERS DEALING IN WHOLE OR IN PART WITH THE GREAT BARRIER REEFS

Agassiz, A. (1898). A visit to the Great Barrier Reef of Australia in the steamer "Croydon" during April and May 1896. *Bull. Mus. Comp. Zool.* **28**, 1–203.

Allen, R. J. (1961). Major developments in the search for petroleum in Queensland 1900–1960. *APEA J.* **1**, 2, 3.

Allen, R. J. (1973). Attractive Queensland on shore basins. *APEA J.* **13**, 26–32.

Allen, R. J., and Hogetoorn, D. J., (1970). Petroleum resources of Queensland. *Geol. Surv. Queensl. Rep.* **43**.

Amos, B. J. (1968). The structure of the palaeozoic sediments of the Mossman and Cooktown areas, north Queensland. *J. Geol. Soc. Aust.* **15**, 195–208.

Anderson, O. (1944). Banks Island wolfram deposits. *Queensl. Gov. Min. J.* **45**, 209–214.

Andrews, E. C., (1902). A preliminary note on the geology of the Queensland coast. *Proc. Linn. Soc. N.S.W.* **27**, 146–185.

Andrews, E. C. (1903). An outline of the Tertiary history of New England. *Rec. Geol. Surv. N.S.W.* **7**, 140–216.

Andrews, E. C. (1910). The geographical unity of eastern Australia in late and post-Tertiary time. *J. Proc. R. Soc. N.S.W.* **44**, 420–480.

Andrews, E. C. (1922). A contribution to the hypothesis of coral reef formation. *J. Proc. R. Soc. N.S.W.* **56**, 10–38.

Andrews, E. C. (1933). The origin of modern mountain ranges with special reference to the Eastern Australian Highlands. *J. Proc. R. Soc. N.S.W.* **67**, 251–350.

Anonymous (1976). Sailing Directions for the Ports and Harbours of Queensland." Gov. Printer, Brisbane.

Australia, Bureau of Mineral Resources, Geology and Geophysics (1957–). Geological Maps, Standard 1:250,000 sheet areas. Explanatory Note Series. Maps are accompanied by a booklet of Explanatory Notes. The maps are identified only by the Australian National Grid reference of each map sheet. The following sheets are pertinent to the present work, but some have not yet been published. SC/54-7, 8, 12, 16; SC/55–5; SD/54-4, 8, 12, 16; SD/55-9, 13; SE/55-1, 2, 5, 6, 10, 15; SF/55-3, 4, 8, 12, 13; SF/56-5, 9, 13, 14; SG/56-1, 2, 3, 6, 7.

Australia, Bureau of Mineral Resources, Geology and Geophysics. (1971, 1972). Annual Report, Vol. 10, pp. 27 and 28.

Australia, Bureau of Mineral Resources, Geology and Geophysics. (1969). Sedimentary basins of Australia and Papua-New Guinea and the stratigraphic occurrence of hydrocarbons. *ECAFE* **1**, NR/PR 4-12-63.

Australia, Geological Society, Queensland Division, (1974). Symposium on The Tasman Geosyncline (A. K. Denmead, G. N. Tweedale, and A. F. Wilson, eds.), University of Queensland, Brisbane.

Australian Government and Government of the State of Queensland. (1974). "Report of the Royal Commissions into Exploratory and Production Drilling for Petroleum in the Area of the Great Barrier Reef," Vol. 1, pp. LII and 548; Vol. 2, pp. V and 549–1052. Aust. Gov. Publ. Serv. (see Section F for list of Transcripts).

Avias, J. (1959). Les récifs corraliens de la Nouvelle-Calédonia et quelques-uns de leur problème. *Bull. Soc. Geol. Fr.* [Ser. 7] **1**, 424–430.

Ball, L. C. (1901). The Hamilton and Coen Gold Fields. *Geol. Surv. Queensl., Publ.* **163**, 1–28.

Ball, L. C. (1902a). The Burrum Coalfield. *Geol. Surv. Queensl., Publ.* **170**.

Ball, L. C. (1902b). *Queensl. Gov. Min. J.* **3**, 130–140.

Ball, L. C. (1904). Certain iron maganese ore and limestone deposits in the central and southern districts of Queensland. *Geol. Surv. Queensl., Publ.* **194**.

Ball, L. C. (1909). The Starkey Goldfield. *Geol. Surv. Queensl., Publ.* **223**, 1–47.

Ball, L. C. (1910). Certain mines and mineral fields in north Queensland. *Geol. Surv. Queensl., Publ.* **222**.

Ball, L. C. (1914). Tertiary oil shales of the Narrows, Port Curtis District. *Queensl. Gov. Min. J.* **25**, 73–76.

Ball, L. C. (1915). Oil shales in the Port Curtis District. *Geol. Surv. Queensl., Publ.* **249**, 1–35.

Ball, L. C. (1916). Lowmead No. 1 bore and the Tertiary oil shales of Baffle Creek. *Queensl. Gov. Min. J.* **17**, 13–16.

Ball, L. C. (1924). Report on oil prospecting near Tewantin. *Queensl. Gov. Min. J.* **70**, 354–362.

Ball, L. C., (1945). Oil shales in Queensland. *Queensl. Gov. Min. J.* **46**, 74–75.

Ball, L. C. (1946). Oil shales of the Narrows, central Queensland, *Queensl. Gov. Min. J.* **47**, 176–179.

Basedow, H. W. (1918). A synopsis of the geology. *In* "Narrative of an Expedition of Exploration in North-western Australia," Proc. South. Aust. Branch R. Geogr. Soc. Aust., pp. 105–295, (esp. pp. 240–249).

Beasley, A. W. (1947). The place of black sand seams in the physiographic history of the south coast region, Queensland. *Aust. J. Sci.* **9**, 208–210.

Beasley, A. W. (1948). Heavy mineral beach sands of southern Queensland. *Proc. R. Soc. Queensl.* **59**, 109–140.

Beddoes, L. R., Jr. (1973). "Oil and Gas Fields of Australia, Papua-New Guinea and New Zealand." Tech. Press, Sydney.

Bennett, I. (1971). "The Great Barrier Reef." Lansdowne Press, Melbourne.

Benson, W. N. (1923). Palaezoic and Mesozoic Seas in Australasia. *Trans. Proc. N.Z. Inst.* **54**, 1–62.

Benson, W. N. (1924). The structural features of the margin of Australasia. *Trans. Proc. N. Z. Inst.* **55**, 99–137.

Best, J. G., Stevens, N. C., and Tweedale, G. W. (1960). Upper Cainozoic: Igneous rocks. *J. Geol. Soc. Aust.* **7**, 419–423.

Bird, E. C. F. (1965). The formation of coastal dunes in the humid tropics: Some evidence from north Queensland. *Aust. J. Sci.* **27**, 258–259.

Bird, E. C. F. (1968). "Coasts," pp. 21, 87, and 190–211. Aust. Natl. Univ. Press, Canberra.

Bird, E. C. F. (1970). Coastal evolution in the Cairns district. *Aust. Geogr.* **11**, 327–335.

Bird, E. C. F. (1971a). Holocene shore features at Trinity Bay north Queensland. *Search* **2**, 27–28.

Bird, E. C. F. (1971b). The fringing reefs near Yule Point, north Queensland. *Aust. Geogr. Stud.* **9**, 107–115.

Bird, E. C. F. (1971c). The origin of beach sediments on the north Queensland Coast. *Earth Sci. J.* **5**, 95–105.

Bird, E. C. F. (1972). Mangroves and coastal morphology in Cairns Bay, north Queensland, *J. Trop. Geogr.* **35**, 11–16.

Bird, E. C. F., and Hopley, D. (1969). Geomorphological features on a humid tropical sector of the Australian coast. *Aust. Geogr. Stud.* **7**, 89–108.

Bock, P. E., and Houston, B. R. (1960). Petrology of cores Nos. 12, 13, 14. *In* "H.B.R. No. 1 Bore Wreck Island, Queensland," p. 15, Bur. Miner. Resour. Geol. Geophys. Aust. Petrol. Search Subsidy Acts.

Brandon, D. E. (1973). Waters of the Great Barrier Reef Province. *In* "Biology and

Geology of Coral Reefs" (O. A. Jones and R. Endean, eds.), Vol. I, pp. 187–232. Academic Press, New York.

British Admiralty, London. (1960–1970). "Australia Pilot," Vol. III and charts. HM Stationery Office, London, (Latest edition is 5th, 1960, with Supplements Nos. 1–6, 1970. A map shows the numbers of the charts covering the area of the Reefs.)

Brooks, J. H. (1960a). Precambrian—central and northern coastal Queensland. *J. Geol. Soc. Aust.* **7**, 87.

Brooks, J. H. (1960b). Silurian—northern coastal Queensland. *J. Geol. Soc. Aust.* **7**, 130–131.

Brooks, J. H. (1960c). Silurian—north of Gladstone. *J. Geol. Soc. Aust.* **7**, 137–138.

Brooks, J. H. (1970). Summary report—iron ore resources of Queensland. *Geol. Surv. Queensl., Rep.* **56**.

Brown, D. A., Campbell, K. S. W., and Crook, K. A. W. (1968). "The Geological Evolution of Australia and New Zealand," pp. 294, 295, 339, and 340. Pergamon, Sydney.

Brown, G. A. (1973). Energy, environment and conservation—Australian case histories and their implications for the future. *APEA J.* **13**, 132–139.

Browne, W. R. (1945). An attempted post-Tertiary chronology for Australia. *Proc. Linn. Soc. N.S.W.* **70**, v–xxiv.

Bryan, W. H. (1926). Earth movements in Queensland. *Proc. R. Soc. Queensl.* **37**, 1–82.

Bryan, W. H. (1928a). The Queensland continental shelf. *Rep. Gt. Barrier Reef Comm.* **2**, 58–69.

Bryan, W. H. (1928b). Metamorphic rocks of Queensland. *Rep. 19th Meet. Aust. N. Z. Assoc. Adv. Sci.* 30–41.

Bryan, W. H. (1930). Physiography of Queensland. *Handb. Queensl., 20th Meet, Aust. Assoc. Adv. Sci.*, pp. 17–22.

Bryan, W. H. (1936). The supposed deepening of the sea floor off Breaksea Spit. *Rep. Gt. Barrier Reef Comm.* **4**, 45–50.

Bryan, W. H. (1944). The relationship of the Australian continent to the Pacific Ocean—now and in the past. *J. Proc. R. Soc. N.S.W.* **78**, 42–62.

Bryan, W. H., and Jones, O. A. (1945). The geological history of Queensland. A stratigraphical outline. *Pap. Dep. Geol. Univ. Queensl.* [N. S.] **2**, 1–78.

Bryan, W. H., and Whitehouse, F. W. (1926). Later palaeogeography of Queensland. *Proc. R. Soc. Queensl.* **38**, 103–114.

Burns, R. E., Andrews, J. E., van der Lingen, C. J., Churkin, M., Jr., Galehouse, J. S., Packham, G. H., Davies, T. A., Kennett, J. P., Dumitrica, P., Edwards, A. R., and von Herzen, R. P. (1972). Glomar Challenger down under: Deep Sea Drilling Project, Leg 21. *Geotimes* **17**, May, pp. 14–16.

Cameron, W. E. (1902). The coal beds of Waterpark Creek, near Port Clinton. *Geol. Surv. Queensl., Publ.* **174**, 1–6.

Cameron, W. E., (1903). Additions to the geology of the Mackay and Bowen districts. *Geol. Surv. Queensl., Publ.* **181**, 1–21.

Cameron, W. E. (1907a). Some goldfields of Cape York Peninsula. *Geol. Surv. Queensl. Publ.* **209**.

Cameron, W. E. (1970b). The Annan River tinfield, Cooktown District. *Geol. Surv. Queensl., Publ.* **210**.

Carey, S. W. (1970). Australia, New Guinea and Melanesia in the current evolution in concepts of the evolution of the earth. *Search* 1, 178–189.

Carruthers, D. S. (1969). Limestone mining. "The future of the Great Barrier Reef." *Spec. Publ. Aust. Conserv. Found.* 3, 47–50.

Chapman, F. (1931). A report on samples obtained by boring into Michaelmas Reef, about 22 miles N.E. of Cairns, Queensland. *Rep. Gt. Barrier Reef Comm.* 3, 32–42.

Chappel, J., and Thom, B. G. (1974). Problems of dating upper Pleistocene sea-levels from coral reef areas. *Proc. Int. Symp. Coral Reefs, 2nd, 1973* Vol. 2, pp. 563–571.

Chu Chong, E. S. (1965). Coal resources—Burrum Coalfield. *Geol. Surv. Queensl., Rep.* 9, 17 pp.

Clarke, D. E., Paine, A. G. L., and Jensen, A. R. (1971). Geology of the Proserpine 1:250,000 sheet area, Queensland. *Aust., Bur. Miner. Resour., Geol. Geophys., Rep.* 144, 98 pp.

Coaldrake, J. E. (1955). Fossil soil hardpans and coastal sandrock in southern Queensland. *Aust. J. Sci.* 17, 132–133.

Coaldrake, J. E. (1960). Upper Cainozoic—Coastal deposits. Quaternary history of the coastal lowlands of southern Queensland. *J. Geol. Soc. Aust.* 7, 403–412.

Coaldrake, J. E. (1961). The ecosystem of the coastal lowlands ("Wallum" of southern Queensland). *Aust., C. S. I. R. O., Bull.* 283, 138 pp.

Coaldrake, J. E. (1962). The coastal sand dunes of southern Queensland. *Proc. R. Soc. Queensl.* 72, 101–116.

Conaghan, P. J. (1966). Sediments and sedimentary processes in Gladstone Harbour, Queensland. *Pap. Dep. Geol. Univ. Queensl.* [N.S.] 6, 1–52.

Condon, M. A., Fisher, N. H., and Terpstra, G. R. (1960). Summary of oil search activities in Australia and New Guinea to June 1959. *Aust., Bur. Miner. Resour., Geol. Geophys., Rep.* 41A, 68 pp.

Connah, T. H. (1958). Summary report of limestone resources of Queensland. *Queensl. Gov. Min. J.* 59, 636–653 and 739–755 (p. 651–Duke group of islands).

Connah, T. H. (1961). Beach sand heavy mineral deposits of Queensland. *Geol. Surv. Queensl., Publ.* 302.

Connah, T. H., and Hubble, G. D. (1960). Laterites. *J. Geol. Soc. Aust.* 7, 373–386.

Connah, T. H., and Newman, P. W. (1960). Upper Cainozoic—beach sands and heavy mineral concentrations. *J. Geol. Soc. Aust.* 7, 409–412.

Cook, Captain James. (1955). "The Journals of Captain James Cook" (J. C. Beaglehole, ed.), Vol. 1. The Voyage of the Endeavour 1769–1771. Cambridge Univ. Press, London and New York (for the Hakluyt Society).

Cook, P. J., and Polach, H. A. (1972). Discovery of recent supra tidal dolomite at Broad Sound, Queensland. *Search* 4, 78.

Cook, P. J., and Polach, H. A. (1973). A chenier sequence at Broad Sound, Queensland and evidence against an Holocene high sea level. *Mar. Geol.* 14, 1–16.

Cotton, C. A. (1947). The pulse of the Pacific. *J. Proc. R. Soc. N.S.W.* 80, 41–76.

Cotton, C. A. (1949). A review of tectonic relief in Australia. *J. Geol.* 57, 280–296.

Crespin, I. (1960). Preliminary note on Palaeontology. *In* "H.B.R. No. 1 Bore, Wreck Island, Queensland," p. 12. Published under Petrol. Search Subsidy Acts, Bur. Miner. Resour., Geol. Geophys., Australia.

Cribb, H. G. S. (1960a). Lower Burdekin Valley. *J. Geol. Soc. Aust.* 7, 147–149.

Cribb, H. G. S. (1960b), Carboniferous—The Pascoe River Beds. *J. Geol. Soc. Aust.* **7**, 181.

Crosslands, C. (1931). The coral reefs of Tahiti compared with the Great Barrier Reefs. *Geogr. J.* **77**, 395–396.

Cushman, J. A. (1942). A report on samples obtained by the boring on Heron Island, Great Barrier Reef, Australia. Appendix to Richards and Hill (1942, pp. 112–119).

Daintree, R. (1872). Notes on the geology of the colony of Queensland. *Q. J. Geol. Soc. London* **28**, 271–317.

Daneš, J. V. (1911). On the physiography of North-Eastern Australia. *Proc. R. Boheme Soc. Sci. Ser.* **2**, 1–18.

Daneš, J. V. (1912). La région des rivières Barron et Russel (Queensland). *Ann. Geog.* **21**, 344–363.

David, T. W. E. (1911). Notes on some of the chief tectonic lines of Australia. *J. Proc. R. Soc. N.S.W.* **45**, 4–60.

David, T. W. E. (1914). Geology of the Commonwealth. *In* "Federal Handbook" (G. H. Knibbs, ed.), pp. 241–325. Br. Assoc. Adv. Sci., London.

David, T. W. E. (1927). Notes on the bore in the Great Barrier Reef of Australia. *Proc. Geol. Soc. London* **83**, vii.

David, T. W. E. (1932). "Explanatory Notes to Accompany a New Geological Map of Australia." Aust. Med. Publ., Sydney.

David, T. W. E., and Browne, W. R. (1950). "The Geology of the Commonwealth of Australia," 3 vols. Arnold, London.

Davies, P. J. (1974a). Subsurface solution unconformities in the Great Barrier Reef. *Proc. Int. Symp. Coral Reefs, 2nd 1973* Vol. 2, pp. 573–578.

Davies, P. J. (1974b). Electrode cation measurements in the Capricorn area. *Proc. Int. Symp. Coral Reefs, 2nd, 1973* Vol. 2, pp. 449–455.

Davies, P. J., and Kinsey, D. W. (1973). Organic and inorganic factors in Recent beach rock formation, Heron Island, Great Barrier Reef. *J. Sediment. Petrol.* **43**, 59–81.

Davis, W. M. (1917). The Great Barrier Reef of Australia. *Am. J. Sci.* [4] **44**, 339–350.

Davis, W. M. (1928). The coral reef problem. *Am. Geogr. Soc., Spec. Publ.* **9**, 596 pp.

Davis, W. M. (1934). Submarine mock valleys. *Geogr. Rev.* **34**, 302–304.

Dear, J. F. (1969). The Permian system in Queensland. *Spec. Publ., Geol. Soc. Aust.* **2**, 1–6.

de Jersey, N. J. (1951). Microspore correlation, Burrum Coalfield. *Queensl. Gov. Min. J.* **52**, 780–789.

de Jersey, N. J. (1960a). Cretaceous—The Plutoville Beds. *J. Geol. Soc. Aust.* **7**, 329–330.

de Jersey, (1960b). Cretaceous—The Styx Coal Measures. *J. Geol. Soc. Aust.* **7**, 330–333.

de Kayser, F. (1963). The Palmerville Fault—a "Fundamental Structure in North Queensland." *J. Geol. Soc. Aust.* **10**, 273–278.

de Kayser, F. (1965). The Barnard Metamorphics and their relation to the Barron River Metamorphics and the Hodgkinson Formation, North Queensland. *J. Geol. Soc. Aust.* **12**, 91–103.

de Kayser, F., and Lucas, K. G. (1968). Geology of the Hodgkinson and Laura Basins, North Queensland. *Aust., Bur. Miner. Resour., Geol. Geophys., Bull.* 84, 254 pp.

Denmead, A. K. (1949). Water search—Olive Vale Station, Laura. *Queensl. Gov. Min. J.* 50, 40–42.

Denmead, A. K. (1971). The Atherton Tableland. *In Geol. Excurs. Handb.* 43rd Congr. Aust. N.Z. Assoc. advnt. Sci., pp. 1–20.

Derrington, S. S. (1960). Completion rep., *In* "H.B.R. No. 1 Bore, Wreck Island, Queensland," pp. 1–11. Published under Petrol. Search Subsidy Acts, Bur. Miner. Resour. Geol. Geophys., Australia.

Derrington, S. S. (1961). The tectonic framework of the Bowen Basin, Queensland. *APEA J* 1, 18–21.

Dooley, J. C. (1965). Gravity surveys of the Great Barrier Reef and adjacent coast, North Queensland, 1954–60. *Aust., Bur. Miner. Resour., Geol. Geophys., Rep.* 73, 26 pp.

Doutch, H. F. (1972). *In* "The Paleogeography of Northern Australia and New Guinea and its Relevance to the Torres Strait Area—Bridge and Barrier: The natural and cultural History of Torres Strait" (D. Walker, ed.), BG/3. Publ. Res. Sch. Pac. Stud. Aust. Natl. Univ. Canberra,

Driscoll, E. M., and Hopley, D. (1968). Coastal development in part of tropical Queensland, Australia. *J. Trop. Geogr.* 26, 17–28.

Dunstan, B. (1898). The geology of Collaroy and Carmilla near Broadsound. *Geol. Surv. Queensl. Publ.* 141, 1–7.

Dunstan, B. (1901). Geological features of Hazledean, west of Mackay. *Geol. Surv. Queensl., Bull.* 17, 1–16.

Dunstan, B. (1902). Geological features of Hazledean, west of Mackay. *Queensl. Gov. Min. J.* 3, 28–33.

Dunstan, B. (1912a). Coal measures and marine beds around Maryborough. *Annu. Rep. Dep. Min. Queensl.* p. 195.

Dunstan, B. (1912b). Geological features of the Maryborough district. *Queensl. Gov. Min. J.* 13, 641–642.

Dunstan, B. (1914). Boring for coal between Takura and Colton, near Maryborough. *Queensl. Gov. Min. J.* 15, 405–407.

Dunstan, B. (1918a). In Walkom (1918, pp. 3–21) (including a geological note, map and section by B. Dunstan).

Dunstan, B. (1918b). Prospects of coal on Fraser Island. *Queensl. Gov. Min. J.* 19, 307–308.

Dunstan, B. (1919). *In* Walkom (1919), pp. 1–5 (including a geological note and sections by B. Dunstan), 64 pp.

Dunstan, B. (1920). Salt. *Geol. Surv. Queensl., Publ.* 268, Art. 1, 1–56.

Dunstan, B., and Cameron, W. E. (1910). The Burrum Coalfield. *Queensl. Gov. Min. J.* 11, 341–342.

Edgell, J. A. (1928a). Some remarks on coral formations. *Rep. Gt. Barrier Reef Comm.* 2, 52–56.

Edgell, J. A. (1928b). Changes at Masthead Island. *Rep. Gt. Barrier Reef Comm.* 2, 57.

Ellis, P. L. (1964). Geological notes on the Maryborough Gayndah area. *Guideb. Maryborough Basin Field Conf., Geol. Soc. Aust., 1964*, 8.

Ellis, P. L. (1966). The Maryborough Basin. *APEA J.* 6, 30–36.

Ellis, P. L. (1968). Geology of the Maryborough 1:250 000 sheet area. *Geol. Surv. Queensl., Rep.* 26.

Ellis, P. L. (1971). Maryborough-Bundaberg area. *In Geol. Excurs. Handb.,* 43rd Meet. Aust. N.Z. Assoc. Adv. Sci. pp. 45–61.

Ewing, M., Hawkins, L. V., and Ludwig, W. J. (1970). Crustal structure of the Coral Sea. *J. Geophys. Res.* 75, 1953–1962.

Ewing, M., Houtz, R., and Ludwig, W. (1970). Sediment distribution in the Coral Sea. *J. Geophys. Res.* 75, 1963–1972.

Fairbridge, R. W. (1950a). Problems of Australian geotectonics. *Scope* 1, 22–29.

Fairbridge, R. W. (1950b). Recent and Pleistocene coral reefs of Australia. *J. Geol.* 58, 330–401.

Fairbridge, R. W. (1953). "Australian Stratigraphy," 2nd ed., pp. xii/64–xii/66. Univ. West. Aust. Text Books Board, Perth.

Fairbridge, R. W. (1967). Coral reefs of the Australian region. *In* "Landform Studies from Australia and New Guinea" (J. N. Jennings and J. A. Mobbott, eds.), pp. 386–417 Aust. Natl. Univ. Press, Canberra.

Fairbridge, R. W. (1968a). Coral reefs—morphology and theories. *In* "Encyclopedia of Geomorphology" (R. W. Fairbridge, ed.), pp. 186–196. Van Nostrand-Reinhold, Princeton, New Jersey.

Fairbridge, R. W. (1968b). Lagoon—coral reef type. *In* "Encyclopedia of Geomorphology" (R. W. Fairbridge, ed.), pp. 594–598. Van Nostrand-Reinhold, Princeton, New Jersey.

Fairbridge, R. W., and Teichert, C. (1947). The rampart system at Low Isles, 1928–45. *Rep. Gt. Barrier Reef Comm.* 6, 1–16.

Fairbridge, R. W., and Teichert, C. (1948). The Low Isles of the Great Barrier Reef: A new analysis. *Geogr. J.* 111, 67–88.

Fairbridge, R. W., and van der Linden, W. J. M. (1963). Coral Sea. *In* "Encyclopedia of Oceanography" (R. W. Fairbridge, ed.), pp. 219–223. Van Nostrand-Reinhold, Princeton, New Jersey.

Falvey, D., and Talwani, M. (1970). Structure of the Coral Sea region from marine gravity data. *Geoexploration* 8, 248–249 (obs.)

Farrow, G. F. (1976). The Great Barrier Reef, geomorphology. *Nature (London)* 259, 528–529.

Findlayson, D. M. (1968). First arrival data from the Carpentaria region upper mantle project (Crump). *J. Geol. Soc. Aust.* 15, 33–50.

Fischer, P. H. (1961). Coup d'oeil sur le Grande-Barrière d'Australie et en particulier sur un recif du group Capricorne. *Cah. Pac.* 3, 52–74.

Fisher, N. H., ed. (1962). "Geological Notes in Explanation of the Tectonic Map of Australia." Bur. Miner. Resour., Geol. Geophys., Australia, 72 pp.

Fleischman, K. R. (1953). Inspection of the Lockerbie (Cape York) area and wolfram workings on Moa, Torres Strait. *Queensl. Gov. Min. J.* 54, 935–937.

Flinders, M. (1814). "A voyage to Terra Australis, . . . in the years 1801, 1802 and 1803 in His Majesty's Ship 'The Investigator.'" G. and W. Nicol, London.

Flood, P. G. (1974). Sand movements on Heron Island—a vegetated sand cay, Great Barrier Reef Province. *Proc. Int. Symp. Coral Reefs, 2nd, 1973* Vol. 2, pp. 387–394.

Folk, R. L. (1967). Sand cays of Alacrán Reef, Yucatán, Mexico: Morphology. *J. Geol.* 75, 412–437.

Folk, R. L. (1972). Electron microscope reconnaissance of muds from the Great Barrier Reef area, Australia. *Search* 3, 171–173.

Forbes, H. O. (1893). Review of "The Great Barrier Reef of Australia" by Saville-Kent. *Geogr. J.* 2, 540–546.

Fosberg, F. R., Thorne, R. F., and Moulton, J. M. (1961). Heron Island, Capricorn Group, Australia. *Atoll Res. Bull.* 82, 1–16.

Frankel, E. (1972). Development of the continental shelf in the Princess Charlotte Bay area, Great Barrier Reef Province. *44th Meeting Aust. New Zealand Assoc. Adv. Sci.*, Section 3—Geology, p. 35 (abstract).

Frankel, E. (1974). Recent sedimentation in the Princess Charlotte Bay area, Great Barrier Reef Province. *Proc. Int. Symp. Coral Reefs, 2nd, 1973* Vol. 2, pp. 355–369.

Gardner, D. E. (1955). Beach-sand heavy-mineral deposits of eastern Australia. *Aust., Bur. Miner. Resour., Geol. Geophys., Bull.* 28, 1–103 pp.

Gardner, J. V. (1970). Submarine geology of the Western Coral Sea. *Geol. Soc. Am. Bull.* 81, 2599.

Geological Society of Australia. (1960). "Tectonic map of Australia, 1:2,534,400." Geol. Soc. Aust., Sydney.

Geological Society of Australia. (1971). "Tectonic Map of Australia and New Guinea, 1:5,000,000." Geol. Soc. Aust., Sydney.

Geological Survey of Queensland. (1960). Occurrence of petroleum and natural gas in Queensland. *Geol. Surv. Queensl. Publ.* 299.

Geological Survey of Queensland and University of Queensland. (1953). "Geological Map of Queensland, 1:2,534,400 (40 Miles to an Inch)." Gov. Printer, Brisbane, Queensland.

Gill, E. D. (1955). Radio-carbon dates for Australian archaeological and geological samples. *Aust. J. Sci.* 18, 49–52.

Gill, E. D. (1961). Changes in the level of the sea relative to the land in Australia during the Quaternary era. *Z. Geomorphol.* [N.S.] 3, 73–79.

Gill, E. D. (1968). Quaternary shoreline research in Australia and New Zealand. *Aust. J. Sci.* 31, 106–111.

Gill, E. D., and Hopley, D. (1972). Holocene sea levels in eastern Australia—a discussion. *Mar. Geol.* 12, 223–233.

Glaessner, M. F. (1952). The geology of the Tasman Sea. *Aust. J. Sci.* 14, 111–114.

Gleghorn, R. J. (1947). Cyclone damage on the Great Barrier Reef. *Rep. Gt. Barrier Reef Comm.* 6, 17–19.

Glomar Challenger (1972). Glomar Challenger down under, deep sea drilling project, leg 21. *Geotimes* 17 (5), May 1972.

Gregory, A. C. (1879a). Geological features of the south-eastern districts of the colony of Queensland. *Votes Proc. Legis. Assem. Queensl.*, 1875 2, 365–375.

Gregory, A. C. (1879b). Report on the Burrum coal mines. *Votes Proc. Legis. Assem. Queensl.*, 1875 2, 1154–1155.

Gregory, J. W. (1912). The structural and petrographic classifications of coast-types. *Scientia, Bologna,* 11, 1–30.

Gregory, J. W. (1930). The geological history of the Pacific Ocean. *Q. J. Geol. Soc.* 86, lxviii et seq.

Guilcher, A. (1971). Mayotte barrier reef and lagoon, Comoro Islands as compared with other barrier reefs. *Symp. Zool. Soc. London* 28, 65–86.

Haddon, A. C., Sollas, W. J., and Cole, G. A. J. (1894). On the geology of the Torres Straits. *Trans. R. Ir. Acad.* **30**, 419–476.

Hawthorne, W. L. (1954–1963). A series of reports on the coal resources in various parts of the Burrum Coalfield, in the Maryborough Basin: *Queensl. Gov. Min. J.* **55**, 749–755 and 824–827 (1954); **56**, 127–128 (1955); **59**, 439–444 and 573–578 (1958); **63**, 465–469, 535–540, 602–608, 609–616, and 668–677 (1962); **64**, 16 (1963).

Hawthorne, W. L. (1960a). The Burrum coalfield. *Geol. Surv. Queensl., Publ.* **296**.

Hawthorne, W. L. (1960b). Triassic—The Maryborough Basin. *J. Geol. Soc. Aust.* **7**, 278–279.

Hawthorne, W. L., Siller, C. W., and de Jersey, N. J. (1960). Cretaceous—The Maryborough Basin. *J. Geol. Soc. Aust.* **7**, 333–340.

Hedley, C. (1906). The mollusca of Mast Head Reef, Capricorn Group, Queensland. Part 1. *Proc. Linn. Soc. N.S.W.* **31**, 453–479.

Hedley, C. (1911). A study of marginal drainage. *Proc. Linn. Soc. N.S.W.* **36**, 13–39.

Hedley, C. (1924). The Great Barrier Reef of Australia. *Nat. Hist.* **24**, 62–67.

Hedley, C. (1925a). The natural destruction of a coral reef. *Trans. R. Geogr. Soc. Aust. Queensl. Branch* **1**, 35–40.

Hedley, C. (1925b). A raised beach at the North Barnard Islands. *Trans. R. Geogr. Soc. Aust. Queensl. Branch* **1**, 61–62.

Hedley, C. (1925c). The Townsville Plain. *Trans. R. Geogr. Soc. Aust. Queensl. Branch* **1**, 63–65.

Hedley, C. (1925d). Coral shingle as a beach formation. *Trans. R. Geogr. Soc. Aust. Queensl. Branch* **1**, 66.

Hedley, C. (1925e). A disused river mouth at Cairns. *Trans. R. Geogr. Soc. Aust. Queensl. Branch* **1**, 69–72.

Hedley, C. (1925f). The surface temperature of Moreton Bay. *Trans. R. Geogr. Soc. Aust. Queensl. Branch* **1**, 149–150.

Hedley, C. (1925g). The Queensland earthquake of 1918. *Trans. R. Geogr. Soc. Aust. Queensl. Branch* **1**, 151–156.

Hedley, C. (1925h). Report of the scientific director for 1924. *Rep. Gt. Barrier Reef Comm.* **1**, 157–160.

Hedley, C. (1926). Recent studies on the Great Barrier Reef of Queensland. *Am. J. Sci.* [5] **11**, 187–193.

Hedley, C., and Taylor, T. G. (1907). Coral reefs of the Great Barrier, Queensland: A study of their structure, life distribution and relation to mainland physiography. *Rep. Aust. Assoc. Adv. Sci.* **11**, 397–413.

Heickel, H. (1972). Pollen and spore assemblages from Queensland Tertiary sediments. *Geol. Surv. Queensl., Publ.* **355**.

Heidecker, E. J. (1973). Structural and tectonic factors influencing the development of Recent coral reefs off northeastern Queensland. *In* "Biology and Geology of Coral Reefs" (O. A. Jones and R. Endean, eds.), Vol. I, pp. 273–298. Academic Press, New York.

Heidecker, E. J. (1976). Structural controls of mineralization in fault troughs, exemplified by the Indooroopilly and Mount Keelbottom ore deposits, Queensland, Australia. *Inst. Mining Metall., trans.* **85**, B47–B52.

Henderson, Commander D. A. (1931a). Morinda shoal. *Rep. Gt. Barrier Reef Comm.* **3**, 46.

Henderson, D. A. (1931b). Subsidence of the continental shelf northward of Sandy Cape. *Rep. Gt. Barrier Reef Comm.* **3**, 43–45.

Henderson, R. A. (1971). Townsville-Charters Towers area. *In Geol. Excurs. Handb., Queensl., 43rd Congr. Aust. N.Z. Adv. Sci.* pp. 79–90.

Highley, E. (1967). Oceanic circulation patterns off the east coast of Australia. *Aust., C.S.I.R.O., Div. Fish. Oceanogr. Tech. Pap.* **23**, 19 pp.

Hill, D. (1951). Geology. *Handb. Queensl., 28th Meet. Aust. N.Z. Assoc. Adv. Sci.* pp. 13–24.

Hill, D. (1960a). Devonian—Boyne Valley, Gladstone, Rockhampton and Clermont districts. *J. Geol. Soc. Aust.* **7**, 143–144.

Hill, D. (1960b). The Great Barrier Reefs. *J. Geol. Soc. Aust.* **7**, 412–413.

Hill, D. (1961). Geology of South-Eastern Queensland. *Handb. Queensl. 35th Meet. Aust. N.Z. Assoc. Adv. Sci.*, pp. 125–135.

Hill, D. (1967). *In* "Elements of the Stratigraphy of Queensland" (D. Hill and W. G. H. Maxwell, eds.), pp. 70–71, and fig. 1 on p. 74. Univ. of Queensl. Press, Brisbane.

Hill, D. (1970). The Great Barrier Reef. *In* "Captain Cook, Navigator and Scientist" (G. M. Badger, ed.), pp. 70–86 Aust. Natl. Univ. Press, Canberra.

Hill, D. (1974). An introduction to the Great Barrier Reef. *Proc. Int. Symp. Coral Reefs, 2nd, 1973* Vol. 2, pp. 723–731.

Hills, E. S. (1955). A contribution to the morphotectonics of Australia. *J. Geol. Soc. Aust.* **3**, 1–15.

Hogan, J. (1925). Record of sea temperatures at Willis Island during the cyclone season of 1922–23. *Trans. R. Geogr. Soc. Aust. Queensl. Br.* **1**, 41–46.

Hopley, D. (1968). Morphology of Curacao Island spit, North Queensland. *Aust. J. Sci.* **31**, 122–123.

Hopley, D. (1970). The geomorphology of the Burdekin Delta, north Queensland. *Dep. Geogr., James Cook Univ., Monogr. Ser.* **1**, 1–66.

Hopley, D. (1971a). Sea level and environment changes in the late Pleistocene and Holocene in north Queensland, Australia. *Quaternaria* **14**, 267–274.

Hopley, D. (1971b). The origin and significance of north Queensland island spits. *Z. Geomorphol.* [N.S.] **15**, 371–389.

Hopley, D. (1972). 1971–72 research on Quaternary shorelines in Australia and New Zealand—a summary report of the ANZAAS Quaternary shorelines Committee. *Search* **3**, 412.

Hopley, D. (1974a). Investigations of sea level changes along the Great Barrier Reef coastline. *Proc. Int. Symp. Coral Reefs, 2nd, 1973* **2**, 551–562.

Hopley, D. (1974b). Coastal changes produced by tropical cyclone Althea in Queensland Dec. 1971. *Austr. Geogr.* **12**, 445–456.

Hopley, D. (1976). The pattern of coastal changes produced by tropical cyclones in Queensland. *Aust. Geogr.* (in press).

Iredale, T. (1942). Report on molluscan content of Heron Island reef boring samples. Appendix to Richards and Hill (1942, pp. 120–122).

Isbell, R. F. (1960). Upper Cainozoic—deposits of the coastal lowlands of northern Queensland. *J. Geol. Soc. Aust.* **7**, 408–409.

Jack, R. L. (1879). Geology and mineral resources of the district between Charters Towers and the coast. *Geol. Surv. Queensl., Publ.* **1**, 1–28.

Jack, R. L. (1881). On explorations in Cape York Peninsula. *Geol. Surv. Queensl., Publ.* **8**, 1–46.

Jack, R. L. (1887). Geological features of the Mackay district. *Geol. Surv. Queensl., Publ.* 39, 1–10.

Jack, R. L. (1899). On some salient points in the geology of Queensland. *Rep. Aust. Assoc. Adv. Sci.* 1, 196–206.

Jack, R. L. (1922a). On explorations in Cape York Peninsula. *Geol. Surv. Queensl., Publ.* 8.

Jack, R. L. (1922b). "Northmost Australia," 2 vols. George Robertson, Sydney.

Jack, R. L., and Etheridge, R. (1893). "Geology and Palaeontology of Queensland and New Guinea." Gov. Printer, Brisbane.

Jackson, C. F. V. (1902). Report on a visit to the west coast of Cape York Peninsula and some islands of the Gulf of Carpentaria; also reports on the Horn Island and Possession Island gold fields and the recent prospecting of the Cretaceous coals of the Cook district. *Geol. Surv. Queensl., Rep.* 180, 1–27.

Jardine, F. (1923). The physiography of the lower Fitzroy basin. *Queensl. Geogr. J.* 38, 1–42.

Jardine, F. (1925a). The physiography of the Port Curtis district. *Trans. R. Geogr. Soc. Aust. Queensl. Branch* 1, 73–110.

Jardine, F. (1925b). The development and significance of benches in the littoral of eastern Australia. *Trans. R. Geogr. Soc. Aust. Queensl. Branch* 1, 111–130.

Jardine, F. (1925c). The drainage of the Atherton tableland. *Trans. R. Geogr. Soc. Aust. Queensl. Branch* 1, 131–148.

Jardine, F. (1928a). The topography of the Townsville littoral. *Rep. Gt. Barrier Reef Comm.* 2, 70–87.

Jardine, F. (1928b). The Broadsound drainage in relation to the Fitzroy River. *Rep. Gt. Barrier Reef Comm.* 2, 88–92.

Jardine, F. (1928c). Bramble Cay, Torres Strait—geological notes. *Rep. Gt. Barrier Reef Comm.* 2, 93–100.

Jardine, F. (1928d). Darnley Island—geological and topographical notes. *Rep. Gt. Barrier Reef Comm.* 2, 101–109.

Jell, J. S., Maxwell, W. G. H., and McKellar, R. G. (1965). The significance of the larger foraminifera in the Heron Island reef sediments. *J. Palaeontol.* 39, 273–279.

Jensen, A. R., Gregory, C. M., and Forbes, V. R. (1966). Geology of the Mackay 1:250,000 sheet area, Queensland. *Aust., Bur. Miner. Resour., Geol. Geophys., Rep.* 104, 58 pp.

Jensen, H. I. (1911). The building of eastern Australia. *Proc. R. Soc. Queensl.* 23, 149–198.

Jensen, H. I. (1918). Notes on the geology of the Gladstone district. *Queensl. Gov. Min. J.* 19, 10–15.

Jensen, H. I. (1923). The geology of the Cairns hinterland and other parts of north Queensland. *Geol. Surv. Queensl., Publ.* 274, 1–75.

Jensen, H. I. (1925). Palaeongeography of Queensland. Part III. *Queensl. Gov. Min. J.* 26, 459–464.

Jensen, H. I. (1941). The manganese deposits of the Cairns district. *Rep. Aeronaut. Surv. N. Aust., Queensl.* 52.

Jensen, H. I. (1960a). Precambrian—Cape York Peninsula. *J. Geol. Soc. Aust.* 7, 77–78.

Jensen, H. I. (1960b). Jurassic—Cape York Peninsula. *J. Geol. Soc. Aust.* 7, 305.

Jones, L. S. (1951). Banks Island and mainland, Cape York. Inspections July 1951. *Queensl. Gov. Min. J.* **52**, 722–726.

Jones, O. A. (1948). The Maryborough earthquake of 1947. *Univ. Queensl. Pap., Dep. Geol.* [N.S.] **3**, 1–10.

Jones, O. A. (1958). Queensland earthquakes and their relations to structural features. *J. Proc. R. Soc. N.S.W.* **92**, 176–181.

Jones, O. A. (1966). Geological questions posed by the Reef. *Aust. Nat. Hist.* **15**, 245–248.

Jones, O. A. (1967). The Great Barrier Reef Committee—its work and achievements, 1922–66. *Aust. Nat. Hist.* **15**, 315–318.

Jones, O. A. (1968). Great Barrier Reefs. *In* "Encyclopedia of Geomorphology" (R. W. Fairbridge, ed.), pp. 492–499. Van Nostrand-Reinhold, Princeton, New Jersey.

Jones, O. A. (1974). The Great Barrier Reef Committee 1922–1973. *Proc. Int. Symp. Coral Reefs, 2nd, 1973* Vol. 2, pp. 733–737.

Jones, O. A., and Endean, R. (1967). The Great Barrier Reefs. *Sci. J.* **3**, 44–51.

Jones, O. A., and Jones, J. B. (1956a). Notes on the geology of some north Queensland Islands. Part I. The islands of Torres Strait. *Rep. Gt. Barrier Reef Comm.* **6**, 31–44.

Jones, O. A., and Jones, J. B. (1956b). Notes on the geology of some north Queensland Islands. Part II. Cairncross Island to Hudson Island. *Rep. Gt. Barrier Reef Comm.,* **6**, 45–54.

Jones, O. A., and Jones, J. B. (1960a). Precambrian—the coastal islands and adjacent coastal strip. *J. Geol. Soc. Aust.* **7**, 78–79.

Jones, O. A., and Jones, J. B. (1960b). Permian?—Rocks of islands between $9\frac{1}{2}°$ and 17°S. *J. Geol. Soc. Aust.* **7**, 243–244.

Jukes, J. Beete. (1847). "Narrative of the Surveying Voyage of H. M. S. Fly . . . . . During the Years 1842–1846," Vols. 1 and 2. T. and W. Boone, London.

Kennet, J. P., Burns, R. E., Andrews, J. E., Churkin, M., Davies, T. A., Dumitrica, P., Edwards, A. R., Galehouse, J. S., Packham, G. H., and van der Lingen, G. J. (1972). Australian-Antartica continental drift, paleocirculation changes and Oligocene deep-sea erosion. *Nature (London), Phys. Sci.* **239**, 51–55.

King, P. P. (1827). "Narrative of a Survey of Intertropical and Western Coasts of Australia, 1819–22." London.

Kirkegaard, A. G. (1971). Rockhampton—Mt. Morgan area. *Geol. Excurs. Handb., 43rd Meet. Aust. N.Z. Assoc. Adv. Sci.* pp. 91–103.

Kirkegaard, A. G., Shaw, R. D., and Murray, C. G. (1970). The geology of the Rockhampton and Port Clinton 1:250,000 sheet area, Queensland. *Geol. Surv. Queensl., Rep.* **38**.

Krause, D. C. (1967). Bathymetry and geologic structure of the northern Tasman Sea—Coral Sea—South Solomon Sea area of the south-western Pacific Ocean. *N.Z. Dep. Sci. Ind. Res., Bull.* **183**.

Ladd, H. S. (1950). Recent reefs. *Bull. Am. Assoc. Pet. Geol.* **34**, 203–214.

Ladd, H. S. (1968). "Preliminary Report to the Minister for Mines, Queensland on Conservation and Controlled Exploitation of the Great Barrier Reef. Gov. Printer, Brisbane.

Ladd, H. S. (1971). Existing reefs—geological aspects. *Proc. North Am. Palaeontol. Conv., 1969* Part J, pp. 1273–1300.

Layton Geophysical Consultants Pty. Ltd., Benbow, D. D., and Lloyd, A. R. (1973). "Compilation and Evaluation, Hydrocarbon Exploration Off-shore North-eastern Australia and the Gulf of Papua" (this is a very valuable proprietary report, but is available only by purchase; no part of the text or illustrations may be reproduced without permission).

Lennox-Conyngham, G., and Potts, F. A. (1925). The Great Barrier Reef. *Geogr. J.* 65, 314–334.

Levingston, K. R. (1971). Mineral deposits and mines of the Townsville 1:250,000 sheet area, North Queensland. (Generalised geological map includes Magnetic Island.) *Geol. Surv. Queensl., Rep.* 61.

Lloyd, A. R. (1967). Foraminifera from H. B. R. Wreck Island No. 1 well and Heron Island bore, Queensland; their taxonomy and stratigraphic significance. Part 1, Lituolacea and Miliolacea. *Aust., Bur. Miner. Resour., Geol. Geophys., Bull.* 92, 69–133.

Lloyd, A. R. (1968). Outline of the Tertiary geology of northern Australia. *Aust., Bur. Miner. Resour., Geol. Geophys., Bull.* 80, 105–132.

Lloyd, A. R. (1970). Neogene foraminifera from H. B. R. Wreck Island No. 1 bore and Heron Island bore, Queensland; their taxonomy and stratigraphic significance. Part 2. Nodosariacea and Buliminacea. *Aust., Bur. Miner. Resour., Geol. Geophys., Bull.* 108, 145–203.

Lloyd, A. R. (1972). Tertiary stratigraphic correlations in the Indo-Pacific region. *Aust. Oil, Gas Rev.,* 18, 11–15.

Lloyd, A. R. (1973). Foraminifera of the Great Barrier Reef bores. *In* "Biology and Geology of Coral Reefs" (O. A. Jones and R. Endean, eds.), Vol. I, pp. 347–366. Academic Press, New York.

McDougall, I. J., and Slessar, G. C. (1972). Tertiary volcanism in the Cape Hillsborough area, north Queensland. *J. Geol. Soc. Aust.* 18, 401–408.

MacGillvray, J. (1852). "Narrative of the Voyage of H. M. S. Rattlesnake . . . . . 1846–50," 2 vols. T. and W. Boone, London.

McLean, R. F. (1974). Geologic significance of bioerosion of beachrock. *Proc. Int. Symp. Coral Reefs, 2nd, 1973* Vol. 2, pp. 401–408.

McNeill, F. (1951). Wealth in coral gravels. *Aust. Mus. Mag.* 10, 190–192.

McNeill, F. (1955a). One Tree Island—remote outpost of the Capricorns. *Aust. Mus. Mag.* 11, 333–337.

McNeill, F. (1955b). Coral paradise of One Tree Island. *Aust. Mus. Mag.* 11, 404–408.

McTaggart, N. (1960). Lower Cainozoic, Bundaberg—Gin Gin. *J. Geol. Soc. Aust.* 7, 351.

Maiklem, W. R. (1967). Black and brown speckled foraminiferal sand from the southern part of the Great Barrier Reef. *J. Sediment. Petrol.* 37, 1023–1030.

Maiklem, W. R. (1968). The Capricorn Reef complex, Great Barrier Reef, Australia. *J. Sediment. Petrol.* 38, 785–798.

Maiklem, W. R. (1970). Carbonate sediments in the Capricorn Reef complex, Great Barrier Reef, Australia. *J. Sediment. Petrol.* 40, 55–80.

Maitland, A. G. (1889). Geological features and mineral resources of the Mackay district. *Geol. Surv. Queensl., Publ.* 53, 1–12.

Maitland, A. G. (1891). The geology of the Cooktown district. *Geol. Surv. Queensl., Publ.* **70.**

Maitland, A. G. (1892a). The physical geology of Magnetic Island. *Geol. Surv. Queensl., Publ.* **75,** 1–8.

Maitland, A. G. (1892b), Geological observations in British New Guinea in 1891. *Geol. Surv. Queensl., Rep.* **85.**

Maitland, A. G. (1905). "The Salient Geological Features of British New Guinea." West. Aust. Natl. Hist. Soc., Perth.

Malone, E. J., (1964). Depositional evolution of the Bowen Basin. *J. Geol. Soc. Aust.* **11,** 263–282.

Malone, E. J. (1966). Fitzroy region, Queensland—geology. *Resour. Surv. Dep. Natl. Dev. Aust.*

Marks, E. O. (1924). Some doubts in Queensland physiography. *Proc. R. Soc. Queensl.* **36,** 3–18.

Marsden, M. A. H. (1972). The Devonian history of northeastern Australia. *J. Geol. Soc. Aust.* **19,** 125–162.

Marshall, C. E., and Narain, H. (1954). Regional gravity investigations in the eastern and central Commonwealth. *Sydney Univ., Dep. Geol. Geophys., Mem.* **2,** 1–101.

Marshall, P. (1912). Pres. address Section C (what is the true structural boundary of the Pacific?). *Rep. Australas. Assoc. Adv. Sci.* **12,** 90–99.

Marshall, P. (1931). Coral reefs—rough-water and calm-water types. *Rep. Gt. Barrier Reef Comm.* **3,** 64–72.

Marshall, P. (1933). Stability of lands in the southwest Pacific. *Rep. Meet. Aust. N. Z. Assoc. Adv. Sci.* **21,** 399–411.

Marshall, P., Richards, H. C., and Walkom, A. B. (1925). Recent emergence at Holbourne Island, Great Barrier Reef. *Trans. R. Geogr. Soc. Aust. Queensl. Br.* **1,** 29–33.

Marshall, S. M., and Orr, A. P. (1931). Sedimentation on Low Isles Reef and its relation to coral growth. *Sci. Rep. Gt. Barrier Reef Exped.* **1,** 93–133.

Maxwell, W. G. H. (1962). Lithification of carbonate sediments in the Heron Island Reef, Great Barrier Reef. *J. Geol. Soc. Aust.* **8,** 217–238.

Maxwell, W. G. H. (1968). "Atlas of the Great Barrier Reef." Elsevier, Amsterdam.

Maxwell, W. G. H. (1969a). Physical geology and geography. *Spec. Publ., Aust. Conserv. Found.* **3,** 5–14.

Maxwell, W. G. H. (1969b). Relict sediments, Queensland continental shelf. *Aust. J. Sci.* **31,** 85–86.

Maxwell, W. G. H. (1969c). Radiocarbon ages of sediment, Great Barrier Reef. *Sediment. Geol.* **3,** 331–333.

Maxwell, W. G. H. (1969d). The structure and development of the Great Barrier Reef. *In* "Stratigraphy and Palaeontology" (K. S. W. Campbell, ed.), pp. 353–374. Aust. Natl. Univ. Press, Canberra.

Maxwell, W. G. H. (1969e). Regional aspects of the Great Barrier Reef Province. *Aust. Oil, Gas Rev.* **15,** 15-17, 20, and 22.

Maxwell, W. G. H. (1970). Deltaic patterns in reefs. *Deep-Sea Res.* **17,** 1005–1018.

Maxwell, W. G. H. (1971). The Great Barrier Reef. 1. Origin. *Aust. Fish.,* **30**(1), 2–7, 35.

Maxwell, W. G. H. (1972). The Great Barrier Reef—past, present and future. *Queensl. Nat.* **20,** 65–78.

Maxwell, W. G. H. (1973a). Geomorphology of Eastern Queensland in relation to to the Great Barrier Reef. *In* "Biology and Geology of Coral Reefs" (O. A. Jones and R. Endean, eds.), Vol. I, pp. 229–345. Academic Press, New York.

Maxwell, W. G. H. (1973b). Sediments of the Great Barrier Reef Province. *In* "Biology and Geology of Coral Reefs" (O. A. Jones and R. Endean, eds.), Vol. I, pp. 299–345. Academic Press, New York.

Maxwell, W. G. H., and Maiklem, W. R. (1964). Lithofacies analysis, southern part of the Great Barrier Reef. *Pap. Dep. Geol. Univ. Queensl.* [N.S.], **5**, 1–21.

Maxwell, W. G. H., and Swinchatt, J. P. (1970). Great Barrier Reef—regional variation in a terrigenous-carbonate province. *Geol. Soc. Am. Bull.* **81**, 691–724.

Maxwell, W. G. H., Day, R. W., and Fleming, P. J. G. (1961). Carbonate sedimentation on the Heron Island Reef. *J. Sediment. Petrol.* **31**, 215–230.

Maxwell, W. G. H., Jell, J. S., and McKellar, R. G. (1963). A preliminary note on the mechanical and organic factors influencing carbonate differentiation, Heron Island Reef, Australia. *J. Sediment. Petrol.* **33**, 962–963.

Maxwell, W. G. H., Jell, J. S., and McKellar, R. G. (1964). Differentiation of carbonate sediments in the Heron Island Reef. *J. Sediment. Petrol.* **34**, 294–308.

Meyers, N. A. (1969). Carpentaria Basin. *Geol. Surv. Queensl., Rep.* **34**.

Moorhouse, F. W. (1933). The recently-formed natural brestwork on Low Isles. *Rep. Gt. Barrier Reef Comm.* 4 (part 1), pp. 35, 36.

Moorhouse, F. W. (1936). The cyclone of 1934 and its effect on Low Isles with special observations on *Porites*. *Rep. Gt. Barrier Reef Comm.* 4, 37–44.

Morgan, W. R. (1960). Petrography of core samples. *In* "H. B. R. No. 1 Bore, Wreck Island, Queensland," Vol. 4, pp. 13-14. Petrol. Search Subsidy Acts, *Bur. Miner. Resour., Geol. Geophys.*, Australia.

Morgan, W. R. (1968). The geology and petrology of Cainozoic basaltic rocks in the Cooktown area north Queensland. *J. Geol. Soc. Aust.* **15**, 65–78.

Morton, C. C. (1924). Geology and mineral occurrences, Pascoe River district, Cape York Peninsula. *Geol. Surv. Queensl., Publ.* **25**, 78–83.

Morton, C. C. (1930). Batavia River gold diggings, Cape York Peninsula. *Queensl. Gov. Min. J.* **31**, 452–455 and 492–498.

Morton, C. C. (1937). Mourilyan harbour iron deposit. *Queensl. Gov. Min. J.* **38**, 427.

Moseley, H. N. (1892). "Notes by a Naturalist. An Account of Observations Made During the Voyage of H. M. S. Challenger." Murray, London.

Mott, C. V. C. (1967). The character and revolution of the Australian continental shelf. APEA *J.* **7**, 44.

Mutter, J. C. (1973). Aspects of structure and tectonic history of the continental margin of northern Queensland. *Int. Conf. Geophys. Earth Oceans* Abstr. No. 2.

Nathan, M. (1927). The Great Barrier Reef of Australia. *Geogr. J.* 70, 541–551.

Officer, C. B. (1935). South-west Pacific crustal structure. *Trans. Am. Geophys. Union* 36, 449–459.

Ogilvie, C., and Weller, N. H. F. (1949). "Report on the Utilization of the Water Resources of the Torres Strait Islands." Queensl. Irrig. Water Supply Comm., Brisbane.

Oppel, T. W. (1970). Exploration of the south-west flank of the Papuan Basin. *APEA J.* **10**, (part 2) 62–69.

Orme, G. R., Flood, P. G., and Ewart, A. (1974). An investigation of the sediments and physiography of Lady Musgrave Reef—a preliminary account. *Proc. Int. Symp. Coral Reefs, 2nd, 1973*, **2**, 371–386.

Orr, A. P. (1933). Physical and chemical conditions in the sea in the neighbourhood of the Great Barrier Reef. *Sci. Rep. Gt. Barrier Reef Exped.* **2**, 37–86.

Orr, A. P., and Moorhouse, F. W. (1933). Variations in some physical and chemical conditions on and near Low Isles Reef. *Sci. Rep. Gt. Barrier Reef Exped.* **2**, 87–98.

Packham, G., and Falvey, D. (1970). A hypothesis for the formation of marginal seas in the western Pacific. *Tectonophysics* **11**, 79–109.

Paine, A. G. L., Gregory, C. M., and Clarke, D. E. (1970). Geology of the Ayr 1:250,000 Sheet area, Queensland. *Aust. Bur. Miner. Resour. Geol. Geophys.*, Rep., 58 pp.

Palmieri, V. (1971). Tertiary subsurface biostratigraphy of the Capricorn Basin. *Geol. Surv. Queensl., Rep.* **52**.

Palmieri, V. (1969). Micropalaeontological report on Tenneco-Signal Anchor Cay No. 1 well. *Geol. Surv. Queensl., Rep.* 18 pp.

Paradice, W. E. J. (1925). The pinnacle or mushroom-shaped coral growths in connection with the reefs of the outer barrier. *Trans. R. Geogr. Soc. Aust. Queensl. Br.* **1**, 52–60.

Paradice, W. E. J. (1928). The divergence of the ends of the Great Barrier Reef from the coast. *Rep. Gt. Barrier Reef Comm.* **2**, 110–111.

Paten, R. J. (1971). Permian geology, central-western Bowen Basin. *In Geol. Excurs. Handb. 43rd Congr. Aust. N.Z. Assoc. Advmt. Sci.*, pp. 21–44.

Pearce, L. G. G. (1963). Coal resources, Burrum coalfield. *Geol. Surv. Queensl., Rep.* **1**.

Penck, A. (1896). Das grosse australische Wallriff. *Vortr. Ver. Vebr. Naturw. Kenntnisse Wien* **36**, No. 13.

Phillips Australian Oil Company, Sunray DX Oil Company and Anacapa Corporation. (1969). Marine seismic survey permit No. 39, Papua. *Bur. Miner. Resour. Geol. Geophys. Aust., Petrol. Search Subsidy Acts* **84**, 1–27.

Phipps, C. V. G. (1967). The character and evolution of the Australian continental shelf. *APEA J.* **7**(2), 44–49.

Pickett, J. W. (with contributions by others). (1972). Correlation of the Middle Devonian formations of Australia. *J. Geol. Soc. Aust.* **18**, 457–466.

Pinchin, J. (1974). A reinterpretation of seismic data in the Laura Basin, Queensland. *J. Geol. Soc. Aust.* **21**, 437–445.

Queensland Government, Department of Lands. (1972). "Strip Map, The Great Barrier Reef and Adjacent Islands. Three Sheets, 16 Miles to an Inch." *Gov. Printer.* Brisbane.

Queensland Littoral Society. (1968). "Ellison Reef Report," Aquatic Conservation Rep. No. 2, Brisbane.

Rade, J. (1973). *Scyphosphaera* in Upper Tertiary offshore, eastern Queensland. *Proc. R. Soc. Queensl.* **84**, 35–41.

Rainford, E. H. (1925). Destruction of the Whitsunday coral reefs, Queensland. *Aust. Mus. Mag.* **2**, 175–177.

Rands, W. H. (1886a). Notes on the Burrum coalfield. *Geol. Surv. Queensl., Publ.* 23.

Rands, W. H. (1886b). On the Burrum coalfield. *Geol. Surv. Queensl., Publ.* 24.

Rands, W. H. (1887). The Burrum coalfield. *Geol. Surv. Queensl., Publ.* 33.

Rands, W. H. (1896). Report on the Horn Island goldfield. *Geol. Surv. Queensl., Rep.* 112.

Rands, W. H. (1902). On the Burrum coalfield. *Geol. Surv. Queensl., Publ.* 170.

Rands, W. H. (1912). Geological features of the Maryborough district. *Queensl. Gov. Min. J.* 13, 641–642.

Rattray, A. (1869). Notes on the geology of Cape York Peninsula, Australia. *Q. J. Geol. Soc. London* 25, 297–305.

Reid, J. H. (1924). Proserpine-Mackay-Styx River. *Queensl. Gov. Min. J.* 25, 384.

Reid, J. H. (1926). Tertiary unity or disunity in Queensland. *Rep. Meet. Aust. N.Z. Assoc. Adv. Sci.* 17, 300–312.

Reid, J. H. (1942). The Tertiary oil shales of the Plevna-Eungella district. *Queensl. Gov. Min. J.* 43, 2–4.

Richards, H. C. (1913). The Cretaceous rocks of Woody Islands. *Rep. Meet. Aust. Assoc. Adv. Sci.* 14, 179.

Richards, H. C. (1922). Problems of the Great Barrier Reef. *Queensl. Geogr. J.* 36–37, 42–54.

Richards, H. C. (1923a). The importance of completing the survey of the Great Barrier Reef . . . . . *Proc. Pan-Pac. Sci. Congr., 1st 1923* Vol. 1, No. 16, pp. 704–707.

Richards, H. C. (1923b). The Great Barrier Reef of Australia. *Proc. Pan-Pac. Sci. Congr., 1st 1923* Vol. 2, No. 82, pp. 1104–1119.

Richards, H. C. (1928). Scientific investigations on the Great Barrier Reef. *Rep. Gt. Barrier Reef Comm.* 2, vii–xvi.

Richards, H. C. (1937). Some problems of the Great Barrier Reef. *J. Proc. R. Soc. N.S.W.* 71, 68–85. [Preprint issued December 1937, complete volume 1938.]

Richards, H. C. (1938). Boring operations at Heron Island, Great Barrier Reef. *Rep. Gt. Barrier Reef Comm.* 4, 135–142.

Richards, H. C. (1941a). Recent sea-level changes in eastern Australia. *Proc. Pac. Sci. Congr., 6th, 1939* pp. 853–856.

Richards, H. C. (1941b). Results of deep-boring operations on the Great Barrier Reef. *Proc. Pac. Sci. Congr., 6th, 1939* p. 857.

Richards, H. C., and Hedley, C. (1923). The Great Barrier Reef of Australia. *Queensl. Geogr. J.* 38, 105–109.

Richards, H. C., and Hedley, C. (1925). A geological reconnaissance in north Queensland. *Trans. R. Geogr. Soc. Aust. Queensl. Br.* 1, 1–26.

Richards, H. C., and Hill, D. (1942). Great Barrier Reef Bores, 1926 and 1937—descriptions, analyses and interpretations. *Rep. Gt. Barrier Reef Comm.* 5, 1–111.

Ridley, W. F. (1957). A sedimentary formation of Tertiary age between Maryborough and Bundaberg, Queensland. *Proc. R. Soc. Queensl.* 68, 11–16.

Ridley, W. F. (1960a). Jurassic—The Maryborough Basin. *J. Geol. Soc. Aust.* 7, 297–301.

Ridley, W. F. (1960b). Geological map of the Cainozoic deposits of the Maryborough basin. *J. Geol. Soc. Aust.* 7, 352.

Ridley, W. F. (1960c). Geological studies in the Maryborough basin, south-east Queensland. *Proc. R. Soc. Queensl.* 72, 83–99.

Roberts, J., Jones, P. J., Jell, J. S., Jenkins, T. B. H., Marsden, M. A. H., McKellar, R. G., McKelvey, B. C., and Seddon, G. (1972). Correlation of the Upper Devonian rocks of Australia. *J. Geol. Soc. Aust.* 18, 467–490.

Rochford, D. J. (1959). The primary external water masses of the Tasman and Coral Seas. *Aust., C.S.I.R.O., Div. Fish. Oceanogr.,* 7, 1–28.

Rochford, D. J. (1960a). The intermediate depth waters of the Tasman and Coral Seas. *Aust. J. Mar. Freshwater Res.* 11, 127–165.

Rochford, D. J. (1960b). Some aspects of the deep circulation of the Tasman and Coral Seas. *Aust. J. Mar. Freshwater Res.* 11, 166–181.

Rochford, D. J. (1968a). The continuity of water masses along the western boundary of the Tasman and Coral Seas. *Aust. J. Mar. Freshwater Res.* 19, 77–90.

Rochford, D. J. (1968b). Origin and circulation of water types on the 26.00 sigma-*t* surface of the south-west Pacific. *Aust. J. Mar. Freshwater Res.* 19, 107–127.

Rod, E. (1966). Clues to ancient Australian geosutures. *Ecol. Geol. Helv.* 59, 849–883.

Rotschi, H., and Lemasson, L. (1967). Oceanography of the Coral and Tasman Seas. *Oceanogr. Mar. Biol.* 5, 49–97.

Saint-Smith, E. C. (1916). Geology and mineral resources of the Cooktown tinfields. *Geol. Surv. Queensl., Publ.* 250, 1–206.

Saint-Smith, E. C. (1918). Limestone at Ben Lomond, Bowen district. *Queensl. Gov. Min. J.* 19, 559–560.

Saint-Smith, E. C. (1919). Rock phosphate deposits on Holbourne Island. *Queensl. Gov. Min. J.* 20, 122–124.

Sandars, D. (1961). The Great Barrier Reef of Australia. *Geol. Excurs. Handb., 35th Meet. Aust. N.Z. Assoc. Adv. Sci.* pp. 109–116.

Saville-Kent, W. (1893). "The Great Barrier Reef of Australia, its Products and Potentialities." Allen, London.

Schuchert, C. (1916). The problem of continental fracturing and diastrophism in Oceania. *Am. J. Sci.* 42, 91–105.

Shepherd. S. R. L. (1936). Portland Roads—preliminary notes. *Queensl. Gov. Min. J.* 37, 336.

Shepherd, S. R. L. (1944). Wolfram on Banks Island. *Queensl. Gov. Min. J.* 45, 209–214.

Shepherd, S. R. L. (1955). Limestone on Duke Island, Broad Sound. *Queensl. Gov. Min. J.* 56, 926.

Siller, C. W. (1961a). The geology and petroleum prospects of the Maryborough Basin, Queensland. *Aust. Oil, Gas J.* 8, 30–36.

Siller, C. W. (1961b). The geology and petroleum prospects of the Maryborough Basin, Queensland. *APEA J.* 1, 30–36.

Simmonds, N. A. H. (1961). The geology of some of the islands of the Cumberland Group. *Proc. R. Soc. Queensl.* 72, 69–73.

Simmonds, N. A. H., and Tucker, R. M. (1961). Limestone inspection, Repulse Island. *Queensl. Gov. Min. J.* 61, 23.

Spender. M. A. (1930). Island Reefs of the Queensland coast. *Geogr. J.* 76, 193–214 and 273–297.

Spender, M. A. (1932). Tidal observations on the Great Barrier Reef expedition. *Geogr. J.* 79, 201–209.

Spender, M. A. (1936). The cyclone at Low Isles. *Geogr. J.* 88, 383–4.

Stack, L. W. (1964). Petroleum potentialities of the continental shelf between Cape York and south-western Papua. *APEA J.* 4, 68–73.

Stanley, G. A. V. (1928a). The physiography of the Bowen district and of the northern islands of the Cumberland Group (Whitsunday Passage). *Rep. Gt. Barrier Reef Comm.*, 2, 1–51.

Stanley, G. A. V. (1928b). Physiographic investigations in Queensland with reference to the Great Barrier Reef. *Am. J. Sci.* 16, 45–50.

Steers, J. A. (1929). The Queensland coast and the Great Barrier Reefs. *Geogr. J.* **74**, 232–257 and 341–370.

Steers, J. A. (1930). A geographical introduction to the biological reports. *Sci. Rep. Gt. Barrier Reef Exped.* **3**, 1–15.

Steers, J. A. (1931). Evidences of recent movements of sea-level on the Queensland coast: Raised beaches and the coral reef problem. *Congr. Geog. Int., C. R.* Vol. 2.

Steers, J. A. (1937). The coral islands and the associated features of the Great Barrier Reefs. *Geogr. J.* **89**, 1–29 and 120–146.

Steers, J. A. (1938). Detailed notes on the islands surveyed and examined by the geographical expedition to the Great Barrier Reef in 1936. *Rep. Gt. Barrier Reef Comm.* **4**, 51–96.

Steers, J. A. (1945). Coral reefs and air photography. *Mon. Rec. Geogr. J.* **106**, 233–235.

Steers, J. A. (1969). Essay review, "The Great Barrier Reef" (review of "Atlas of the Great Barrier Reef," W. G. H. Maxwell). *Geol. Mag.* **106**, 217–220.

Stephenson, P. J. (1970a). The Townsville-Charters Towers area, physiography. *Field Conf. Handb., Geol. Soc. Aust., Queensl. Div., 1970* pp. 5–7.

Stephenson, P. J. (1970b). Geology of the Townsville area. *Field Conf. Handb. Geol. Soc. Aust., Queensl. Div., 1970* pp. 23–27.

Stephenson, P. J., Stevens, N. C., and Tweedale, G. W. (1960). Lower Cainozoic—igneous rocks. *J. Geol. Soc. Aust.* **7**, 355–369.

Stephenson, T. A., Stephenson, A., Tandy, G., and Spender, M. (1931). The structure and ecology of Low Isles and other reefs. *Sci. Rep. Gt. Barrier Reef Exped.* **3**, 17–112.

Stevens, N. C. (1961). Upper Cainozoic volcanism near Gayndah, Queensland. *Proc. R. Soc. Queensl.* **72**, 75–82.

Stevens, N. C. (1968). Triassic volcanic rocks of Agnes Water, Queensland. *Pap. Dep. Geol. Univ. Queensl.* [N.S.] **6**, 147–154.

Stevens, N. C. (1969). The volcanism of southern Queensland. *Spec. Publ., Geol. Soc. Aust.* **2**, 193–202.

St. John, B. E. (1974). Heavy metals in the skeletal carbonate of Scleractinian corals. *Proc. Int. Symp. Coral Reefs, 2nd, 1973,* Vol. 2, pp. 461–469.

Stoddart, D. R. (1965). British Honduras Cays and the Low Wooded Island problem. *Trans. Inst. Br. Geogr.* **64**, 131–147.

Stoddart, D. R. (1969a). Ecology and morphology of Recent coral reefs. *Biol. Rev. Cambridge Philos. Soc.* **44**, 433–498.

Stoddart, D. R. (1969b). Reef morphology (review of "Atlas of the Great Barrier Reef," W. G. H. Maxwell). *Nature (London)* **221**, 978.

Stoddart, D. R. (1971). The unquiet landscape; variations on a coral theme. *Geogr. Mag., London* pp. 610–615.

Stokes, J. L. (1846). "Discoveries in Australia . . . . . The voyage of H. M. S. Beagle, 1837–43 . . . .," 2 vols. T. and W. Boone, London (Vol. 1, pp. 323–380 deals with the Reefs and adjacent coast and islands).

Strusz, D. L. (with contributions by others). (1972). Correlation of the Lower Devonian rocks of Australasia. *J. Geol. Soc. Aust.* **18**, 427–455.

Stubbings, H. G. (1938). Marine sediments from islands and reefs of the Great Barrier Reef. (Appendix to Steers, 1938, q.v.) *Rep. Gt. Barrier Reef Comm.* **4**, 97–104.

Sussmilch, C. A. (1930). Post-Pliocene tectonic movements in eastern Australia. *Proc. Pac. Sci. Congr., 4th, 1929* Vol. 2, p. 657.

Sussmilch, C. A. (1938). The geomorphology of eastern Queensland. *Rep. Gt. Barrier Reef Comm.* 4, 105–134.

Swinchatt, J. P. (1967). Marine geology and sedimentology, Arlington Reef Complex and adjacent areas northern Great Barrier Reef. *Abstr. Geol. Soc. Am. Annu. Meet. Program* p. 219.

Swinchatt, J. P. (1968). Sedimentary facies in areas of mixed carbonate-terrigenous clastic deposition, Arlington Reef complex and adjacent areas, northern Great Barrier Reef. *Am. Assoc. Pet. Geol. Bull.* 52, 551 (abstr.).

Taylor, T. (1942). Movement of Sand Cays. *Queensl. Geogr. J.* 39, 38.

Taylor, T. G. (1911). Physiography of eastern Australia. *Bull. Bur. Met. Aust.* 8, 1–17.

Taylor, T. G. (1914). Physiography and general geography of Australia. *In "Handb. Aust., Bt. Assoc. Adv. Sci.* (G. H. Knibbs, ed).

Taylor, T. G. (1927). The topography of Australia. *Yearb. Commonw. Aust.* 20, 75–86.

Tenison-Woods, J. E. (1880). On a fossiliferous bed at the mouth of the Endeavour River. *Proc. Linn. Soc. N.S.W.* 5, 187–189.

Tenison-Woods, J. E. (1881). Geology of northern Queensland. *Trans. Queensl. Philos. Soc.* 3, 189–238.

Thom, B. G., Hails, J. R., and Martin, A. R. H. (1969). Radiocarbon evidence against higher post-glacial sea-levels in eastern Australia. *Mar. Geol.* 7, 161–168.

Thom, B. G., Hails, J. R., and Martin, A. R. H. (1972). Post-glacial sea-levels in eastern Australia—a reply. *Mar. Geol.* 12, 233–242.

Thompson, J. E. (1967). A Geological history of eastern New Guinea. *APEA J.* 7, 83–93.

Thomson, W. C. (1905). Upheavals and depressions in the Pacific and on the Australian coast. *Queensl. Geogr. J.* 20, 1–8.

Thyer, R. F. (1970). Progress in petroleum prospecting on Australia's continental shelf *in* mining and petroleum geology. *Proc. Commonw. Min. Metall. Congr., 9th, 1969* Vol. 2.

Traves, D. M. (1960). Lower Cainozoic—Wreck Island, subsurface. *J. Geol. Soc. Aust.* 7, 369–371.

Twidale, C. R. (1956). A physiographic reconnaissance of some volcanic provinces in north Queensland. *Bull. Volcanol.* 18(2), 3–23.

Vaughan, T. W. (1916). Some littoral and sublittoral physiographic features of the Virgin and northern Leeward Islands and their bearing on the coral reef problem. *Geol. Soc. Am. Bull.* 27, 41–45 (abstr.).

Vaughan, T. W. (1919a). Corals and the formation of coral reefs. *Smithson. Inst. Annu. Rep.* pp. 189–276.

Vaughan, T. W. (1919b). Great Barrier Reef of Australia. *In* Vaughan (1919, pp. 306–329). Vaughan, T. W. (1919). Fossil corals from Central America, Cuba and Porto Rico, with an account of the American Tertiary, Pleistocene and Recent coral reefs. *U.S., Natl. Mus., Bull.* 103, 189–524.

Vaughan, T. W., in collaboration with J. A. Cushman, M. I. Goldman, M. A. Howe, and others. (1918). Some shoal water bottom samples from Murray Island, Australia and comparisons of them with samples from Florida and the Bahamas. *Carnegie Inst. Washington Publ.* 213, 235–316.

Vaux, D. (1970). Surface temperatures and salinity for Australian waters, 1961–65. *Atlas, C.S.I.R.O., Div. Fish. Oceanog.* **1.**

Veeh, H. H., and Veevers, J. J. (1970). Sea-level at −175 metres off the Great Barrier Reef 13,600 to 17,000 years ago. *Nature (London)* **226,** 536–537.

Verstappen, H., T. (1968). Coral reefs—wind and current growth control. *In* "Encyclopedia of geomorphology" (R. W. Fairbridge, ed.), pp. 197–202. Van Nostrand-Reinhold, Princeton, New Jersey.

Vind, E. W., and Harwood, C. R. (1965). Geophysical exploration in Torres Strait. *APEA J.* **5,** 188–190.

Walkom, A. B. (1918). Mesozoic floras of Queensland. Part 2. The flora of the Maryborough (marine) series. *Geol. Surv. Queensl. Publ.* **262.**

Walkom, A. B. (1919). Mesozoic floras of Queensland. Parts 3 and 4. The floras of the Burrum and Styx River Series. *Geol. Surv. Queensl., Publ.* **263.**

Walkom, A. B. (1928). Fossil Plants from Plutoville, Cape York Peninsula. *Proc. Linn. Soc. N.S.W.* **53,** 145–150.

Watkins, J. R. (1967). The relationship between climate and the development of landforms in the Cainozoic rocks of Queensland. *J. Geol. Soc. Aust.* **14,** 153–168.

Webb, A. W., and McDougall, I. (1964). Granites of Lower Cretaceous age near Eungella, Queensland. *J. Geol. Soc. Aust.* **11,** 151–153.

Webb, A. W., and McDougall, I. (1968). The geochronology of the igneous rocks of eastern Queensland. *J. Geol. Soc. Aust.* **15,** 313–346.

Webb, A. W., Cooper, J. A., and Richards, J. R. (1963). K-Ar ages on some central Queensland granites. *J. Geol. Soc. Aust.* **10,** 317–324.

Weber, J. N., and Woodhead, P. M. J. (1968). Factors affecting the carbon and oxygen isotopic composition of marine carbonate sediments. II. Heron Islands, Great Barrier Reef, Australia. *Geochim. Cosmochim. Acta* **33,** 19–38.

Weeks, L. G. (1959). Geologic architecture of the circumpacific. *Bull. Am. Assoc. Pet. Geol.* **43,** 350–380.

White, D. A. (1961). Geological history of the Cairns-Townsville hinterland, northern Queensland. *Aust., Bur. Miner. Resour., Geol. Geophys., Rep.* **59.**

Whitehouse, F. W. (1926). The correlation of the marine Cretaceous deposits of Australia. *Rep. Meet. Aust. N.Z. Assoc. Adv. Sci.* **18,** 275–280.

Whitehouse, F. W. (1930). The geology of Queensland. *Handb. Aust. Assoc. Adv. Sci., Brisbane Meeting.* pp. 23–29.

Whitehouse, F. W. (1947). A marine early Cretaceous fauna from Stanwell (Rockhampton District). *Proc. R. Soc. Queensl.* **57,** 7–20.

Whitehouse, F. W. (1951). Physiography. *Handb. Queensl., Aust. Assoc. Adv. Sci., Brisbane* pp. 5–12.

Whitehouse, F. W. (1956). Australian and New Zealand research in eustasy 11. *Aust. J. Sci.* **19,** 54.

Whitehouse, F. W. (1963). The sandhills of Queensland, coastal and desert. *Queensl. Nat.* **17,** 1-10.

Whitehouse, F. W. (1964). Sediments forming. *Queensl. Nat.* **17,** 51–58.

Whitehouse, F. W. (1967). Wallum country. *Queensl. Nat.* **18,** 64–70.

Wilkinson, C. S. (1877). Notes on a collection of geological specimens . . . from the coasts of New Guinea, Cape York and neighbouring islands. *Proc. Linn. Soc. N.S.W.,* **1,** 113–117.

Willman, P., and McDougall, I. (1974). Cainozoic igneous activity in Eastern Australia. *Tectonphysics* **23,** 49–65.

Willmott, W. F. (1972). Late Cainozoic shoshonitic lavas from Torres Strait, Papua and Queensland. *Nature* (*London*), *Phys. Sci.* **235**, 33–35.

Willmott, W. F., Whitaker, W. G., Pafreyman, W. D. and Trail, D. S. (1973). Igneous and metamorphic rocks of Cape York Peninsula and Torres Strait. *Aust. Bur. Miner. Resour. Geol. Geophys.*, *Bull.* **135** (viii) + 145 pp.

Wilson, T. C. (1968). Exploration—Great Barrier Reef area. *APEA J.* **7**, 33–39.

Wilson, T. C. (1969). Geology of southern Swain Reef area, Queensland, Australia. *Am. Assoc. Pet. Geol. Bull.* **53**, 750 (abstr.).

Winterer, E. L. (1968). Submarine valley systems around the Coral Sea Basin. *In* Falvey, D. A. (1972). On the margins of marginal plateaux. Thesis (unpublished) University of New South Wales.

Wolf, K. H., and Ostlund, G. (1967). ^{14}C dates of calcareous samples, Heron Island, Great Barrier Reef. *Sedimentology* **8**, 249–251.

Woodhead, P. M. J. (1968). Sea surface drift between central Queensland and New Zealand. *Aust. J. Sci.* **31**, 195–196.

Woods, J. T., and Wood, P. (1973). Offshore petroleum and mineral resources. *Proc. R. Soc. Queensl.* **84**, 87–97.

Wyatt, D. H. (1968). Explanatory notes on the Townsville, Queensland, 1:250,000 geological series map. *Aust., Bur. Miner. Resour., Geol. Geophys.* SE/55-1.

Wyatt, D. H., and Webb, A. W. (1970). Potassium-argon ages of some northern Queensland basalts and an interpretation of late Cainozoic history. *J. Geol. Soc. Aust.* **17**, 39–51.

Wyrtki, K. (1960). The surface circulation in the Coral and Tasman Seas. *Aust., C.S.I.R.O., Div. Fish. Oceanogr., Tech. Pap.* **8**, 1–44.

Wyrtki, K. (1962a). The subsurface water masses in the western South Pacific Ocean. *Aust. J. Mar. Freshwater Res.* **13**, 18–47.

Wyrtki, K. (1962b). Geopotential topographies and associated circulation in the western Pacific Ocean. *Aust. J. Mar. Freshwater Res.* **13**, 89–105.

Yonge, C. M. (1930). "A Year on the Great Barrier Reef." Putnam, New York.

Yonge, C. M. (1931). The Great Barrier Reef Expedition 1928–1929. *Rep. Gt. Barrier Reef Comm.*, **3**, 1–25.

Yonge, C. M. (1951). The form of coral reefs. *Endeavour* **39**, 136–144.

B. Unpublished Theses Available for Perusal at Various Libraries

Conaghan, P. I. (1963). Terrestrial and marine geology of the Gladstone Area, Queensland (Univ. Qd).

Conaghan, P. I. (1967). "Marine Geology of the Southern Tropical Shelf, Queensland." Ph.D. Thesis, University of Queensland, Brisbane.

Frankel, E. (1971). Recent sedimentation in the Princess Charlotte Bay and Edgecumbe Bay Areas, Great Barrier Reef Province. Ph.D. Thesis, University of Sydney (unpublished).

Jell, J. S. (1961). "Stratigraphy, Structure and Palaeontology of the Mt. Larcom District, Central Queensland. B.Sc. Thesis, University of Queensland, Brisbane.

Lloyd, A. R. (1961). "Foraminifera from the Subsurface Miocene of Wreck Island, Queensland." M.Sci. Thesis, of University of Adelaide.

Mutter, J. C. (1972). "A Recent Geophysical Reconnaissance of the Gulf of Papua and N. W. Coral Sea," *Symp. University of New South Wales.*

Slessar, G. C. (1970). Geology of Cape Hillsborough, Mackay Region. B.Sc. Hons. Thesis (unpublished).

St. John, B. E. (1970). "Trace Elements in Corals of the Coral Sea. An Investigation Relating their Distribution to Oceanographic Factors." Ph.D. Thesis, University of Queensland, Brisbane.

Tanner, J. J. (1969). The ancestral Great Barrier Reef in the Gulf of Papua. *ECAFE Conf., Canberra* pp. 1–5.

Thomas, B. M. (1966). "The Marine Geology of the Whitsunday Passage Area, Queensland." Hons. Thesis, University of Queensland, Brisbane.

Winterer, E. L. (1968). Submarine valley systems around the Coral Sea Basin (in Falvey, 1972). Thesis, University of New South Wales.

C. Records of the Bureau of Mineral Resources, Geology and Geophysics, Australia. Unpublished, but Available at University and Geological Survey Libraries

Best, J. G. (1960). Some Cainozoic basaltic volcanoes in north Queensland. *Aust., Bur. Miner. Resour., Geol. Geophys., Rec.* **78.**

Burgis, W. A. (1972). Cainozoic history of the peninsula east of Board Sound, Queensland. *Aust., Bur. Miner. Resour., Geol. Geophys., Rec.* **78.**

Chamberlain, N. G., and Maddern, C. A. (1959). Review of geophysical exploration for oil in Australia and New Guinea to the end of 1958. *Aust., Bur. Miner. Resour., Geol. Geophys. Rec.* **15.**

Clarke, D. E. (1969). Geology of the Ravenswood 1—mile sheet area, Queensland. *Aust., Bur. Miner. Resour., Geol. Geophys., Rec.* **117.**

Clarke, D. E., Paine, A. G. L., and Jensen, A. R. (1968). The geology of the Proserpine 1:250,000 sheet area, Queensland. *Aust., Bur. Miner. Resour., Geol. Geophys., Rec.* **22.**

Cook, P. J. (1972), The supra tidal environment and recent dolomitization, Broad Sound, Queensland. *Aust., Bur. Miner. Resour., Geol. Geophys., Rec.* **125.**

de Kayser, F. (1961). Geology and mineral deposits of the Mossman 1:250,000 sheet area, North Queensland. *Aust., Bur. Miner. Resour., Geol. Geophys., Rec.* **110.**

de Kayser, F. (1964). The Barnard metamorphics and their relation to the Barron River metamorphics and the Hodgkinson Formation, North Queensland. *Aust., Bur. Miner. Resour., Geol. Geophys., Rec.* **122.**

de Kayser, F., Fardon, R. S., and Cuttler, L. G. (1964). Geology of the Ingham 1:250,000 geological sheet area, Queensland. *Aust., Bur. Miner. Resour., Geol. Geophys., Rec.* **78.**

Dooley, J. C. (1959). Preliminary report on underwater gravity survey, Hervey Bay, Queensland. *Aust., Bur. Miner. Resour., Geol. Geophys., Rec.* **68.**

Dooley, J. C. (1963). Gravity surveys of the Great Barrier Reef and adjacent coast, north Queensland. *Aust., Bur. Miner. Resour., Geol. Geophys., Rec.* **163.**

Dooley, J. C., and Goodspeed, M. J. (1959). Preliminary report on underwater gravity survey, Great Barrier Reef area—Rockhampton to Gladstone. *Aust., Bur. Miner. Resour., Geol. Geophys. Rec.* **69.**

Evans, P. R. (1966). Contribution to the palynology of Northern Queensland and Papua. *Aust., Bur. Min. Resour., Geol. Geophys., Rec.* **198.**

Gibson, D. L. (1972). Clay mineralogy of Recent sediments of the Broad Sound area, Queensland. *Aust., Bur. Miner. Resour., Geol. Geophys., Rec.* **65.**

Goodspeed, M. J., and Williams, L. W. (1959). Preliminary report on underwater gravity survey, Great Barrier Reef area, Thursday Island to Rockhampton. *Aust., Bur. Miner. Resour., Geol. Geophys., Rec.* **70**.

Jensen, A. R. (1963). The geology of the southern half of the Proserpine 1:250,000 sheet area. *Aust., Bur. Resour., Geol. Geophys. Rec.* **65**.

Jensen, A. R., Gregory, C. M., and Forbes, V. R. (1963). Regional geology of the Mackay 1:250,000 sheet area, Queensland. *Aust., Bur. Miner. Resour., Geol. Geophys., Rec.* **71**.

Jongsma, D., and Marshall, J. F. (1971). BMR marine geology cruise in the southern Barrier Reef and northern Tasman Sea, 12/9/70 to 14/12/70. *Aust., Bur. Miner. Resour., Geol. Geophys., Rec.* **17**.

Konecki, M. C., and Blair, K. (1970). Preliminary analyses of natural gases encountered in exploration drilling in Australia and Papua-New Guinea. *Aust., Bur. Miner. Resour., Geol. Geophys., Rec.* **78**.

Lucas, K. G. (1962). The geology of the Cooktown 1:250,000 sheet area, SD 55/13, north Queensland. *Aust., Bur. Miner. Resour., Geol. Geophys., Rec.* **149**.

Lucas, K. G. (1964). The geology of the Cape Melville 1:250,000 sheet area, SD 55/9, north Queensland. *Aust., Bur. Miner. Resour., Geol. Geophys., Rec.* **93**.

McKay, B. A., and Loban, V. (1970). Some characteristics of formation waters obtained from petroleum exploration and development wells in Australia and Papua-New Guinea including the continental shelf. *Aust., Bur. Miner. Resour., Geol. Geophys., Rec.* **78**.

Malone, E. J., Mollan, R. G., Jensen, F., Gregory, A. R., Kirkegaard, A. G., and Forbes, V. R. (1963). Geology of the Duaringa and St. Lawrence 1:250,000 sheet areas, Queensland. *Aust., Bur. Miner. Resour., Geol. Geophys., Rec.* **60**.

Marshall, J. F. (1972). Morphology of the east Australian continental margin between 21°S and 23°S. *Aust., Bur. Miner. Resour., Geol. Geophys., Rec.* **70**.

Paine, A. G. L., Clarke, D. E., and Gregory, C. M. (1970). Geology of the northern half of the Bowen 1:250,000 sheet area, Queensland (with additions to the geology of the southern half). *Aust., Bur. Miner. Resour., Geol. Geophys., Rec.* **50**.

Paine, A. G. L., Gregory, C. M., and Clarke, D. E. (1966). The geology of the Ayr 1:250,000 sheet area, Queensland. *Aust., Bur. Miner. Resour., Geol. Geophys., Rec.* **68**.

Pinchin, J. (1973). A geophysical review of the Carpentaria, Laura and Olive River Basins. *Aust., Bur. Miner. Resour., Geol. Geophys., Rec.* **132**.

Reynolds, M. A. (1965). The sedimentary basins of Australia and the stratigraphic occurrence of hydrocarbons. *Aust., Bur. Miner. Resour., Geol. Geophys., Rec.* **196**.

Reynolds, M. A. *et al.* (1963). The sedimentary basins of Australia and New Guinea. *Aust., Bur. Miner. Resour., Geol. Geophys., Rec.* **159**.

Trail, D. S., Willmott, D. F., Palfreyman, W. D., Spark, R. F., and Whitaker, W. G. (1969). The igneous and metamorphic rocks of the Coen and Cape Weymouth 1:250,000 sheet areas, Cape York Peninsula. *Aust., Bur. Miner. Resour., Geol. Geophys., Rec.* **64**.

Whitaker, W. G., and Willmott, W. F. (1968). The nomenclature of the igneous and metamorphic rocks of Cape York Peninsula, Qld. Part 1. The southern area. *Aust., Bur. Miner. Resour., Geol. Geophys., Rec.* **48**.

Williams, L. W., and Waterlander, S. (1958). Preliminary report of an underwater gravity survey. Bramble Cay to Cape Arnhem. *Aust., Bur. Miner. Resour., Geol. Geophys., Rec.* **102**.

Wilmott, W. F., Palfreyman, W. D., Trail, D. S., and Whitaker, W. G. (1969). The igneous rocks of Torres Strait, Queensland and Papua. *Aust., Bur. Miner. Resour., Geol. Geophys., Rec.* **119.**

Wyatt, D. H., Paine, A. G. L., Harding, R., and Clarke, D. E. (1965). The regional geology of the Townsville 1:250,000 sheet area. *Aust., Bur. Miner. Resour., Geol. Geophys., Rec.* 1965, **159.**

Wyatt, D. H., Paine, A. G. L., Clarke, C. M., and Harding, R. R. (1967). The geology of the Charters Towers 1:250,000 sheet area, Queensland. *Aust., Bur. Miner. Resour., Geol. Geophys., Rec.* **104.**

D. Record of the Geological Survey Branch of the Mines Department, Queensland. Unpublished, but Available at University and Geological Survey Libraries

Kirkegaard, A. C., Shaw, R. D., and Murray, C. G. (1966). The geology of the Rockhampton and Port Clinton 1:250,000 sheet areas. *Geol. Surv. Queensl., Rec.* 1.

E. Unpublished Company Reports on "Open File" in the Library of Queensland Geological Survey Branch

Afflick, J., and Landau, J. F. (1965). Interpretation of an aeromagnetic survey, Swain Reefs area, Concession 90P; Report M—Australia—168. Gulf Research and Development Co., for Australian Gulf Oil Co.

American Overseas Petroleum Ltd. (1968). Final report, Warrior Reef seismic survey A. to P. 133P. Queensland.

Ampol Exploration (Queensland) Pty. Ltd. (1963). Interpretation report of airborne magnetometer survey over concessions 93P and 94P. Mackay Area, Queensland.

Ampol Exploration (Queensland) Pty. Ltd. (1964a). Seismic survey report on Authority to Prospect 94P. Queensland, Australia.

Ampol Exploration (Queensland) Pty. Ltd. (1964b). Final report offshore area 93P and 94P. Australia, marine seismic survey.

Ampol Exploration (Queensland) Pty. Ltd. (1965). Proserpine No. 1 well. Completion report.

Ampol Exploration (Queensland) Pty. Ltd. (1966). Final report of the sparker survey, Broad Sound area, A to P 103P. Queensland.

Australian Aquitane Petroleum Pty. Ltd. (1964). An airborne magnetometer survey of Winlock River (Cape York Peninsula).

Australian Aquitane Petroleum Pty. Ltd. (1965). Pascoe "B"—William Thompson, Field report 12.

Australian Gulf Oil Co. (1965a). "Interpretation of an Aeromagnetic Survey Swain Reef Area," Concession 90P. Great Barrier Reef, Australia.

Australian Gulf Oil Co. (1965b). Preliminary marine seismic project Swain Reef area, authority to prospect 90P.

Australian Gulf Oil Co. (1968). Completion report seismic survey, 90. Swain Reefs Area.

Australian Gulf Oil Co. (1969). Aeromagnetic survey, Townsville area.

Australian Oil and Gas Corp. Ltd. (1962). Preliminary interpretation of airborne magnetometer profiles over Barrier Reef, Queensland, Australia.

Broken Hill Proprietary Company Limited. (1962). Final report on A. to P. 86M. Cape York, Queensland.

Bruce, I. D. (1965). Final report, Bunker Group marine seismic survey, Maryborough Basin, Queensland. A. to P. 70P. Shell Development (Australia) Pty. Ltd.

Carlson, C. T., and Wilson, T. C. (1968a). Gulf—A.O.G., Aquarius No. 1 well, completion report. Australian Gulf Oil Co.

Carlson, C. T., and Wilson, T. C. (1968b). Capricorn No. 1A well, completion report. Australian Gulf Oil Co.

Corbett Reef Ltd. (1968). Completion report, Cooktown aeromagnetic survey.

de Jong, H. (1969). Final report, Bligh Entrance marine seismic survey, Gulf of Papua, PNG/IP, 2P, 3P. Philips Petroleum Co.

Endeavour Oil Co. N. L. (1969). Q/9P, Offshore Laura Basin seismic survey.

Evans, P. R. (1957). A to P Cape York Peninsula summary report—Reconnaissance Survey, 1956-7. Aust. Min. and Smelting Co.

Exoil N. L. (1970). Q/8P Princess Charlotte Bay (Offshore) seismic survey.

Fowle, R. A. (1968). Final report, northern Great Barrier Reef (A to P 88P), detail Triangle Reef (A to P 134P), marine seismic survey, Queensland, Australia. Tenneco Australia Inc.

Gough, K. (1961). Final report on Queensland A to P. No. 77P. Reef Oil Co. Ltd.

Gulf Interstate Overseas Ltd. (1963). Aerial magnetic report. Authority to Prospect 88P. Queensland, Australia.

Gulf Interstate Overseas Ltd. (1965). Final report marine seismograph survey, off-shore Australia, Torres Strait, Princess Charlotte Bay areas.

Gulf Interstate Overseas Ltd. (1967). Aeromagnetic survey, Torres Strait area. Gulf Interstate Overseas Co.

Hartman, R. R. (1962). A preliminary interpretation of airborne magnetometer profiles over Barrier Reef, Queensland, Australia. For Australian Oil and Gas Corporation Ltd.

Hightower, W. E., and Thralls, H. M. (1958). Seismic survey report, the Pialba area, country of March, Queensland, Australia.

Hill, D. (1956). The geology of the Great Barrier Reef in relation to oil potential, 150 pp. Aust. Min. and Smelt Co. Ltd.

Jannick, A. (1968). Marble Island iron prospect, A. to P. 490M, Progress Report No. 1. Metals Exploration N.L.

Jenny, W. P. (1963). Aerial magnetic survey report, A. to P. 88P, Queensland, Australia. Gulf Interstate Overseas Ltd.

Jenny, W. P. (1964). An airborne magnetometer survey of Wenlock River (Cape York Peninsula), part of A to P 95P. Australia Aquitaine Petroleum Pty. Ltd.

Laing, A. C. M. (1959). Reconnaissance geology of the northern section of the Bowen Basin and Styx Basin. Mines Administration Pty. Ltd. Rep. Q/56P/63.

Lloyd, A. R. (1968). Palaeontological report, Aquarius No. 1 well, Queensland. (Appendix to Carlson and Wilson, 1968a.)

McKeague, E., and Paterson, O. D. (1957a). Final report on authority to prospect, Yeppoon to Cape Clinton. Mineral Deposits Pty. Ltd.

McKeague, E., and Paterson, O. D. (1957b). Final report on authority to prospect, Bustard Head to Burnett River. Mineral Deposits Pty. Ltd.

Madden, J. J. (1958). Interim report, A to P 35P, Flinders Group and Princess Charlotte Bay. Aust. Mining and Smelting Co.

Marathon Petroleum Aust. Ltd. (1962). Cabot-Blueberry Marina No. 1, Qld., well completion report.

Marathon Petroleum Aust. Ltd. (1963). Marina Plains reflection seismograph survey, A to P 61P, Queensland, Australia.

Mead (1955). The Maryborough Basin, Queensland, Australia. Western Petroleum Explor. Co.

Middle States Petroleum Company. (1965). Torres Strait—Princess Charlotte Bay marine seismic survey. Gulf Interstate Overseas Ltd.

Mines Administration Pty. Ltd. (1962). Cabot-Blueberry Marina No. 1, well completion report.

Mines Administration Pty. Ltd. (1965). A to P 61P Queensland, final report.

Namco (1967). Final report, North Bundaberg seismic survey, Maryborough Basin. Exoil N.L.

Nicholls, C. W. (1967). Final report, seismic and magnetic survey, northern Great Barrier Reef area, Queensland, Australia. Tenneco Australia Inc.

Nicholls, C .W. (1969). Final report, Pearce Cay marine seismic survey, petroleum exploration permit Q/10P, Queensland, Texaco Overseas Petroleum Co.

North Australian Cement Ltd (1967). Report on investigation of marine carbonate sand deposits near Townsville for possible use as a portland cement raw material. Unpublished (open file) report to Dept. of Mines.

Oppel, T. W. (1969). Anchor Cay No. 1 offshore Queensland. Well completion report. Tenneco Australia Inc.

Paterson, O. D. (1957). Final report on authority to prospect, Cape York Peninsula. Mineral Deposits Pty. Ltd.

Ranneft, T. S. M. (1968). Review of A to P 135P, Laura Basin, Queensland. Exoil N.L.

Roosjen, B. (1966). Final report of the Sparker survey, Broad Sound area, A. to P. 103P, Queensland, by United Geophysical Corp. for Ampol Exploration (Queensland) Pty. Ltd.

Shell Development (Aust.) Pty. Ltd. (1964). Final report Pialba-Fraser Island seismic survey, Maryborough Basin, Queensland, A to P 70P.

Shell Development (Aust.) Pty. Ltd. (1967). Gregory River No. 1 well, Queensland. Well completion report.

Shell Development (Aust.) Pty. Ltd. (1969). Final Report, Hervey Bay R—1 seismic survey, Maryborough Basin.

Shell Development (Aust.) Pty. Ltd. (1970). Final report Hervey Bay seismic survey, Maryborough Basin.

Siller, C. W. (1955). Final report on L.S.D. No. 1 Cherwell Well 6P. Lucky Strike Drilling Co.

Siller, C. W. (1956). Final report on L.S.D. No. 2 Susan River Well, 6P. Lucky Strike Drilling Co.

Siller, C. W. (1957). Final report on A. to P. 6P Lucky Strike Drilling Co.

Smith, R. (1965). Hervey Bay regional seismic survey, Maryborough Basin, Queensland, A. to P. 70P. Shell Development (Aust.) Pty. Ltd.

Swindon, V. G. (1960). Report on the Laura Basin area of A to P 61/P, Queensland, Australia. Plymouth Oil Company.

Tenneco Australia Inc. (1966). Final report, seismic and magnetic survey, 88P and IIIP northern Great Barrier Reef.

Tenneco Australia Inc. (1968). Final report north Great Barrier Reef (AP88P), detail Triangle Reef (AP 134P), marine seismic survey, Queensland, Australia.

Tenneco Australia Inc. (1969). Completion report Anchor Cay No. 1 well, QI/P.

Fig. 1. The eight major areas of the Queensland Shelf. A, Torres Strait; B, Warrior Reef; C, Northern Great Barrier Reefs; D, Cape York–Lloyd Bay; E, Lloyd Bay–Cape Melville; F, Cape Melville–Townsville–Mackay; G, Mackay–Swain Reefs; H, Bunker–Capricorn Reefs.

geological history and tectonic framework of the Queensland Shelf and briefly outlined the geology of isolated areas. More recent geophysical and geological studies of the Queensland Shelf and adjacent areas permit a better understanding of the geology of the Queensland Shelf and the development of the Great Barrier Reefs.

The geology of the Shelf is complex and will be divided into eight areas, as shown in Fig. 1. The term "basement" is used in the sense of economic basement and includes indurated sediments, and metamorphic and igneous rocks.

II. Torres Strait (Area A)

The islands of the Torres Strait consist of granetic rocks, schists, and volcanics (Jones and Jones, 1956; Willmott *et al.*, 1973). The granites are Late Carboniferous in age, as determined by radiometric dating. The schists are considered to be Proterozoic in age, based on the results of the Weipa No. 1, Wyabba No. 1, Normanton No. 1, and Karumba No. 1 wells drilled along the western coast of Cape York Peninsula (Myers, 1969) and onshore outcrop. The volcanics include ignimbrites, which were considered by Jones and Jones (1956) to be possibly Early Triassic in age, and olivine nephelinite, which is Pleistocene in age (Willmott *et al.*, 1973).

These islands mark the Oriomo Ridge which extends from Cape York to Mabaduan on the southwestern coast of Papua New Guinea. Paleozoic, Mesozoic, and Tertiary sediments are absent in Area A. This area was therefore a part of a land mass throughout the Paleozoic, Mesozoic, and Tertiary. It was not transgressed by the seas until the Middle Pleistocene when the reefs began to develop. The reefs of Area A have therefore formed directly on Proterozoic and Late Carboniferous rocks of the Oriomo Ridge.

III. Warrior Reefs (Area B)

The Warrior Reefs are controlled by a basement high which is the eastern part of the Oriomo Ridge. The rocks of this high are thought to consist of Proterozoic and possible Late Carboniferous rocks similar to those in Area A. From geophysical data, the basement at the Warrior Reefs appears to be overlaid directly by Early to early Middle Miocene and Pliocene limestones similar to those penetrated in the Anchor Cay No. 1 Well. The Warrior Reefs have therefore developed directly on

Tertiary sediments and reef conditions may have been continuous in this area since the Early or Late Pliocene.

IV. Northern Great Barrier Reefs (Area C)

The Northern Great Barrier Reefs (Area C) are also controlled by a basement ridge which extends northward from the vicinity of Lloyd Bay in Queensland. The nature of this ridge is not known but it is thought to consist of Proterozoic sediments or low-grade metamorphics similar to those outcropping on the Cape York Peninsula. Mesozoic sediments are absent over the ridge which is directly overlaid by Tertiary sediments thought to be similar to those in Area B. The reefs in Area C are therefore developed directly on Tertiary sediments and reef conditions may have been continuous in this area since the Early or Late Pliocene.

The reefs around Darnley Island and between the Warrior Reefs and the Outer Barrier Reefs are associated with Pleistocene vulcanism. This is an area of thick Mesozoic and Tertiary sediments as intersected in Anchor Cay No. 1 Well (Lloyd, 1973).

V. Cape York–Lloyd Bay (Area D)

Area D extends from just north of Cape York to the south of Lloyd Bay. The basement in this area is believed to be composed of Proterozoic sediments or low-grade metamorphics similar to those outcropping on the Cape York Peninsula and to be an extension of these rocks. Middle or Late Paleozoic intrusives could also be present. The basement in Area D is overlaid by only a thin veneer of sediments which may not be older than Pleistocene. The reefs of this area therefore appear to have developed directly on basement rocks during the Pleistocene.

VI. Lloyd Bay–Cape Melville (Area E)

South from Lloyd Bay to the vicinity of Cape Melville, the reefs are developed on thin Tertiary sediments (possibly Pliocene only) which overlie Jurassic to Middle Cretaceous (Albian) sediments of the Laura Basin. Onshore these Mesozoic sediments attain a thickness of 3200 ft and consist of Jurassic to Neocomian freshwater sandstone and claystone and Late Neocomian to Albian marine silt stone and clay stone. The Mesozoic sediments overlie Late-Permian-indurated sediments or possible Middle Paleozoic volcanics and granite.

VII. Cape Melville–Townsville–Mackay (Area F)

From Cape Melville to Townsville and Mackay (Area F) there appears to be only a thin sequence of Tertiary strata (possibly Early to early Middle Miocene and Pliocene) overlying Middle Paleozoic sediments and Late Cretaceous volcanics. Late Paleozoic intrusions could also be present. The reefs in Area F therefore appear to overlie Tertiary sediments but are controlled by basement highs.

VIII. Mackay–Swain Reefs (Area G)

The Swain Reefs and the reefs to the north are controlled by a basement high, the Swain Ridge, which is thought to be composed of Late Paleozoic to Early Triassic mildly metamorphosed sediments similar to those intersected in the Aquarius No. 1 Well. These basement rocks are overlaid directly by Tertiary sediments which are thought to be Early to early Middle Miocene and Pliocene marls and limestones similar to those intersected in the Aquarius No. 1 and Capricorn No. 1A Wells (Carlsen and Wilson, 1968a,b).

The Pleistocene to Recent reefs of Area G are therefore possible continuations of reef conditions which began in the Early Pliocene.

IX. Bunker–Capricorn Reefs (Area H)

The Bunker and Capricorn Reefs are controlled by the Bunker Ridge which is composed of Late Jurassic to Early Neocomian volcanics as intersected in the Wreck Island No. 1 bore (Lloyd, 1973). The Ridge is overlaid at Wreck Island by 835 ft of Early to early Middle Miocene sandstone, siltstone, and limestone and 960 ft of Early Pliocene to Recent reefal limestone with minor sandstone. The Bunker and Capricorn Reefs are therefore a continuation of reef conditions which have existed in this area since Early Pliocene.

X. Capricorn Channel

Lloyd (1973) stated that the Early Tertiary sediments in the Aquarius No. 1 and Capricorn No. 1A Wells were possibly Early or Middle Oligocene in part. It is now believed that they are entirely Late Oligocene in age, based on the findings of Heickel (1972). Lloyd (in Carlsen and Wilson, 1968a,b; and Lloyd, 1973) placed a part of the sequences in the

Aquarius No. 1 and Capricorn No. 1A Wells in the Late Miocene. These sections are now placed in the early Middle Miocene, based on the results of Palmieri (1971). Late Miocene sediments are therefore absent from the Capricorn Channel area as well as from the Swain and Bunker Ridges or Highs.

Acknowledgments

Thanks are extended to Dr. O. A. Jones, of the Great Barrier Reef Committee, for the invitation to write this paper and to Messrs. D. D. Benbow, A. J. Flavelle, and B. E. Sealy for their assistance with geophysical and geological interpretations.

References

Carlsen, C. T., and Wilson, T. C. (1968a). *Rep. Aust. Gulf Oil* (unpublished).
Carlsen, C. T., and Wilson, T. C. (1968b). *Rep. Aust. Gulf Oil* (unpublished).
Heickel, H. (1972). *Geol. Surv. Queensl., Publ.* **335.**
Heidecker, E. (1973). *In* "Biology and Geology of Coral Reefs" (O. A. Jones and R. Endean, eds.), Vol. 1, pp. 273–298. Academic Press, New York.
Jones, O. A., and Jones, J. B. (1956). *Rep. Gt. Barrier Reef Comm.* **6,** 31–44.
Lloyd, A. R. (1973). *In* "Biology and Geology of Coral Reefs" (O. A. Jones and L. Endean, eds.), Vol. 1, pp. 347–366. Academic Press, New York.
Maxwell, W. G. H. (1973). *In* "Biology and Geology of Coral Reef" (O. A. Jones and R. Endean, eds.), Vol. 1, pp. 233–272. Academic Press, New York.
Myers, N. A. (1969). *Geol. Surv. Queensl., Rep.* **34.**
Palmieri, V. (1971). *Geol. Surv. Queensl., Rep.* **52.**
Willmott, W. F., Whitaker, W. G., Palfreyman, W. D., and Trail, D. S. (1973). *Aust., Bur. Miner. Resour., Geol. Geophys., Bull.* **135.**

Supplementary References

Australasian Petroleum Company. (1961). *J. Geol. Soc. Aust.,* **8.**
Doutch, H. F. (1972). *Publ. Res. Sch. Pac. Stud. Dep. Biogeogr. Geomorphol., Aust. Natl. Univ.* Bb/3.
Hill, D., and Denmead, A. K., eds. (1960). *J. Geol. Soc. Aust.* **7.**
Hill, D., and Maxwell, W. G. H. (1967). *Univ. Queensl. Pap.*
Layton Geophysical Consultants Pty, Ltd., Benbow D. D., and Lloyd, A. R. (1973). Unpublished report.
Oppel, T. W. (1969). Report. Tenneco Inc. (unpublished).
Oppel, T. W. (1970). *J. Aust. Petrol. Explor. Assoc.* **10** (2), 62–69.
St. John, V. P. (1970). *J. Aust. Petrol. Explor. Assoc.* **10** (2), 41–45.
Trail, D. S., Pontifex, I. R., Palfreyman, W. D., Willmott, W. F., and Whitaker, W. G. (1968). *Aust., Bur. Miner. Resour., Geol. Geophys., Rec.* 1968/26 (unpublished).

10

THE CORAL SEA PLATEAU—A MAJOR REEF PROVINCE

G. R. Orme

I. Introduction

The Coral Sea Plateau (latitude, 13°30′S–18°10′S; longitude, 146°15′E–152°30′E) is a large mid-ocean plateau (see Sheridan and Fairbridge, 1966, p. 876) lying adjacent to the Queensland continental slope. The region is sometimes called the "Queensland Plateau", Carte Internationale, sheet A′111; Krause, 1967; Gardner, 1970; Winterer, 1970; and is referred to by Maxwell (1968) as the "Coral Sea Platform." It covers an area of approximately 177,400 km². 85,800 km² lie above the 1000 m isobath, with extensive shoals and shallows less than 100 m deep, and 13,000 km² rise almost to sea level. Its maximum lateral dimension is 675 km, between Osprey Reef in the northwest and Lihou Reefs and Cays in the southeast. From Flinders Reefs it extends eastward for approximately 420 km to the Plateau Slope, and it narrows gradually northward to a width of approximately 25 km at Osprey Reef. Extensive reefs punctuate the surface of the Coral Sea Plateau, which is a major coral reef province.

Until recently the geology of this region was little known, even in the most general terms, and nothing more than educated guesses could be made regarding its nature and origin. Bathymetric data, reference to isolated sediment samples, and some deep-sea photographs of bottom sediments were included in an account of the bathymetry and geological structure of part of the South-Western Pacific Ocean by Krause (1967). However, confirmation of its continental character, information regarding its general structure, and a broad assessment of its geological history were the outcome of more recent geophysical investigations (J. I. Ewing *et al.,* 1970; M. Ewing *et al.,* 1970; Gardner, 1970; Andrews *et al.,* 1973).

It is the purpose of this paper to present an account of the Coral Sea Plateau drawing on the limited published material and incorporating original, previously unpublished observations made during the Japanese/ Australian Oceanographic Expedition of 1969 and during brief, marine geological expeditions from the University of Queensland in 1968 and 1970.

GEOLOGICAL SETTING

The geographical location and general bathymetric characteristics of the Coral Sea Plateau region have a significance in relation to the geological evolution of the continental margin of Queensland (see Heidecker, 1973), and bear a relationship to the regional tectonic trends of the South-West Pacific region. The tectonic grain of Australia is imparted by two intersecting elements (Maxwell, 1968) namely, the latitudinal trend of ancient crustal swells and depressions, e.g., the Forsayth (Euroka Ridge) and Longreach Swells and the Amadeus Trough, and the newer meridional trend of geosynclinal belts, e.g., the Tasman Geosyncline which, in Palaeozoic times extended from Tasmania to Cape York peninsula (see Hill and Denmead, 1960; Brown *et al.,* 1968) (Fig. 1). On the continent where the swells are transected by geosynclines, extensive exposures of

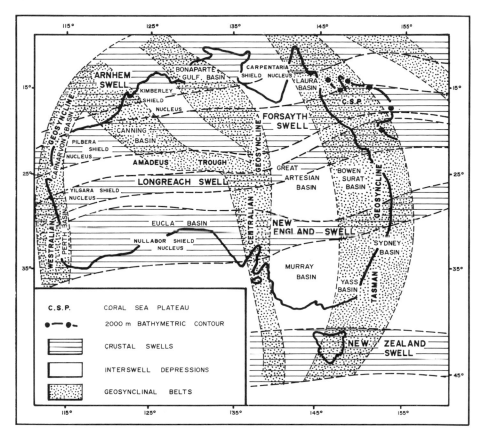

Fig. 1. Sketch map showing regional tectonic trends of Australia (after Maxwell, 1968).

granitic and metamorphic rocks commonly occur and extensive, deep basins are formed where the geosynclinal belts intersect the latitudinal depressions. As pointed out by Maxwell (1968), it could be inferred that the swells and depressions are ancient features which still react epeirogenically. Thus the Coral Sea Plateau could be regarded as a positive element lying at the intersection of the Forsayth Swell and the Tasman geosynclinal belt. The site of the axis of the Tasman Geosyncline lies in the Eastern Highlands (one of the three physiographic regions of Australia) and its eastern part is believed to lie beneath the Coral Sea in the Queensland area (Bryan, 1944).

Recent gravity and aeromagnetic studies of Eastern Queensland have demonstrated the northward extension, beneath the sea, of the meridional

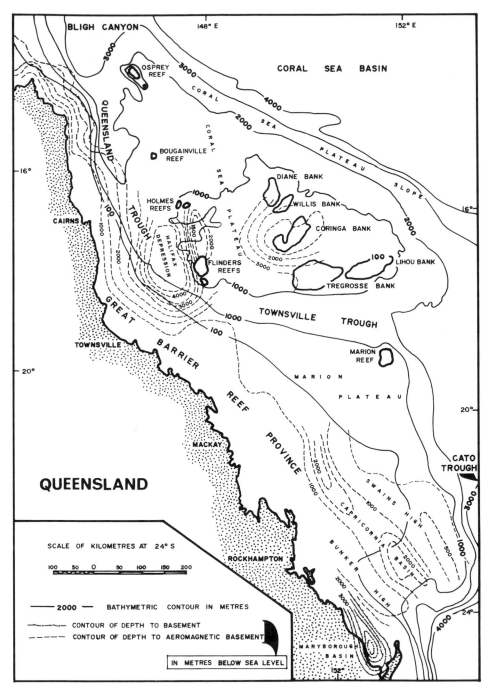

Fig. 2. Sketch map showing basement contours (after Geological Society of Australia, 1971) and bathymetric provinces.

geologic grain; eastward from the Maryborough Basin the Bunker High, Capricorn Basin (expressed bathymetrically by the Capricorn Channel), and Swain's High have been proven (see Geological Society of Australia, 1971; Maxwell, 1968). Further north, this meridional grain is a feature of the Coral Sea Plateau and adjacent marine provinces but the influence of the older latitudinal trend on the configuration of the basement is also evident. Thus the Halifax Depression corresponding approximately with the Queensland Trough lies between the Queensland Continental Shelf and the eastern margin of the Coral Sea Plateau, and the basement contours (Fig. 2) indicate a "high" beneath Flinders and Holmes reefs also with a meridional trend. East of this high a depression occurs separating it from the next high which lies beneath the South Central Coral Sea Plateau. The influence of a latitudinal component is expressed by the "high" beneath the junction of the Queensland and Townsville troughs, a trend which may be present also beneath Tregrosse and Lihou Reefs. Highs beneath the Western Coral Sea Plateau, the Swain's high, and the Guyots of the North Tasman Sea are in line and probably of similar composition, viz., metamorphic and volcanic.

The similarity between the gross geological features of the Coral Sea Plateau and of adjacent Queensland is therefore immediately apparent and leads to anticipation of a similar geological history. A cordilleran system was formed by the uplift of the Tasman Geosyncline in late Permian time; further movements occurred during the Mesozoic. Along the Queensland coast prominent basins received terrestrial sediments during the Triassic and Jurassic, as did the larger interior basins, and terrestrial and marine sedimentation followed in the Cretaceous. Widely scattered deposits of terrestrial Tertiary sediments interbedded with basaltic and rhyolitic lavas occupy depressions in the post-Mesozoic landscape in the coastal areas.

An episode which is considered to be of great significance in the interpretation of the geological history of the Coral Sea Plateau is an extensive planation of the Australian continent which occurred in early Tertiary time. This was followed in late Tertiary time by the arching of the Eastern Highlands and subsidence along major normal faults of adjacent areas lying to the east.

II. General Climatic and Oceanographic Factors

A salient characteristic of the Coral Sea Region is the profound change in prevailing winds and water circulation patterns between the winter and summer months. An indication of meteorological conditions over the Coral Sea Plateau is provided by the records of the manned weather station on Willis Island (latitude 16°18′S; longitude, 149°59′E; see British Admiralty, 1962, p. 54), and an insight into the oceanographic characteristics of the general region is given by Rotschi and Lemasson (1967).

A. Winds and Rainfall

Over the Coral Sea region, from May to November the southeast trade winds are fairly constant in intensity and direction, forming a stable dominant system between 0° and 25°S. Observations at Willis Island show that the southeast wind direction dominates and reaches maximum importance in May, when the highest mean wind speed (16 knots) was recorded.

During summer (December–April) the southeast trade wind belt is displaced southward by the southward movement of the Intertropical Convergence Zone, or Thermal Equator, until it lies between 10° and 30°S (Wyrtki, 1960). The resulting low-pressure cells which develop create cyclonic disturbances which pass over the Coral Sea and subject the area to monsoonal northeast winds. The Northeasterlies become important in December, January, and February, and the lowest mean wind speed (9 knots) recorded was in December.

The rainfall records for the Willis Island station show that the wettest period is from January to April and the lowest average rainfall figures are for August. The highest average over 26 years is 251 points of rain.

B. Sea Surface Temperatures

Mean sea surface temperatures over the Coral Sea Plateau region vary between 28°C in January to 22°C in July. During January there is a wide separation of the isotherms and uniform temperature conditions under the influence of the South Equatorial water mass. In July and October temperature gradients are more marked over the plateau. Figure 3 shows that during July a mean temperature of 21°C prevails at the extreme south of the plateau region.

C. Tides and Tidal Currents

The Coral Sea tidal oscillations are predominantly lunar, semidiurnal in character, and co-tidal and equirange lines for the region have been

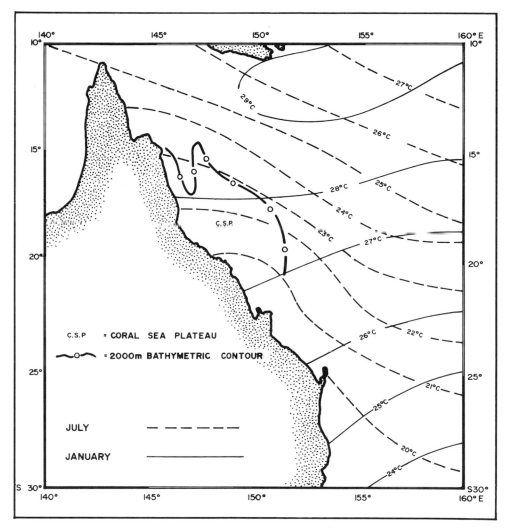

Fig. 3. Mean Sea Surface temperatures (°C) (after British Admiralty, 1962).

computed by Dietrich (1963; see also Maxwell, 1968. Chapter 4). Over the Coral Sea Plateau the tidal range indicated is 6 ft increasing to 8 ft toward the Queensland coast. Because of the relatively large tidal ranges in the Coral Sea, powerful tidal currents are commonplace. Currents greater than 1 knot flowing over the plateau surface adjacent to Coringa Bank were encountered during submersible dives, and are considered to be probably tidal (Professor M. Uda, personal communication). The effects of such currents on the surface of the plateau are referred to in subsequent sections of this chapter. Variable or complex 0.5–1-knot currents flow along the outer slopes of reefal banks; others move downslope and assist in the dispersal of sediment from the banks. However, the analysis of the complex current regime in the vicinity of the reefs and banks of the Coral Sea Plateau awaits detailed systematic investigation.

D. General Oceanic Circulation

The Coral Sea comes under the influence of the counterclockwise gyre formed by the South Equatorial Current to the north, the East Australian Current to the west, the West Wind Drift to the south, and the Trade Wind Drift to the east. This constitutes the western cell of a major South Pacific oceanic gyral formed by the South Equatorial Current, East Australian Current, West Wind Drift, and the Peru Current, and basically controls the hydrological character of the Coral Sea Plateau and adjacent areas.

1. The Trade Wind Drift

The Southeast Trades induce a wide surface current, the Trade Wind Drift which varies in velocity according to season. It is most stable between April and December, and reaches maximum strength between August and October. The flow is generally divergent in the Coral Sea, the main branch turning south to become part of the East Australian Current. A weaker drift continues westward into the Western Coral Sea, crossing the northern tip of the Coral Sea Plateau to meet the Queensland Shelf as far south as 15°S latitude, and another branch passes northwestward into the Solomon Sea. Torres Strait offers only a restricted exit for water masses flowing into the Coral Sea. These water masses consequently sink or flow away southward along the coast and become integrated into the East Australian Current (Figs. 4a and 4b).

Under the monsoonal influence this pattern changes from January to March when South Equatorial water masses enter the region from the north and northeast and flow southwest between the Solomon Islands and the New Hebrides, forming the East Australian Current, and extending to the Great Barrier Reef in the Townsville–Cairns region (Rochford, 1968), after crossing the Coral Sea Plateau. In the Western Coral Sea during January the prevailing northwest winds cause a general surface flow to the southeast parallel to the coast. At this time water from the Arafura Sea enters the region through the Torres Strait.

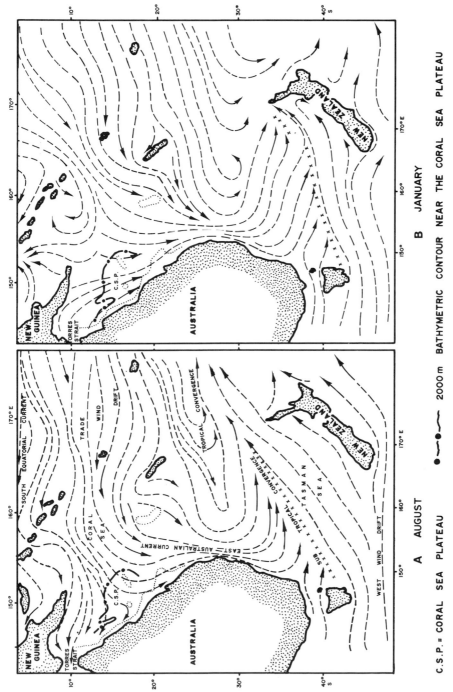

C.S.P. = CORAL SEA PLATEAU   ●━━● 2000 m BATHYMETRIC CONTOUR NEAR THE CORAL SEA PLATEAU

Fig. 4. Streamlines of the surface oceanic circulation in the Coral Sea-Tasman Sea region (after Wyrtki, 1960).

2. The East Australian Current

The tendency to build up water in the Coral Sea, a result of the influence of the trade winds and the obstruction to water movement provided by the land masses of Australia and New Guinea, is relieved by a strong narrow southward flow of water known as the East Australian Current, which is formed at about 20°S between Chesterfield Reef and the Great Barrier Reef. It is supplied with equatorial water masses driven west-southwest by the monsoonal winds from January to March, and by subtropical water masses from the Central South Pacific under the influence of the Trade Wind Drift from April to December.

3. Surface Water Flow over the Coral Sea Plateau

Changes in the pattern of general surface flow over the Coral Sea Plateau region, indicated in the diagrams presented by Wyrtki (1960), reflect the waxing and waning influence of the Trade Wind Drift which is moved from 5° to 10° further south during the summer period so that its southern limit reaches approximately 30°S between December and May. Thus, between May and October there is a marked south-westerly to westerly flow which diverges northwest and south as it approaches Australia. Contrasting strongly with this is the January flow pattern which shows a general south-southeastward movement. More complex flow patterns are depicted for the intervening months, February–April, and November–December (see Wyrtki, 1960).

E. WATER MASSES

The surface waters of the South-West Pacific are formed basically from three primary types. These are: (1) the South Equatorial Water (minimum salinity, 34.69‰; temperature range, 28.2°C–28.8°C) to the north; (2) the Subantarctic Surface Water (minimum salinity, 34.69‰; temperature range, 9°C–12°C) to the south; and (3) the Central South Pacific Water or Subtropical Lower Water (maximum salinity, 36.00‰; temperature, 26°C) (Rotschi and Lemasson, 1967).[1] Mixing of Central South Pacific Water and the Equatorial Water produces Coral Sea Water (salinity range, 35.41‰–35.59‰; temperature range, 20°C–26°C; Rochford, 1957), which extends beneath the South Equatorial and Central South Pacific Masses. Coral Sea Water probably rises in the region of di-

[1] Brandon (1973) states that the South Equatorial Water Mass has a temperature greater than 24°C (salinity range 34.0‰–35.6‰), and that the West Central South Pacific Water Mass (Subtropical Lower Water) may have temperature range of 18°C–25°C, with a salinity maximum characteristically between 35.5‰ and 36‰.

vergence where the main Trade Wind Drift divides, forming the East Australian Current, and sending branches into the Solomon Sea and the Western Coral Sea. When the Trade Wind Drift reaches maximum stability during the winter, the Coral Sea Water reaches its maximum salinity and minimum temperature owing to the large component of Central South Pacific Water. In the summer, when a branch of the South Equatorial Current enters the Coral Sea (see Fig. 4b), lower salinities and higher temperatures prevail. Also in summer a water mass with low salinity (34.14‰) and high temperatures (28°C–29°C), the Gulf-Arafura and Coastal water, affects the northwestern part of the Coral Sea (see Brandon, 1973).

Coral Sea Water is considered to be of great importance to reef development (Maxwell, 1968) because of its temperature and salinity characteristics, both of which favor high carbonate ion concentration. Under the influence of the Trade Wind Drift, part of the Coral Sea Water is carried westward from the region of divergence toward the Continental Slope of North Queensland while part turns southward along the Queensland Trough and over the Coral Sea Plateau, where it gradually intermixes with South Equatorial Water and Central South Pacific Water, surface waters which reduce its salinity and density, eventually destroying its identity (Maxwell, 1968, p. 16).

The characteristics of the water column at a station "32 km from Second 3 Mile Opening in the outer barrier" (Great Barrier Reef) which is just northwest of the Coral Sea Plateau have been recorded by Brandon (1973, p. 222, Fig. 14). Brandon (1973) also investigated temperature, density, and PO_4–P characteristics of the upper 250 m of the water column in the vicinity of the Great Barrier Reef and Coral Sea Plateau. He concluded that a southward surface flow along the Queensland continental margin and northward-flowing currents further offshore are indicated by his calculations. Still further out, over the southern limit of the Coral Sea Plateau, his data indicate southward-flowing currents. The distribution of PO_4–P values indicates uplifting of water near the continental margin, and correlation of his station locations (Brandon, 1973, p. 225, Fig. 15) with the bathymetry of the region further suggests uplifting of water at the northeastern margin of the Coral Sea Plateau and also near its southern boundary. Limited upwelling, probably bringing nutrient rich waters into contact with living reefs, is hypothesized by Brandon, as opposed to the deep-seated upwelling proposed by Maxwell (1968). This is implied by the small-scale cellular structure in the upper few hundred meters of the water column demonstrated by Brandon.

Temperatures in the upper 280 m of the water column were measured at five sites during January, 1969, from the *Yomiuri-Go* (submersible) and its mother vessel, the *Yamato*, during the Japanese/Australian Oceanographic Expedition, four near the central bank or platform (here named Coringa Bank) which supports the Coringa Islets, and one at Flinders Reefs (see Figs. 7 and 11). From sea surface temperatures of between 27°C–28°C there was a fall to 18°C at 240 m below stations 2 and 3. Figures provided by Professor Uda show a sharp thermocline (23°C–20°C) at approximately 200 m below station 3 and a homogeneous colder layer (18°C) below 240 m extending over the interbank sea floor lying 32 km due south of Chilcott Island, down to the limit of the dives (280 m). However, the lowest temperature measured below station 1 was 19°C at a depth of 274 m, and, at Flinders Reef (station 5), at 250 m a temperature of 20.5°C prevailed, thereby indicating that the 18°C isotherm at these stations lies at a much lower level. The data also indicate a closer spacing of isotherms in the channel south of Coringa Bank compared with stations 1 and 5. A local inversion of isotherms is suggested by an anomalously high reading (24.2°C) at station 3, which coincides with a strong downslope current encountered at 162 m depth. Furthermore, near station 4, some upwelling and divergence is suspected, which may account for a zone of "oily" reddish water observed to fringe the bank (Professor M. Uda, personal communication). Therefore, small-scale cellular motion may be important over parts of the Coral Sea Plateau, especially near bathymetric prominences such as Coringa Bank.

F. PHOSPHATES, NITRATES, AND TRACE ELEMENTS

Relevant to a consideration of the condition of the coral reefs is an assessment of certain chemical characteristics of the water masses in which they occur. Rochford (1959) suggested that the South Equatorial Water Mass was identifiable by its high total phosphate content as compared with that of the Central South Pacific Water Mass. He also showed that, although total phosphorus tends to be a conservative property in the Coral Sea at surface and intermediate depths, in regions where water depths do not exceed 1000 m, including the Coral Sea Plateau, anomalously high total phosphate content in the surface water occurred. He concluded that bottom currents picked up phosphates from sedimentary interstitial waters in such areas.

Since this property in part determines the fertility of the water for the support of the phytoplanktonic community, phosphate abundance in waters of the Coral Sea might be anticipated to bear some relationship to the condition of reefs. The distribution charts prepared by Kirkwood

(1967), who summarized the previous work on inorganic and organic phosphorus and on nitrates, show an increase in both inorganic and organic phosphate abundances in the Coral Sea surface waters during the Trade Wind regime in winter, but there is no apparent relationship between the winter organic phosphate abundances and the distribution of reef areas. Brandon's data on PO_4–P content favoring limited upwelling in this region have been discussed previously.

Nitrate concentrations in the surface waters of the Coral Sea are sparse (Kirkwood, 1967), with the notable exception of localized areas around the Coral Sea Plateau reefs.

Studies carried out by St. John (1971), of the geologically significant trace-element abundances in living corals of the Coral Sea region, and their relationship to oceanographic factors included observations on some Coral Sea Plateau reefs.

III. Bathymetry

The Coral Sea Plateau forms a distinct bathymetric province which is flanked by the deep Coral Sea Basin, the Queensland Trough, and the Townsville Trough (see Fig. 2). Despite its present isolation from the Great Barrier Reef Province, a glance at a bathymetric chart of the Coral Sea Region is sufficient to suggest an affinity between the Coral Sea Plateau and the Queensland Continental Shelf. The intervening Queensland Trough (approximately 75 km wide) descends from a point just southeast of Flinders Reefs at a depth of approximately 1100 m towards the north-northwest, reaching 2800 m at the latitude of Osprey Reef. A continuation of the Queensland Trough, now known as the Townsville Trough, which is aligned east-west between latitudes 18°S and 19°S, separates the southern margin of the Coral Sea Plateau from the Marion Plateau, and descends from approximately 1100 m where it joins the Queensland Trough (148°50' and 150°E), to a depth of 2500 m just to the northeast of Marion Reef.

The nature of the Queensland Continental Slope and that of the western slope of the Coral Sea Plateau is illustrated in Figs. 5 and 6.[2] The former, due west of Osprey Reef, descends very steeply from the Great Barrier Reef to 1600–2000 m before a gradient decrease gives place to a continental rise. Both the gradient and depth range of the former decrease southward, which is also generally true for the western slope of the Coral

[2] See foldout for Figs. 5 and 6.

Sea Plateau, although the latter begins at greater depth and is less steep except where it is modified by reefs.

To the northeast the slope at the margin of the Coral Sea Plateau eventually reaches the abyssal plain (4600 m) of the Coral Sea Basin, which is connected with the Queensland Trough via the Bligh Canyon (Winterer, 1970) lying to the north of Osprey Reef.

The following description is basically an interpretation of information obtained from Ocean Sounding Plotting Sheets (see Fig. 5).

A. Nature of the General Plateau Surface and Marginal Slopes

Although the gently undulating plateau surface deepens in all directions from a point just south of Coringa Bank, it is generally tilted towards the north and descends from 600 m to 1500 m near Osprey Reef (see Figs. 5 and 6). There is a decrease of gradient just to the northwest of Herald Cays (gradient change at 149°E, 16°32′S) but north of this point the gradient increases progressively until an equivalent to a continental slope, the Coral Sea Plateau Slope, is encountered at about 2000 m. The linear nature of this northeastern slope of the plateau has promoted speculation regarding its significance in the structural evolution of this region (Heidecker, 1973). The steeper gradient of the Coral Sea Plateau Slope begins in some areas at shallower depths (e.g., 1500 m). However, at approximately 3500 m, the gradient is reduced to give a gentler, usually more irregular slope which continues to the abyssal plain of the Coral Sea Basin. The Coral Sea Plateau Slope is dissected by submarine valleys, and its general complexity, particularly evident between 148°–151°E and 13°40′–15°S, and between 152°–153°E, and 16°–17°S, may reflect the effects of submarine slumping and erosion. At 16°23′S and 152°42′E there is a gradient decrease at 2300 m and the slope divides, the lower part passing into a sill which lies between the Coral Sea Basin and the Cato Trough, the upper part continuing southward to the eastern end of the Townsville Trough.

The southern margin of the Coral Sea Plateau descends rapidly from the line of reefs—Malay, Abington, Tregrosse, and Lihou—to the Townsville Trough. This slope becomes very steep in places, e.g., between Malay Reef and Tregrosse Bank, and may be of structural significance. Elsewhere it is crossed by conspicuous submarine valleys, e.g., between 17°–18°S, and 152°–152°40′E.

The steepness of the western slope in the vicinity of Osprey Reef rivals that of the opposing Queensland Continental Slope, and is probably the consequence of a combination of sustained upward reef growth, concomitant with slow subsidence of the sea floor, and the structural characteristics of this area. Steep sectors are evident on the western slope of the plateau and the margin of the Queensland Trough, viz., at 146°10′E, between 13°50′–14°10′S (2300–2700 m), between 14°30′–14°55′S (1800–2400 m), and between 14°50′–15°5′S (1800–2400 m). Similar gradients occur northwest, west, and south of Bougainville Reef from 1400 m to 1800 m (146°45′–147°E, and 15°–15°35′S), and from 1500 m to 1700 m at 147°E (16°–16°30′S). Such linear features suggest the influence of faults, and indeed, normal faults in this area have been detected (J. I. Ewing *et al.*, 1970).

B. Conspicuous Relief Features

Winterer (1970) has described the major submarine valley systems around the Coral Sea Basin, but smaller, yet striking features are associated with the Coral Sea Plateau. A major valley, between 16°–17°S, and 152°–153°E, descends from the eastern side of the plateau and extends east and then northeast towards the Coral Sea Basin. In this area isobaths about the 1500-m and 3000-m level anastomose to define steep slopes which may reflect local fault control. Another conspicuous feature which may be of structural significance is the linear valley trending southwest to northeast between 14°–14°45′S, and 148°12′–149°10′E. This valley is 65 km in length and 3 to 6 km in width, widening towards the northeast and descending from −1400 m to −4500 m.

The area enclosed by the 1000 m isobath, i.e., south of approximately 15°35′S, is bathymetrically more diversified than the northern part of the plateau. Here several small prominences occur, e.g., in the region between Holmes Reefs and Herald Cays, rising 100 m (approximately 5 km across); a larger feature (13 km diameter) rises more than 200 m at 17°23′S, 148°50′E, approximately 35 km east of Herald's Surprise.

Prominent bathymetric features rising from the plateau surface which support viable reefs are considered separately in the following section, but an observation may be made here regarding their apparent relationship to the general plateau surface. In some places their boundaries exhibit discordance with the pattern of contours on the plateau surface, e.g., 17°20′S, 152°E, where the southern face of the Lihou Bank transects the head of a tributary valley leading southward towards the axis of the Townsville Trough. In others, bank margins and plateau surface contours are concordant, e.g., southwest Tregrosse Bank parallels a scarplike feature evident on the plateau surface, and in this case some significance may be attached to the influence of fault control in the siting and configuration of the bank.

C. SMALL-SCALE RELIEF FEATURES

Locally the plateau surface in the vicinity of reefs has a complex relief promoted by slumping. Exposed rock ridges and large ripples also diversify the surface in some interbank areas. Because of parabolic side reflections recorded in seismic profiles at the sea bed/water interface, larger relief features are suspected in some areas.

IV. The Coral Reefs and Their Foundations

A. DISTRIBUTION AND VARIETY

The reefs of the Coral Sea Plateau are oceanic in character, lying seaward of the Great Barrier Reef Province. Their geographical distribution is shown in Fig. 5. They form two groups, one trending north-northwest to south-southeast along the western margin of the Plateau, the other occurring on the southeastern (highest) part of the Plateau and extending almost to the margin of the Townsville Trough. The extensive reefal areas have the general form of large atolls, and although in some cases living reef rising to sea level is only sporadically distributed around the fringes of extensive, relatively shallow, bathymetric features, in others, e.g., Osprey Reef, it forms an almost continuous ring surrounded by deep water and enclosing a lagoon. Their biota is rich, many aspects of which are similar to that described by Maxwell (1968) for the Great Barrier Reef Province.

Flourishing reefs are located on a variety of bathymetric features. These include narrow, steep-sided features (Malay and Abington reefs), broader, steep-sided cones (Osprey, Bougainville, and Holmes reefs), and extensive, steep-sided features which rise from the plateau surface to approximately 60 m below sea level (Lihou, Tregrosse, Diane, Willis, Coringa, and Flinders). Some of the latter have been officially designated banks and are so named on hydrographic charts, others previously unidentified are similarly designated in the present work, their names being derived from those of prominent reefs and/or islets occurring on them. Two smaller, but morphologically similar bathymetric features lie respectively, 37 km (150°E, 17°19'S), and 57 km (150°E, 17°31'S) due south of Chilcott Islet.

Those reefs developed over conical bathymetric features were believed to overlie volcanic cones (Krause, 1967), but more recent work (Gardner, 1970) does not support this view. Their distribution along the western edge of the plateau coincides with a zone of tensional faulting (M. Ewing *et al.*, 1970) where basement highs appear to have influenced the location of such reefs. However, the largest reefal areas are the "banks" which characterize the southern half of the plateau. Some of these banks have been investigated seismically, and also inspected visually and sampled from the *Yomiuri-Go* and a more detailed account follows.

B. THE BANKS

1. Physiography

Without exception, the slopes of the outer margins increase very sharply from relatively shallow bank tops and fall away to the plateau surface at steep angles. The outline of the banks is defined in Figs. 5 and 7 by the 100-m bathymetric contour, but the steep gradients of the outer slopes begin at an average depth of 60 m. On the western side of Diane Bank and Willis Bank the "drop-away" begins at approximately 55 m and on the western side of Coringa Bank the 65-m isobath appears to mark the beginning of the very steep slopes. However, steep gradients begin at shallower depths, viz., 35–45 m on the northeastern, eastern, and southeastern margins of some banks where they are influenced by marginal bank-top features, and, indeed, may start at sea level where flourishing reefs maintain a steep bathymetric slope.

It is convenient to consider the physiography of the banks broadly in terms of the bank top (upper surface), the outer slope, and the immediately adjacent interbank sea floor.

(a) *Bank-Top (upper surface)*. In all cases the viable reefs which approach sea level occur mainly at or near the bank margins. The Lihou, Tregrosse, Coringa, and Willis banks are arranged en echelon with their long axes oriented between northeast to southwest and east-northeast to west-southwest. In some cases large areas are as yet unsurveyed; nevertheless, the available charts indicate that the most favored sites for reef development on the banks varies from the southeastern to the northeastern and northern edges. Which of these positions has the most extensive reef development probably depends on the proximity and position of adjacent banks. Thus there is conspicuous reef development on the southeastern edge of Lihou Bank, which is completely open to the southeast and to the effects of the Southeast Trades. On the other hand, where open sea conditions to the southeast are restricted by a neighboring bank, conspicuous reef development occurs on the northeastern and northern margins, e.g., the Diane, Willis, and Coringa banks.

Although there is a general correlation between the banks in terms of

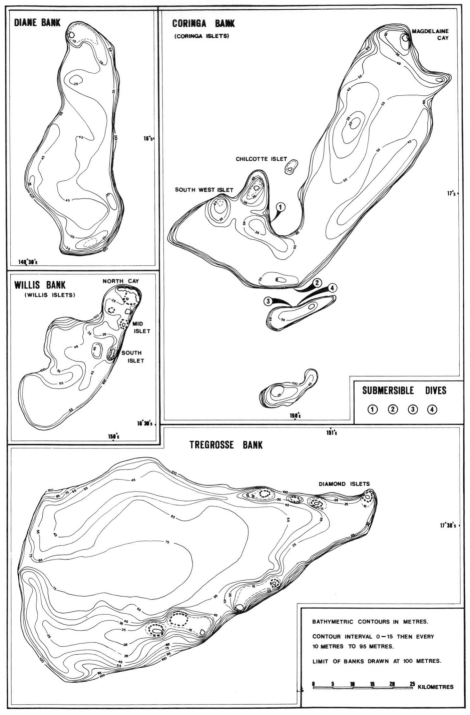

Fig. 7. Interpretation of general bank-top relief based on data from Ocean Sounding Plotting Sheets (see Fig. 6). Submersible dive locations are indicated.

average depth of surface below sea level, some diversity in detail is indicated by available soundings (Fig. 7). Thus soundings over the Tregrosse Bank indicate a central depression reaching approximately 80 m in depth with a pass to the west. The rim of this depression rises to 45 m to the north and 15 m to the southeast. At this southern edge also, viable reefs which reach sea level have developed and sand and gravel cays are present on some of these. The bank top profile is therefore generally asymmetrical with a much higher rim in the southeast.

Soundings over Diane Bank likewise indicate a depression in the south central part opening to the west, but over the Coringa and Willis Banks a series of mounds and ridges occurs and no central depression is evident. The ridges nevertheless lie toward the bank margins.

Bank top profiles of Coringa and adjacent banks show a general surface level at 55 m (180.4 ft)–58 m (190 ft) which rises locally to 37 m (121 ft) and falls to 73 m (239.5 ft) at some margins where the steep descent to the plateau surface begins. In places a marked terrace is evident at this lower level. In the profile of the northwestern part of Coringa Bank (Fig. 8) the bank top rises near its northwestern margin to a depth of 50 m over an irregular surface caused by an extension of the reef on which South-West Islet stands. Pinnacles or knolls representing living coral heads with a 9 m (29.5 ft)–27 m (88.5 ft) relief rise above this general reef surface and also occur peripherally to the northwest and southeast. These profiles invite comparison with those of Alexa and Turpie banks, Melanesian Border Plateau (see Fairbridge and Stewart, 1960). The lower terrace levels which Fairbridge and Stewart (1960, p. 108) suggest may correlate with the last Glacial (Wisconsin/Wurm) correspond to the level of the general surface of Coringa Bank. Furthermore, the Alexa and Turpie Bank profiles show pinnacles of similar dimensions to those on Coringa and adjacent banks, and although some were shown to be living coral heads, for others of larger dimensions an alternative explanation was posed, viz., that they "may in part be erosional remnants of karst solution" active during periods of bank top exposure through low stands of sea level.

(b) *Outer Slopes.* The longest outer slopes extend from sea level to a depth of 900–1000 m at the northeastern tip of Willis Bank and on the western side of Flinders Reefs. An examination of the shorter outer slopes of Coringa Bank, the subsidiary bank immediately to the south in a line due south of Chilcott Islet, and the adjacent interbank sea floor areas was carried out from the *Yomiuri-Go* down to a depth of 285 m. The outer slopes of Flinders Reefs, which afford a somewhat different aspect,

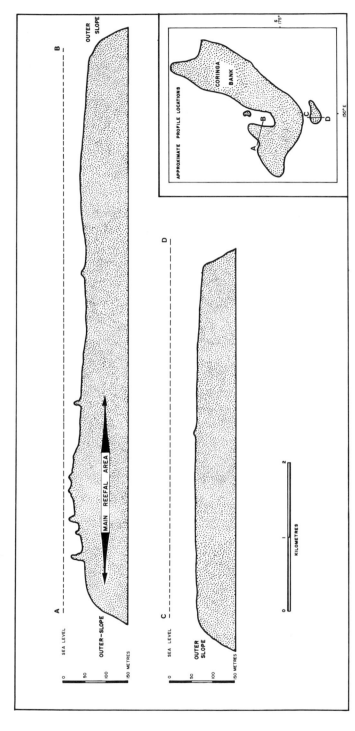

Fig. 8. Bank-top profiles of part of the Coringa Bank and an adjacent subsidiary bank. Profile A-B (N.W. Coringa Bank) shows uneven surface with pinnacles over reef area, and general bank-top level at 55 m (180.4 ft)–37 m (121 ft). The outer-slopes plunge from a depth of approximately 55 m. Profile C-D (subsidiary bank south of Coringa Bank) shows a smooth profile with outer slopes plunging from approximately 73 m (239.5 ft) to 55 m (180.4 ft).

were inspected at 148°27′E, 17°45′S. Together, these two areas provide an indication of the general conditions likely to prevail on the outer slopes of other "banks" in this province.

On Coringa Bank, outward-facing slopes above 55–73 m occur only where reefs or mounds are present near bank margins. Elsewhere, the outer slopes descend abruptly from the bank-top level at 55 m (180.4 ft) or 73 m (239.5 ft). The general inclination of the outer slopes is steep. However, closer examination shows that they consist of 20°–50° rock and scree slopes with less steeply inclined (5°–15°, or less) intervening sections.

Differences in the abundance and variety of the biota, and in the general aspect of the sediments and topography of the bank slopes, with depth permit the recognition of three significant levels, viz., 55 m, 140–150 m, and 200–210 m (see Fig. 9).

Reef-building corals were observed down to 55 m. Slopes above this level vary according to whether or not reefs occur at the bank margins. Thus, where active reef growth reaches sea level, the outer slope is generally steeply inclined, but elsewhere slopes of 15°–20°, increasing toward the 55–73 m level are more usual over what appear to be bank-edge mounds of sediment and old reefs. These have surface deposits of shell gravel and shell sand mixed with sponge and algae encrusted reef rubble.

Below 55 m slopes ranging between 10° and 40° are common, and a marked change in their appearance is correlated with a general decrease in the abundance and variety of the benthos, and with a difference in the nature of the surface sediment. Deposits of white, shell gravel and sand with some small blocks and boulders of reef debris give way at approximately 100 m to a "boulder" zone which extends to the 140-m level. This is a submarine slope of commonly 20°–30°, composed of dark, algae- and sponge-encrusted boulders which decrease in size in a downslope direction. They are mixed with white, shell sand and gravel which only becomes important as a surface deposit on lesser slopes. For instance, they form conspicuous deposits on small ledges which are particularly evident between 100 m and 120 m.

Below 140–150 m rocky slopes, pebbly carbonate sand, and carbonate sand are characteristic and the outer slope continues steeply at angles of up to 50°, decreasing to 10° or less near the junction with the interbank sea floor. Complex relief near this junction is the result of slumping, e.g., large masses of limestone (up to 8 m in height) resting on the rocky sea bed were observed on the northern side of Coringa Bank (Dive No. 1)—see Fig. 10.

Flinders Reefs are comprised of two parts, both forming separate bathymetric features which descend to the general plateau surface at 800–

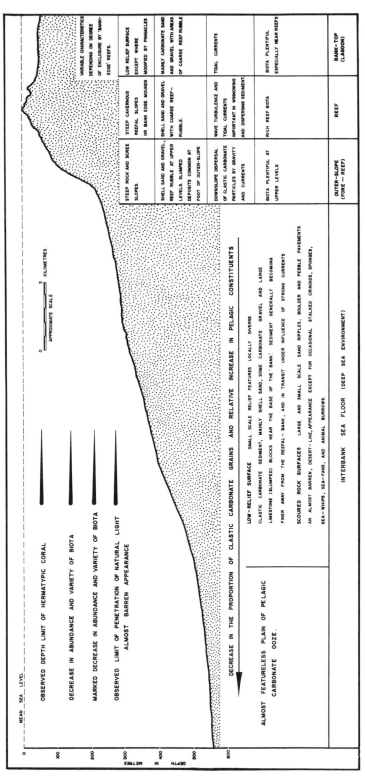

Fig. 9. Sketch profile of an area near the Coringa Bank illustrating the general arrangement of sedimentary environments and indicating their salient characteristics. (This model is based on an area which was investigated between the Coringa Bank and Herald Cays.)

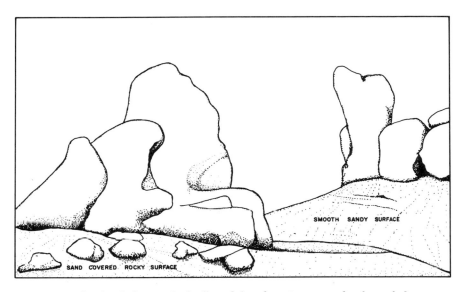

Fig. 10. Sketch of the sea bed affected by slumping near the foot of the outer slope on the northern side of Coringa Bank (Dive No. 1, see Fig. 8), showing large (7–8 m high) limestone blocks resting on a sand-covered surface. Fine carbonate sand partially veneers the limestone especially where crevices and sheltered declivities occur.

900 m. They are here referred to as North Flinders Reef and South Flinders Reef which are separated by a deep, narrow channel. South Flinders Reef features an almost continuous reef encircling a lagoon, and North Flinders Reef, at its southern end, also has a fairly continuous viable reef especially in the southeast. The southern, outer slope of North Flinders Reef is considered here, and an impression of the upper 300 m is presented in Fig. 11. This outer slope descends more abruptly than any of those examined in the Coringa Bank area, there being fewer lower angle slopes and three levels of almost vertical coralline rock faces which have overhanging sections. These limestone cliffs occur at 40–70 m, approximately 90–115 m and 135–150 m below sea level.

The present reef front continues downwards as a craggy surface, and a rocky slope with a surface deposit of carbonate sand and gravel with boulders of reef rubble extends from about 30 m to 40 m. It is inclined outward at about 15° to 20° and may represent a partially sediment-covered terrace on which both hard and soft corals occur, although its lateral extent was not investigated during the dive.

At 40 m the top of the first clifflike slope, which extends to approximately 70 m, is encountered. Between the clifflike sections of coralline

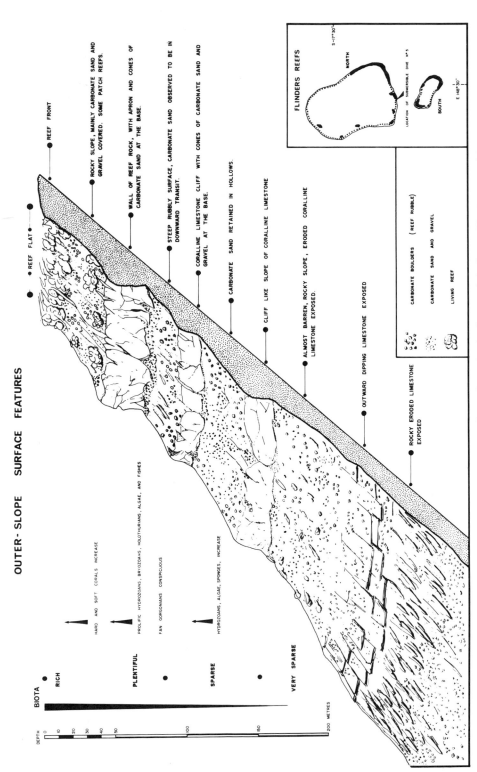

OUTER-SLOPE SURFACE FEATURES

REEF FLAT
REEF FRONT

ROCKY SLOPE, MAINLY CARBONATE SAND AND GRAVEL COVERED. SOME PATCH REEFS.

WALL OF REEF ROCK, WITH APRON AND CONES OF CARBONATE SAND AT THE BASE.

STEEP RUBBLY SURFACE, CARBONATE SAND OBSERVED TO BE IN DOWNWARD TRANSIT.

CORALLINE LIMESTONE CLIFF WITH CONES OF CARBONATE SAND AND GRAVEL AT THE BASE.

CARBONATE SAND RETAINED IN HOLLOWS.

CLIFF LIKE SLOPE OF CORALLINE LIMESTONE

ALMOST BARREN, ROCKY SLOPE, ERODED CORALLINE LIMESTONE EXPOSED.

OUTWARD DIPPING LIMESTONE EXPOSED

ROCKY ERODED LIMESTONE EXPOSED

CARBONATE BOULDERS (REEF RUBBLE)

CARBONATE SAND AND GRAVEL

LIVING REEF

FLINDERS REEFS

S-17°30′

NORTH

LOCATION OF SUBMERSIBLE DIVE N° 5

SOUTH

E 148°30′

BIOTA

RICH

HARD AND SOFT CORALS INCREASE

PLENTIFUL

PROLIFIC HYDROZOANS, BRYOZOANS, HOLOTHURIANS, ALGAE, AND FISHES

FAN GORGONIANS CONSPICUOUS

SPARSE

HYDROZOANS, ALGAE, SPONGES, INCREASE

VERY SPARSE

DEPTH
0
10
20
30
40
50

100

150

200 METRES

Fig. 11. An impression of the upper 300 m of the outerslope of North Flinders Reef (submersible Dive No. 5).

limestone the intervening slopes (70–90 m and 105–135 m) are generally fairly steep with a very sporadic surface deposit of carbonate sand and gravel which tends to fill in hollows in the rocky surface. Cones of carbonate sand occur at the base of the cliffs and a continuous downward movement of shell sand over the steeper rock surfaces was observed to be in progress. Below 150 m the slope is steep (30°–45° being common) and rocky, with a sporadic surface deposit of mainly white carbonate sand. The uneven rocky surface has an eroded appearance and displays small-scale downslope grooving. From approximately 190–230 m on a slightly steeper section, seaward-dipping limestone beds are exposed which could well be ancient lithified talus deposits of sand and gravel.

(c) *The Interbank Sea Floor.* The interbank sea floor examined near the southern margin of Coringa Bank begins at 210–254 m and has the appearance of a "desert landscape." It appears to be almost barren.

Areas of carbonate shell sand alternate with areas of carbonate sand with limestone pebbles, pebble and boulder pavements, slipped limestone masses, limestone boulders, and eroded limestone ridges (see Plate I). Areas of carbonate sand display both small- and large-scale ripples (Plate I, C and D). Simple current ripples are common, but cross ripples occur locally. These sedimentary structures are produced by downslope currents from the banks and by currents which flow over the adjacent interbank sea floor. The cross ripples testify to a complex current regime near the junction of the outer slope with the interbank sea floor.

2. Observations on the Distribution of Organisms

On the outer slope of the Coringa Bank there is a sporadic distribution of bank-edge reefs, with corresponding lateral variations in the biota. Hermatypic corals have been noted down to a depth of 55 m, and a rich biota occurs above this level. Encrusting coralline algae, soft corals, holothurians, sea urchins, gastropods, bivalve molluscs, polyzoans, and Foraminifera are abundant and are conspicuous members of the benthos; clumps of green algae, including *Halimeda* also occur. Sea whips and sea fans become particularly conspicuous toward the 40–55-m level, and in places algal "meadows" dominate the slope down to 75 m.

The epifauna associated with the boulder zone is commonly plentiful except where there are patches of unstable sand. Red, orange, and yellow encrusting sponges are very conspicuous; red and white fan gorgonians and hydroids are anchored to boulders, shells, and exposed rock surfaces. Crustose red algae also occur.

Below the level of the boulder zone the biota is very sparse; however,

80–90 m. Both hard and soft corals increase above 40 m. Hydrozoans which are still plentiful at 100 m, rapidly decrease below 125 m, and below approximately 150 m there are few signs of life on the outer slope except for crustose red algae, sponges, occasional hydrozoans, and some whip and fan gorgonians.

3. Bank Structure and Development

The preliminary results of analysis of continuous seismic reflection profiles indicate that the reefal banks are composed essentially of massive reef rock, steeply inclined fore reef (off reef) talus deposits, and horizontally bedded or gently inclined "lagoonal" sediments.

The general character of the banks revealed in seismic profiles is of steep-sided bathymetric features with planed upper surfaces on which variations of relief have been imposed by reef growth and sedimentation. The steepest inclination of their outer slopes occurs in the middle section, the upper edge of the banks being beveled by erosion, and the lower slopes being modified by an accumulation of talus deposits. Some seismic profiles show outer slopes which have been disrupted by slumping, the effects of which extend for some distance into the interbank areas creating a very irregular sea-floor relief (topography). Below the lower part of the outer slope, in areas where slumping has not occurred, the internal structure consists of steeply dipping talus deposits forming thick, wedging beds which pass laterally into thinner, horizontal units of the interbank areas. There is clearly a stratigraphic link between bank deposits and those of the interbank areas.

At a depth of approximately 100 m below sea level, and buried beneath 30 m of sediment, there is a pronounced terrace. Some of the large pinnacles referred to earlier rise from this surface, extending upward through bedded deposits to become prominent topographic features of the bank top. It is likely that this terrace is significant with regard to eustatic sea level changes and that it probably represents a low sea level of the last Glacial (Wisconsin/Würm). Beneath this horizon, massive lenticular reeflike bodies, which are enclosed by bedded deposits, occur at successive levels. Their greater lateral extent at lower levels within the banks may in part be a consequence of slow rates of subsidence of the sea floor favoring lateral reef extension rather than upward growth.

The banks appear to have no volcanic foundations and the uneven, strongly reflecting surfaces detected beneath banks and interbank areas in some seismic profiles are probably significant in terms of the erosion surface described by J. I. Ewing et al. (1970) and Gardner (1970). Indeed, they may represent high sections of this surface on which platform reefs became established. The development of the enormous reefal banks from ancestral platform-type reefs was facilitated by subsidence of the sea bed at a rate compatible with upward reef growth. Variations in the form and distribution of reeflike masses recorded in seismic profiles may, in part, reflect fluctuations in the relative subsidence rate during subsequent phases of bank development.

V. Sediments and Sedimentation

Previous records have indicated that the Coral Sea Plateau is at present a region of pelagic sedimentation, the deposits covering the plateau surface being described as calcareous ooze. While this is true for the northern half of the region and the deeper interbank areas, in the south sedimentation patterns are more diversified due to the influence of the reefal banks, from which appreciable quantities of clastic carbonate sediment are derived. These components and the processes involved in their dispersal result in a range of sedimentary deposits which dominate the plateau surface locally.

A. Composition and Texture

Forming the textural extremes of the considerable range of carbonate sediments of the Coral Sea Plateau are the slumped deposits occurring at the base of the reefal banks and the carbonate ooze covering the deeper parts of the plateau surface. Grade sizes of carbonate components range from the extremes of huge blocks and boulders of limestone to the fine debris of microscopic tests of Foraminifera. Between these extremes are the sediments more characteristically associated with the reefal banks and the immediately adjacent interbank areas, their components being boulders and pebbles of largely coralline algae and coral, which are usually dark in color where coated by living algae and other encrusting organisms, and the debris of molluscs, *Halimeda,* benthic Forminifera, corals with contributions from other organisms of the bank-top, reefal, and outer-slope environments.

B. Provenance and Dispersal

There are at present three main sources of sediment, namely reef and bank organisms, pelagic organisms, and the limestones exposed both on the outer slopes and locally on the plateau surface in some interbank areas.

Erosion of the reefs by wave action maintains a supply of boulders of reef rubble and may, during storms, assist in the detachment, breakup, and removal of material from the outer slopes. Skeletal material is broken down by wave action and corrasion and comminuted by organisms, and the breakdown of limestone blocks may be induced by storm waves and by slumping, which appears to have been locally effective. Biological erosion is active in limestone blocks dredged from considerable depths. The dispersal and sorting of these sediments is achieved by wave action, currents and gravity. Gravity alone is an important factor in sediment dispersal over the outer slopes: its effect was observed in the downward trickle of shell sand and gravel which is in progress on steeper slopes— particularly over the steep rocky surfaces of North Flinders Reef, in settling of the tests of pelagic organisms, and in slumping.

However, currents appear to be the most effective means of dispersing bank-derived sediment. A complex current regime disperses the sediment over outer slopes and interbank areas and transports part of it towards the deeper levels of the plateau surface. In the Coringa Bank area, at least, strong currents descend the outer slopes and assist in the separation of sand from gravel which permits the differentiation of surface features referred to in Section IV. Thus carbonate boulder and cobble gravels tend to form submarine screes at higher levels of the outer slopes while carbonate sand is carried to lower levels. Under the influence of strong tidal currents some shell sand is probably moved to lower parts of the plateau surface and may ultimately reach the deeper bathymetric provinces which surround the Coral Sea Plateau. The strength of such currents and the volume of sediment being transported by them is implicit in the large scale ripples and scoured rock surfaces exposed on the sea floor. Sediment is retained in hollows and elsewhere appears to be in transit. It therefore seems that at present no thick deposits of carbonate sand are accumulating on the higher parts of the interbank areas.

Although no evidence of turbidity flows was observed, this means of sediment transport may be locally important when unstable sedimentary deposits resting on the steep outer slopes are disturbed. Erosion of the banks by wave action and slumping may have been more effective in the past during lower stands of sea level.

The resulting general distribution pattern of sedimentary facies is for carbonate sand and gravel with sporadic patches of coarse reef rubble on bank tops, boulder and cobble screes with some finer shell gravel and shell sand on the upper parts of outer slopes, and slump deposits with carbonate sand and limestone blocks on the lower slopes. Carbonate sand and carbonate sand with pebbles are characteristic of the higher interbank areas, while a deposit of carbonate ooze occupies the surface in the deeper regions.

VI. Cays and Islets

Unvegetated and vegetated cays and islets are associated with some bank-edge reefs. They are widely separated and do not cover a very large total area but they are of considerable geological interest. Sand cays may occur on reefs at the northern, southern, and western bank margins, some of the larger ones being vegetated to form stable, more extensive islets.

Beach rock, which is common on the Coral Sea Plateau, is of considerable thickness, and has stabilized even relatively small cays. It is well developed at Chilcott Islet (Plate II, A) where it displays a variety of interesting structures. The westward migration of the sand cay at North Flinders Reef (17°44′S, 148°26′E) is testified by the eastward extension of beach rock, which defines its former position. For notes regarding anchorages, cays, and islets see British Admiralty (1960).

VII. Structure and Geological History

Seismic reflection measurements (M. Ewing et al., 1970) verified the continental nature of the crust beneath the Coral Sea Plateau and demonstrated a sequence of layers overlying a deeper crustal layer with velocities of 7.3–7.6 km/sec. Such velocities are common beneath the orogenic belts in a "transition layer" representing changes between oceanic and continental crust, and M. Ewing et al. (1970, p. 1958) point out that this supports the idea that the plateau may represent the northeastern part of the Tasman geosyncline. An average thickness of 800 m of sediment overlies a 2-km basement layer of probably lithified sediments, low-grade metamorphics and carbonates, and igneous rocks (5.1–5.7 km/sec). Beneath this is the main crustal layer (5.9–6.3 km/sec) of Palaeozoic and/or Pre-Cambrian age (Fig. 12). Below the main crustal layer, a subcrustal layer (6.9 km/sec) was detected beneath the western side of the Queensland Trough at approximately 18 km, which may correlate with a similar layer reported for Eastern Queensland.

The offset of velocity lines along the eastern margin of the trough has been interpreted as due to normal faulting which lends credence to the view that the Queensland Trough is a deeply submerged graben (see Section III). Block faulting, bringing basement closer to the surface along the boundary of the Queensland Trough and Coral Sea Plateau, has been revealed in seismic reflection profile records (J. I. Ewing et al., 1970; Gardner, 1970); see Fig. 12.

Attention was drawn (M. Ewing et al., 1970, p. 1957) to the striking

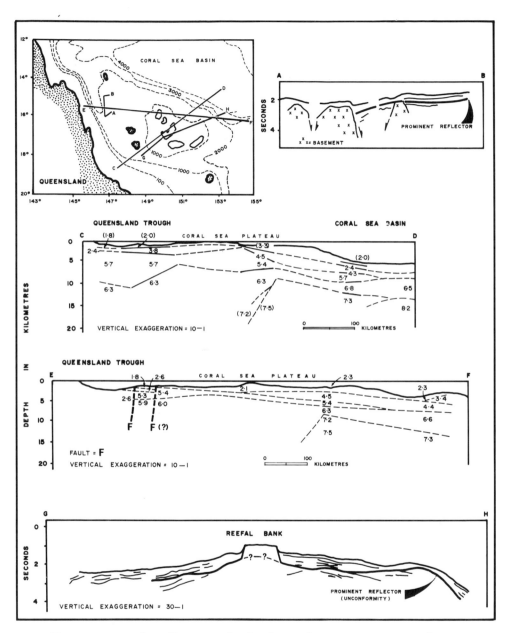

Fig. 12. Structural profiles across the Coral Sea Plateau. A–B: interpreted seismic reflection profile (after J. I. Ewing *et al.*, 1970); C–D and E–F: Seismic refraction profiles (after Gardner, 1970); and G–H: an interpreted seismic reflection profile (after Gardner, 1970; J. I. Ewing, 1970, see page 299, this chapter).

similarity between the velocity characteristics of the Plateau's sedimentary sequence and those of the sequence drilled in the Capricorn Channel at the southern end of the Great Barrier Reef. A late Tertiary change from nonmarine to marine facies occurs in the latter sequence and therefore, by comparison, may imply subsidence of the plateau as well as the shelf.

Within the sediments of the Coral Sea Plateau, occurring approximately 500 m below the sediment/water interface, is a relatively flatlying prominent reflector, which in places is near or at the top of the 5.1–5.7 km/sec basement layer of the plateau (horizon S, J. I. Ewing *et al.*, 1970). This pronounced reflector was believed to represent an erosion surface formed during the great Australian denudation cycle which is generally considered to be of Eocene-Miocene age. Further, J. I. Ewing *et al.* (1970) reported evidence of an Upper Oligocene date for this feature, and postulated that erosion of the plateau sediments, and even of the basement, which would result in its formation may account for a sequence of turbidites which lies just above the basal sediment of the Coral Sea Basin (J. I. Ewing *et al.*, 1970, p. 1968). This strong reflector, which terminates against one of the platform-type reefs (see Fig. 12), called Magdelaine Bank (Coringa Bank) by J. I. Ewing *et al.* (1970), and illustrated as Lihou Bank by Gardner (1970), was therefore believed to be an erosional surface which was formerly at or just below sea level, in order to allow the reef-building processes to begin.

Gardner (1970, p. 2607), citing several lines of evidence, maintains that this surface is a Lower Miocene datum and concludes that, if this is so, large-scale subsidence of the Coral Sea Plateau has occurred at least since Lower Miocene times. However, an initial description of results obtained during Leg 21 of the Deep Sea Drilling Project (Scientific Staff, 1972) indicates that in a core taken from Site 209 on the northeastern margin of the Coral Sea Plateau, the major stratigraphic break, though a less prominent seismic reflector (approximately 150 m below the sediment/water interface) occurs between late Eocene and late Oligocene deposits. Van der Lingen *et al.* (1973) confirmed this stratigraphic break, which occurs where a deposit of silicified chalk and chert in unlithified foraminiferal ooze with a large sand-silt detrital component was succeeded by calcareous biogenic ooze. The break is considered to represent an unconformity of regional importance, having been detected at other sites drilled in the Coral/Tasman Sea Region. A disconformity above this level represents a late Miocene hiatus which is of local importance only.

Further, although the presence of a prominent reflector at 0.25 sec subbottom depth, which was previously interpreted by J. Ewing *et al.*

(1970) as a regional unconformity representing an extensive erosion surface, was confirmed (Andrews *et al.*, 1973), it has been correlated with Eocene chert horizons between 187 m and 250 m subbottom depth (Van der Lingen *et al.*, *op. cit.* p. 309), and was shown to crop out on the Coral Sea Plateau Slope at a depth of 2000 m.

Nevertheless, Gardner (1970, p. 2606) states that this prominent reflector "is a surface which truncates underlying reflectors and upon which overlying reflectors onlap," and therefore there seems to be coincidence of this strong reflector, at least in some areas of the plateau, with an unconformity. Furthermore, since this strong seismic reflector (horizon S) has been shown to terminate against one of the reefal-banks, such reefal structures may have developed on a surface of Eocene deposits. It is conceivable that continuous sedimentation, represented by the Eocene chert bearing deposits in the core from site 209, was in progress at the northeastern margin of the plateau while the central and southern parts were at or near sea level, and being eroded. The latter areas later provided a site for reef development.

From the western margin of the Coral Sea Plateau horizon S was traced (J. Ewing *et al.*, 1970) into the Queensland Trough where it appears to have been affected by faulting. On the northern margin of the plateau it bends down, the down bending roughly following the downwarping of the overlying sediments, and the apparent termination of the reflector supports the idea of fault control, which may be inferred from the linear nature of the Coral Sea Plateau Slope (see Section III, page 280, and Gardner, 1970, p. 2609).

The slight dip of the plateau surface was interpreted by Fairbridge and Van der Linden (1966) as tilting that results from stretching and thinning of the continental crust. Gardner (1970) draws attention to the fact that the dip of the beds coincide with the northward tilt of the plateau, and suggests that if the dip does result from tilting it would be post unconformity (horizon S) in age, which would suggest that the same tectonic activity had resulted in the tilting of the plateau, the block faulting that formed the Queensland Trough, and the general subsidence of the region.

On the north-eastern margin of the Coral Sea Plateau a sequence of calcareous fossil ooze (biogenic ooze) extends upward from the Eocene-Oligocene regional unconformity through the Pleistocene (see Scientific Staff, 1972, and Van der Lingen *et al.*, 1973). The sediments above this unconformity in the core from site 209 indicate a deepening of this depositional area with time from late Oligocene to Recent; below the unconformity an increasing terrigenous content with depth occurs in a late Eocene to late-middle-Eocene sequence of pelagic ooze.

A thickness of 40 cm of Holocene pelagic calcareous ooze overlies Pleistocene glacial pelagic sediments (Gardner, 1970, p. 2602) on the plateau. Basing his calculations on the recognition of a change from a temperate-sub polar foraminiferal fauna to a tropical water fauna at depth of 40 cm in sediment cores (representing the climatic change from Wisconsin glacial to Holocene at 11,000 years B.P.), Gardner (*op. cit.,* p. 2605) estimated a Holocene sedimentation rate over the Coral Sea Plateau of 3.6 cm/1000 years.

The evidence outlined in the foregoing suggests that the Coral Sea Plateau was once continuous with the mainland and that after the formation of a widespread horizon which forms a prominent seismic reflector (horizon S), the plateau gradually subsided carrying this horizon down to −1500 m. The plateau was separated from the Queensland Continental Shelf by a tensional trough, and tilted. Subsidence of the northeastern margin may have begun earlier, allowing more continuous sedimentation there.

To account for the decrease of terrigenous (detrital) sediment in deposits beneath site 209 after late Middle Eocene, Andrews *et al.* (1973) postulated that the Queensland and Townsville Troughs may have developed as early as the Middle Eocene, isolating the plateau from the mainland and thereby forming an obstacle to the transport of terrigenous material, or alternatively, bottom transport of terrigenous sediments may have been blocked during Middle Eocene by reefs and cays developed on the central part of the plateau.

Throughout the subsidence, which began in the Tertiary and continued through the Pleistocene, the plateau has remained a reef province. The emergence and erosion of the tops of large reefal structures, which occurred during the Pleistocene low stands of sea level, resulted in considerable morphological changes.

The development of Holocene beach rock, which helped to stabilize the sand and shingle cays of the Coral Sea Plateau, is the most recent event in the geological history of this area.

VIII. Summary and Conclusions

Results of geophysical investigations tend to support the idea that the Coral Sea Plateau represents the northeastern part of the Tasman Geosyncline, and indicate that its present isolation and basic characteristics are legacies of Tertiary tectonics and differential subsidence, which separated the plateau from the Queensland continental shelf. A prominent seismic reflector, which corresponds with the upper surface of late Eocene chert (Van der Lingen *et al.*, 1973) below site 209 (Deep Sea Drilling Project) on the northeastern margin of the Coral Sea Plateau, but which elsewhere appears to represent an unconformity, may herald a significant palaeogeographic change. If this low relief seismic reflector is indeed widespread as an unconformity it may generally correspond in time with the period of early Tertiary planation recognized on the Australian Mainland. That it was formed very close to sea level is implied by the reefal structures which have gradually developed upward from it. Siting of the large reefal structures may have been influenced by structural "highs" (Gardner, 1970, p. 2609). The elongate nature of some banks might suggest that they are built on large anticlinal upwarps of the plateau (Fairbridge, 1950) but low topographic elevations produced by erosion of the plateau surface may also have provided suitable foundations (see p. 294). There is a curious absence of either living or extinct reefs on the northeastern margin of the plateau, which Gardner (*op. cit.*) suggests may be due to an initially rapid submergence, which carried the plateau surface below the ecological limit of hermatypic coral growth. A subsequent decrease in the subsidence rate would then have allowed the establishment of reefs on higher parts of the plateau surface. Further changes in subsidence rates during the development of the reefal banks may, to some extent, account for structural features such as the variations in the morphology and dimensions of reeflike masses within the banks (p. 294).

Palaeontological and lithological data from the sedimentary record of the core from site 209 indicate that the Coral Sea Plateau shows "a general deepening from outer neritic depths in the Eocene to present-day depths (approximately 1400 m)" (Van der Lingen *et al.*, 1973, p. 309), and that "as deepening progressed, oceanic pelagic sediments began to predominate during latest Middle Eocene as oceanic circulation established itself over the shelf" (Andrews *et al.*, 1973, p. 196).

An important phase in the geological history of the plateau is related to Pleistocene eustatic sea level changes which brought about considerable modifications to "bank-top" morphology. The main terrace so far identified on Coringa Bank lies at approximately 100 m below present sea level, and from this, pinnacles rise which may represent modern reefs on Pleistocene foundations. It is probable that karst-like features are concealed beneath Holocene sediment and reefs. Further, the saucerlike depression noted on Tregrosse Bank (p. 285) is probably in part a reflection of solution effects due to subaerial exposure during low sea level stands (see e.g. MacNeil, 1954; Fairbridge and Stewart, 1960). Indeed, a number of Pleistocene topographic remnants may exist beneath present bank-top mounds and ridges. The cycle of reef morphology proposed by

Wiens (1962, p. 110) probably applies here although further seismic studies will be necessary to define precisely the morphological changes that have taken place.

Contemporary climatic and oceanographic factors probably largely account for the distribution of modern reef growth on the reefal banks. Thus small-scale cellular motion and local upwelling (see pp. 277 and 278) bringing nutrients to reefs, as well as exposure to wave turbulence, may account for the bank-edge siting of reefs, and their preference for the southeast, east, and northeastern localities may be explained by the predominant wind direction from the southeast, and the general surface water flow from the east and northeast.

The general decrease in abundance and variety of organisms with depth and accompanying light and temperature conditions have been noted (page 291), and striking variations related to substrate conditions have been observed on both the outer-slopes and the interbank sea floor. The steep rock faces (e.g. North Flinders Reef) especially where sheltered from sedimentation are more favorable substrates for many forms of sessile benthonic organisms, than either the sediment covered slopes or the rocky surfaces over which sediment is in transit. Evidence of burrows was observed at lower levels only where apparently thicker accumulations of sediment had become relatively stable.

The Coral Sea Plateau is a carbonate province of both pelagic and clastic sedimentation, the latter being associated primarily with the extensive reefal banks of the southern area, and there is a broad differentiation into bank-top (lagoon), reef, outer-slope (fore reef), and interbank (deep sea) environments. Dispersal and "sorting" of sediment derived from bank tops and outer slopes is largely accomplished by wave action and by currents which descend outer slopes and flow over interbank areas. On the interbank sea floor adjacent to the Coringa and neighboring banks, sediment in transit to lower regions is rippled in both simple and complex patterns. Slumping on the outer slope of reefal banks, the effects of which have been revealed in seismic profiles and verified by observations made during dives, has occurred and is probably still an active process on the flanks of reefal structures and on the marginal slopes of the Coral Sea Plateau. Turbidity currents, which locally supplied the Coral Sea Basin with calcareous turbidites during the Pleistocene, are probably still locally generated on the Coral Sea Plateau Slope and may occasionally develop on the flanks of reefal "banks."

The bank-top environment extends to an average depth of 60 m and, being unprotected from the open sea in many areas, it is a high-energy environment characterized by deposits of coarse skeletal sand and gravel with patches of reef rubble. The adjacent reef environment with its rich

biota extends to a depth of 50–55 m, the observed depth limit of hermatypic coral in this area. Reef rubble, carbonate sand, and gravel form
patchy deposits, the finer grades of sediment being removed by wave
turbulence.

A considerable bathymetric range is covered by the outer slope and a
wide range of ecological conditions is included, but the most marked
changes in abundance and variety of the biota were observed to occur
in the Coringa Bank area at 140–150 m and 200–210 m, the latter coin-
ciding with the position of a thermocline (page 278). Sediments vary from
blocks and boulders through pebbly carbonate sand, to carbonate sand,
and are dispersed under the influence of gravity and strong downslope
currents.

The interbank environment is a cold, dark zone, apparently almost
barren of life. Two conditions occur according to the depth of the plateau
surface and proximity of reef supporting structures. The seabed relief in
the more shallow areas adjacent to the banks shows many small-scale
relief features in a region of sandy deposits which are largely in transit to
lower levels of the plateau surface. The deeper areas have little sea floor
relief except for occasional valleys which have been cut into sediment
which is essentially pelagic ooze.

Seismic reflection profiles indicate that the present range and arrangement of sedimentary environments have prevailed for a long time, which
may extend back at least into Oligocene times. However, it is apparent
that further extensive, systematic studies, particularly of the reefal areas,
are necessary in order to establish more precisely the geological history
of this major reef province.

Acknowledgments

The author wishes to express his gratitude to the Bureau of Mineral
Resources, Canberra, and to the late Mr. M. Shoriki and the Yomiuri
Committee who jointly sponsored the Japanese/Australian Oceanographic
Expedition. The officers and crew of the *Yamato* (mother ship) and the
Yomiuri-Go (submersible) are also gratefully acknowledged for their
assistance and cooperation during an important phase of the Coral Sea
Plateau studies.

A discussion of aspects of oceanography and the provision of temperature data by Professor M. Uda, were invaluable.

The Australian Aquitaine Oil Company generously provided financial
aid for the University of Queensland Expeditions in the Coral Sea region.

References

Andrews, J. E., Burns, R. E., Churkin, M., Jr., Davies, T. A., Dumitrica, P., Edwards, A. R., Galehouse, J. S., Kennett, J. P., Packham, G. H., and Van der Lingen, G. J. (1973). Deep Sea Drilling Project: Leg 21; Tasman Sea—Coral Sea. *In* "Oceanography of the South Pacific 1972" (R. Fraser, comp.), pp. 185–199. New Zealand National Commission for UNESCO, Wellington.

Brandon, D. E. (1973). Waters of the Great Barrier Reef Province. *In* "Biology and Geology of Coral Reefs" (O. A. Jones and R. Endean, eds.), Vol. 1, pp. 187–232. Academic Press, New York.

British Admiralty. (1960). "Australia Pilot," 5th ed., Vol. III, Sailing Directions No. 15. Hydrographic Department, Admiralty, London.

British Admiralty. (1962). "Australia Pilot," 5th ed., Vol. IV, Sailing Directions No. 16. Hydrographic Department, Admiralty, London.

Brown, D. A., Campbell, K. S. W., and Crook, K. A. W. (1968). "The Geologic Evolution of Australia and New Zealand." Pergamon, Oxford.

Bryan, W. H. (1944). The relationship of the Australian continent to the Pacific Ocean—now and in the past. *J.R. Soc. N.S.W.* **78**, 42–62.

Dietrich, G. (1963). "General Oceanography." Wiley, New York.

Ewing, J. I., Houtz, R. E., and Ludwig, W. J. (1970). Sediment distribution in the Coral Sea. *J. Geophys. Res.* **75**, 1963–1972.

Ewing, M., Hawkins, L. V., and Ludwig, W. J. (1970). Crustal structure of the Coral Sea. *J. Geophys. Res.* **75**, 1953–1962.

Fairbridge, R. W. (1950). Recent and Pleistocene coral reefs of Australia. *J. Geol.* **58**, 330–401.

Fairbridge, R. W., and Stewart, H. B., Jr. (1960). Alexa Bank, a drowned atoll on the Melanesian Border Plateau. *Deep-Sea Res.* **7**, 100–116.

Fairbridge, R. W., and van der Linden, W. J. M. (1966). Coral Sea. *In* "The Encyclopedia of Oceanography" (R. W. Fairbridge, ed.), pp. 219–224. Van Nostrand-Reinhold, Princeton, New Jersey.

Gardner, J. V. (1970). Submarine geology of the Western Coral Sea. *Geol. Soc. Am. Bull.* **81**, 2599–2614.

Geological Society of Australia. (1971). "Tectonic Map of Australia and New Guinea 1:5,000,000." Geol. Soc. Aust., Sydney.

Heidecker, E. (1973). Structural and tectonic factors influencing the development of Recent Coral Reefs off North-eastern Queensland. *In* "Biology and Geology of Coral Reefs" (O. A. Jones and R. Endean, eds.), Vol. 1, pp. 273–298. Academic Press, New York.

Hill, D., and Denmead, A. K., eds. (1960). "The Geology of Queensland," *J. Geol. Soc. Aust.*, **7** 1–474.

Kirkwood, L. F. (1967). Inorganic phosphate, organic phosphorous, and nitrate in Australian waters. *Aust., CSIRO, Div. Fish. Oceanogr. Tech. Pap. No.* 25.

Krause, D. C. (1967). Bathymetry and geologic structure of the Northwestern Tasman Sea-Coral Sea-South Solomon Sea area of the South-western Pacific Ocean. *N. Z. Dep. Sci. Ind. Res., Bull.* **183**, 1–50.

MacNeil, F. S. (1954). The shape of atolls: An inheritance from subaerial erosion forms. *Am. J. Sci.* **252**, 402–427.

Macurda, D. B., Jr., and Meyer, D. L. (1974). Feeding posture of modern stalked Crinoids. *Nature (London)* **247**, 394–396.

Maxwell, W. G. H. (1968). "Atlas of the Great Barrier Reef." Elsevier, Amsterdam.

Rochford, D. J. (1957). The identification and nomenclature of the surface water masses in the Tasman Sea (data to the end of 1954). *Aust. J. Mar. Freshwater Res.* **8**, 369–413.

Rochford, D. J. (1958). Total phosphorus as a means of identifying East Australian water masses. *Deep-Sea Res.* **5**, 89–110.

Rochford, D. J. (1959). The primary external water masses of the Tasman and Coral Seas. *Aust., CSIRO, Div. Fish. Oceanogr. Tech. Pap. No. 7.*

Rochford, D. J. (1968). The continuity of water masses along the western boundary of the Tasman and Coral Seas. *Aust. J. Mar. Freshwater Res.* **19**, 77–90.

Rotschi, H., and Lemasson, L. (1967). Oceanography of the Coral and Tasman Seas (H. Barnes, ed.). *Oceanogr. Mar. Biol.* **5**, 49–97.

Scientific Staff. (1972). Glomar Challenger down under. Deep sea drilling project—Leg. 21. *Geotimes* **17**, No. 5, 14–16.

Sheridan, R. E., and Fairbridge, R. W. (1966). Submarine plateaus. *In* "The Encyclopedia of Oceanography" (R. W. Fairbridge, ed.), Vol. 1, pp. 872–877. Van Nostrand-Reinhold, Princeton, New Jersey.

St. John, B. E. (1971). Trace elements in corals of the Coral Sea—an investigation relating their distribution to oceanographic factors. Ph.D. Thesis, University of Queensland, Brisbane, Australia.

Van der Lingen, G. J., Andrews, J. E., Burns, R. E., Churkin, M., Jr., Davies, T. A., Dumitrica, P., Edwards, A. R., Galehouse, J. S., Kennett, J. P., and Packham, G. H. (1973). Lithostratigraphy of eight drill sites in the South-West Pacific—preliminary results of leg 21 of the Deep Sea Drilling Project. *In* "Oceanography of the South Pacific 1972" (R. Fraser, comp.), pp. 299–313. New Zealand Commission for UNESCO, Wellington.

Wiens, H. J. (1962). "Atoll Environment and Ecology." Yale Univ. Press, New Haven, Connecticut.

Winterer, E. L. (1970). Submarine Valley systems around the Coral Sea Basin (Australia). *Mar. Geol.* **8**, 229–244.

Wyrtki, K. (1960). The surface circulation in the Coral and Tasman Seas. *Aust, CSIRO, Div. Fish. Oceanogr., Tech. Pap. No. 8.*

AUTHOR INDEX

Numbers in italics refer to the pages on which the complete references are listed.

LOCALITY INDEX

A

Abington Reef, 280, 282
Addu Atoll, 46, 60, 66, 73, 81, 83, 100
Admiralty Islands, 12
Ailuk Atoll, 41, *see also* Subject Index, Dredging
Aitutaki Atoll, 29, 60, 72, 81
Aiwa Island, 12
Alacran Reef, 14, 38, 63, 98, 100, 147, 157, 164, 176
Albany Island, 224
Alberta, 108, 111, 114, 116, 117, 122
Aldabra Atoll, 51, 71
Alexa Bank, 30, 285
Amirante Islands, 12
Anchor Cay, 212
Andes, 23
Andros Island, 41, 77, *see also* Subject Index, Bores
Artic Islands, 116
Argo Reefs, 12
Ant Atoll, 29
Arno Atoll, 60, 72, 82, 95
Atauro, 196
Atlantic Ocean, 4, 33, 34, 68
Aua Reef, 44, *see also* Subject Index, Bores
Austral Islands, 26, 52
Australia, 7, 13, 34, 276
Austria, 125

B

Bacon Isles, 11
Bahama Banks, 41, 44, 46, 49, 50, 131, 141, 150
Bahamas, 60, 61, 69, 71, 77, 194
Barataria Bay, 170
Barbados, 196
Barnard Islands, 223
Bashaw, 119
Belgium, 127

Bermuda, 4, 22, 30, 42, 50, 51, 71, 150, 151, 194, *see also* Subject Index, Bores
Bewich Island, 70, 225
Bikini Atoll, 39, 46, 60, 65, 79, 83, 139, 147, 201, *see also* Subject Index, Bores
Bismarck Archipelago, 202
Bligh Canyon, 280
Bligh Water, 7
Bogue Islands, 69
Borabora Island, 25, 87
Borneo, 43
Bougainville Reef, 281, 282
Bramble Cay, 206, 223, 224
Breaksea Spit, 206
British Columbia, 117
British Honduras, 28, 37, 42, 46, 47, 60, 61, 62, 67, 68, 69, 71, 74, 79, 85, 87, 94, 95, 96, *see also* Subject Index, Bores
British Honduras Shelf, 138
Broad Sound, 206, 213
Bunker Group, 89, 213, 218, 265

C

Cairns, 209, 274
Canada, 108, 109, 112, 113, 116, 122, 125, 126, 127
Cape Clinton, 206
Cape Melville, 206, 26, 265
Cape York, 213, 220, 223, 224, 263, 264
Cape York Peninsula, 263, 264, 268
Capricorn Channel, 20, 212, 217, 228, 265, 271, 299, *see also* Subject Index, Capricorn Basin and Southern Shelf Embayment
Capricorn Group (Reefs), 89, 155, 212, 213, 217, 224, 265
Caribbean Sea, 9, 12, 13, 34, 35, 38, 42, 46, 60, 67, 68, 77, 84, 85, 91, 97, 99, 101, 139, 146, *see also* Subject Index, Bores

SYSTEMATIC INDEX

325

SUBJECT INDEX

A

Absolute age determinations, *see* Radiometric geochronology of coral reefs
Abyssal deposits, 141
Accordant platforms, 32
Accretion, 100
Acropora thickets, 138
Aggradation of lagoon floors, 32
Ailuk Atoll, 41
Alberta basin, 119
Albertite, 115
Algal mounds, 138
Algal ridge, 133, 134, 221–223
Algal rim, *see* Algal ridge
Amadeus Trough, 268
Ancient reef complexes, 143
Andros Island, 42
Arafura Sea, 220, 274
Aragonite *see* Eniwetok bore core
Atlantic fauna, 35
Atoll(s), *see* Reefs, types of
Aua reef, 44

B

Bahama reef, 150
Bahamas, 42
Barrier reef(s), *see* Reef, types of, *also see* Locality Index
Barrier stage, 12
Basin, *see* specific area
Basin deposits, 141
Bathyal deposits, 141
Beach rock, *see* Reef features
Beach sediments, analyses of, 147
Beaverhill Lake Formation, 119
Bermuda, 42
Bermuda patch reef, 150
Bikini Atoll, 39
Bioherm(s), 2, 130, *see also* Coral reef, definition
Biosomes, 115

Biostrome(s), 2, 112
Bligh Canyon, 280
Bores into reefs, *see also* Great Barrier Reef province, Sediment, Stratigraphic basins
Aua reef, 44
Andros Island, 42
British Honduras, 42
Bahamas, 42
Bermuda, 42
Bikini Atoll, 39
Borneo, 43
Caribbean, 42
Elevated Atolls, 42–43
Eniwetok, 39
Florida, 44
Funafuti, 39
Glovers Reef, 42
Hawaiian Islands, 42
Hao, 39
Kita-Daite-Jima, 42, 43
Mahé, Seychelles, 44
Maldives, 39
Maratoea Atoll, Borneo, 43
Marshall Islands, 39
Midway, 42
Mururoa Atoll, 41–42
New Caledonia, 44
Pacific, 42
Pago Pago, Tutuila, Samoa, 44
Samoa, 44
Tahiti, 44
Tuamotus, 39, 42
Turneffe Atoll, British Honduras, 42
Tutuila, 44
Utelei Reef, Tutuila, Samoa, 44
Borneo, 43
British Honduras reef, 140

C

Calcite, *see* Eniwetok bore core
Canning basin, 111, 122
Capricorn basin, 213, 271, 299

327